LONDON MATHEMATICAL SOCIETY LECTURE NOT

Managing Editor: Professor M. Reid, Mathematics Institute,
University of Warwick, Coventry CV4 7AL, United Kingdom

D1388543

The titles below are available from booksellers, or from Cambridge ....
www.cambridge.org/mathematics

London Mathematical Society Lecture Notes series: 379

Finite and Algorithmic Model Theory

Edited by

JAVIER ESPARZA
Technische Universität München, Germany

CHRISTIAN MICHAUX
Université de Mons, Belgium

CHARLES STEINHORN
Vassar College, New York, USA

CAMBRIDGE
UNIVERSITY PRESS

CAMBRIDGE UNIVERSITY PRESS
Cambridge, New York, Melbourne, Madrid, Cape Town,
Singapore, São Paulo, Delhi, Tokyo, Mexico City

Cambridge University Press
The Edinburgh Building, Cambridge CB2 8RU, UK

Published in the United States of America by Cambridge University Press, New York

www.cambridge.org
Information on this title: www.cambridge.org/9780521718202

© Cambridge University Press 2011

First published 2011

Printed in the United Kingdom at the University Press, Cambridge

A catalogue record for this publication is available from the British Library

Library of Congress Cataloguing in Publication data
Finite and algorithmic model theory / edited by Javier Esparza, Christian Michaux,
Charles Steinhorn.
p.   cm. – (London Mathematical Society lecture note series ; 379)
Includes index.
ISBN 978-0-521-71820-2 (pbk.)
1. Finite model theory.   2. Computer science – Mathematics.   I. Esparza, Javier.
II. Michaux, Christian.   III. Steinhorn, Charles.   IV. Title. V. Series.
QA9.7.F565   2011
511.3′4 – dc22   2010048062

ISBN 978-0-521-71820-2 Paperback

Contents

Preface

This volume is based on the satellite workshop on *Finite and Algorithmic Model Theory* that took place at the University of Durham, January 9–13, 2006, to inaugurate the scientific program *Logic and Algorithms* held at the Isaac Newton Institute for Mathematical Sciences during the first six months of 2006. The goal of the workshop was to explore the emerging and potential connections between finite and infinite model theory, and their applications to theoretical computer science. The primarily tutorial format introduced researchers and graduate students to a number of fundamental topics. The excellent quality of the tutorials suggested to the program organizers, Anuj Dawar and Moshe Vardi, that a volume based on the workshop presentations could serve as a valuable and lasting reference. They proposed this to the workshop scientific committee; this volume is the outcome.

The *Logic and Algorithms* program focused on the connection between two chief concerns of theoretical computer science: (i) how to ensure and verify the correctness of computing systems; and (ii) how to measure the resources required for computations and ensure their efficiency. The two areas historically have interacted little with each other, partly because of the divergent mathematical techniques they have employed. More recently, areas of research in which model-theoretic methods play a central role have reached across both sides of this divide. Results and techniques that have been developed have found applications to fields such as database theory, complexity theory, and verification.

Some brief historical remarks help situate the context for this volume. The study of the model-theoretic properties of finite structures emerged initially as a branch of classical model theory, with its focus primarily on first-order logic. Beginning in the late 1980s, however, research concerning logics on finite structures diverged sharply from work in classical model theory. Classical model theory, with its emphasis on infinite structures, had made dramatic advances both theoretically and in applications to other areas of mathematics. Work on finite structures focused on connections with discrete complexity theory and verification. Indeed, the connections between finite model theory, descriptive complexity theory, parameterized complexity, and state machine verification are now so strong that boundaries between them are hard to distinguish.

The methods employed in these two facets of model theory also grew apart during this period. Probabilistic techniques and machine simulations have played a prominent role in the study of finite structures, and stand in contrast to the geometric, algebraic, and analytic methods that pervade classical (infinite) model theory. Although both classical and finite model theory deal with restricted classes of structures, the conditions by which such classes are delimited also have been quite different. Finite model theory and verification typically concentrate on classes linked to particular computing formalisms, or to which decomposition methods from finite graph theory can be applied. In contrast, infinitary model theory usually places restrictions on combinatorial or geometric properties of the definable sets of a structure.

Yet, there are recent indications of a re-convergence of classical model theory and logical aspects of computer science. This has resulted both from the interest of computer scientists in new computing and specification models that make use of infinitary structures, and from the development of powerful model-theoretic techniques that provide insight into finite structures. If there is an overarching theme, it is how various "tameness" hypotheses used to delimit classes of structures *and* logics have deeply impacted the study of those aspects of theoretical computer science in which model-theory naturally comes into play. The chapters that comprise this volume survey many of the common themes that have emerged and gained attention, and point to the significant potential for wider interaction.

The chapter of Bárány, Grädel, and Rubin, *Automata-based presentations of infinite structures* develops what the authors call *algorithmic model theory*. The authors direct their attention to the "tame" class of *automatic structures*, that is structures that have a presentation in a precise sense by automata operating on finite or infinite words or trees. The goal of this work, to extend algorithmic and logical methods from finite structures to finitely presented infinite structures, has been a focal point for research in computer science, combinatorics, and mathematical logic. This point of view allows structures to be viewed alternately from both a finite and infinite model theoretic perspective. The theory that has emerged makes use of techniques both from classical model theory and theoretical computer science, and has found appealing applications to several areas, including database theory, complexity theory and verification.

Classical model theory by and large concentrates on the analysis of the first-order definable sets over a structure, that is, those sets of n-tuples of the universe whose definition is given by a first-order formula. This analysis has predominantly taken two forms. The first is based on the "structural complexity" of the formula, e.g., the number of alternations of blocks of existential and universal quantifiers appearing in its prenex normal form. This theme is best illustrated by *quantifier elimination*, in which definable sets over a structure are shown to have quantifier-free definitions. The second involves assigning a dimension (with a corresponding notion of independence) to the definable sets that is combinatorially, algebraically, or geometrically motivated. Stability theory, with its combinatorial/algebraic account of dimension and independence, is perhaps the most widely known and longest-studied exemplar, its development traceable to Morley's seminal work in the 1960's and to Shelah's deep and extensive work in the 1970's. More recently, o-minimality, and in particular its focus on o-minimal expansions of the ordered field of real numbers, provides another important class of examples. The imposition of "tameness" assumptions in classical model theory such as stability and o-minimality – often verified in examples by quantifier elimination – make the analysis of the structures satisfying these hypotheses not only tractable but also amenable to applications in mathematics outside of logic.

Tarski's quantifier-elimination for real-closed fields which thereby (effectively) equates the first-order definable sets over the field of real numbers with the semialgebraic sets, has long proved a fertile ground for framing and addressing computational issues. Kuijpers and Van den Bussche, in their chapter, *Logical aspects of spatial databases*, model spatial data via semialgebraic subsets of n-dimensional Euclidean space, and investigate the expressive power of several logic-based languages to query these databases. They first characterize the topological properties of planar spatial databases that are first-order expressible over the usual language for the ordered field of real numbers – of interest from the point of view of geographical information systems, for example – in terms of the query language "cone logic". The second half of their chapter deals with query languages that extend first-order logic over the real field by some form of recursion, including spatial Datalog, and first-order logic extended with a while loop or with a transitive closure operator.

Koponen, in her chapter, *Some connections between finite and infinite model theory*, discusses how stability theoretic considerations, as well as other properties and techniques from classical model theory such as smooth approximation, can be imported successfully into the study of finite structures by restricting to bounded variable logic, that is, first-order logic under the restriction that there is a fixed value k such that only formulas in which no more than k variables occur. In particular, Koponen investigates when a theory in bounded variable logic with an infinite model has arbitrarily large finite models and isolates conditions for effectively determining least upper bounds for the size of the smallest such finite model.

The chapter of Macpherson and Steinhorn, *Definability in classes of finite structures*, contains two distinct threads that draw their motivation from classical model theory. The first, inspired by the model theory of finite and pseudofinite fields, concerns asymptotic classes of finite structures. These are non-elementary classes of finite structures whose first-order definable sets asymptotically satisfy cardinality constraints that permit the assignation of a dimension and measure, and have an intimate connection in classical model theory to so-called simple theories. The second theme concerns so-called *robust classes* of finite structures, whose origin lies in attempting to "finitize" classical model-theoretic tameness conditions, such as o-minimality, that are provably excluded in asymptotic classes. Robust classes consist of directed systems of finite structures in which the truth value of a formula requires "looking ahead" into a larger structure in the system.

For the model theory of finite structures that has been developed with great success within theoretical computer science, "tameness" assumptions do not apply only to isolate classes of structures that are well-behaved with respect to a preferred logic, such as first-order logic. Research has prospered by striking a balance between appropriate logics or fragments thereof and classes of finite structures: that is, tame logics matched with tame classes. This theme appears already in Koponen's chapter, with its emphasis on bounded variable logic combined with classical tameness assumptions, and strongly emerges in the chapters of Otto and Kreutzer. As these chapters furthermore show, this point of view can furnish significant computational insights.

Kreutzer's chapter, *Algorithmic meta-theorems*, discusses how constraining both classes of (finite) structures and logics yields a wealth of algorithmic results. An algorithmic meta-theorem has the form that every computational problem that can be expressed in some logic can be solved efficiently on every class of structures that satisfy certain constraints. This is usually accomplished by showing that the model-checking problem for formulas in some logic – typically first-order or monadic second-order – is what is called *fixed-parameter tractable* for a class of structures, typically based on graphs with well-behaved tree decompositions. This point of view goes back to well-known work of Courcelle and his collaborators.

Otto takes as the focus of his chapter the application of game-oriented methods and explicit model constructions in the analysis of fragments of first-order logic restricted to well-behaved (non-elementary) classes of structures, particularly finite structures. Whereas the model-theoretic compactness theorem plays an essential role in the classical setting, paradigmatically in proving *expressive completeness* results such as the Łos-Tarski theorem characterizing those formulas preserved under extensions as the existential formulas, its failure for restricted classes of structures, e.g., classes of finite structures, motivates the introduction of the methods and techniques that Otto places at the center of the chapter. The chapter also surveys how by restricting to classes of finite structures defined by tree-width and locality considerations, expressive completeness results that fail for the class of finite structures can be regained.

The workshop organizer was Professor Iain Stewart (Durham). The members of the Scientific Committee for the workshop included : Michael Benedikt (Oxford), Javier Esparza (Munich), Bradd Hart (McMaster), Christian Michaux (Mons-Hainaut), Charles Steinhorn (Vassar), and Katrin Tent (Münster). Financial support from the Newton Institute and EPSRC is gratefully acknowledged. We also wish to express our appreciation to the staff at Cambridge University Press, in particular Clare Dennison, our maths/computer science editor, and Sabine Koch, our production editor, for their remarkable thoughtfulness, patience, and efficiency throughout the process of bringing this volume into print.

Javier Esparza
Christian Michaux
Charles Steinhorn

1

Automata-based presentations of infinite structures

VINCE BÁRÁNY[1], ERICH GRÄDEL[2] AND SASHA RUBIN[3]

1.1 Finite presentations of infinite structures

The model theory of finite structures is intimately connected to various fields in computer science, including complexity theory, databases, and verification. In particular, there is a close relationship between complexity classes and the expressive power of logical languages, as witnessed by the fundamental theorems of descriptive complexity theory, such as Fagin's Theorem and the Immerman-Vardi Theorem (see [78, Chapter 3] for a survey).

However, for many applications, the strict limitation to finite structures has turned out to be too restrictive, and there have been considerable efforts to extend the relevant logical and algorithmic methodologies from finite structures to suitable classes of infinite ones. In particular this is the case for databases and verification where infinite structures are of crucial importance [130]. *Algorithmic model theory* aims to extend in a systematic fashion the approach and methods of finite model theory, and its interactions with computer science, from finite structures to finitely-presentable infinite ones.

There are many possibilities to present infinite structures in a finite manner. A classical approach in model theory concerns the class of *computable structures*; these are countable structures, on the domain of natural numbers, say, with a finite collection of computable functions and relations. Such structures can be finitely presented by a collection of algorithms, and they have been intensively

[1] Oxford University Computing Laboratory
Wolfson Building, Parks Road, Oxford OX1 3QD, United Kingdom
vbarany@logic.rwth-aachen.de
[2] Mathematical Foundations of Computer Science
RWTH Aachen, D-52056 Aachen, Germany
graedel@logic.rwth-aachen.de
[3] Department of Mathematics and Applied Mathematics
University of Cape Town, Private Bag, Rondebosch 7701, South Africa
srubin@math.cornell.edu

studied in model theory since the 1960s. However, from the point of view of algorithmic model theory the class of computable structures is problematic. Indeed, one of the central issues in algorithmic model theory is the effective evaluation of logical formulae, from a suitable logic such as first-order logic (FO), monadic second-order logic (MSO), or a fixed point logic like LFP or the modal μ-calculus. But on computable structures, only the quantifier-free formulae generally admit effective evaluation, and already the existential fragment of first-order logic is undecidable, for instance on the computable structure $(\mathbb{N}, +, \cdot)$.

This leads us to the central requirement that for a suitable logic L (depending on the intended application) the model-checking problem for the class $\mathcal{C}$ of finitely presented structures should be algorithmically solvable. At the very least, this means that the L-theory of individual structures in $\mathcal{C}$ should be decidable. But for most applications somewhat more is required:

> **Effective semantics:** There should be an algorithm that, given a finite presentation of a structure $\mathfrak{A} \in \mathcal{C}$ and a formula $\psi(\bar{x}) \in L$, expands the given presentation to include the relation $\psi^{\mathfrak{A}}$ defined by ψ on $\mathfrak{A}$.

This also implies that the class $\mathcal{C}$ should be closed under some basic operations (such as logical interpretations). Thus we should be careful to restrict the model of computation. Typically, this means using some model of *finite automata* or a very restricted form of rewriting.

In general, the finite means for presenting infinite structures may involve different approaches: logical interpretations; finite axiomatisations; rewriting of terms, trees, or graphs; equational specifications; the use of synchronous or asynchronous automata, etc. The various possibilities can be classified along the following lines:

> **Internal:** a set of finite or infinite words or trees/terms is used to represent the domain of (an isomorphic copy of) the structure. Finite automata/ rewriting-rules compute the domain and atomic relations (eg. prefix-recognisable graphs, automatic structures).
>
> **Algebraic:** a structure is represented as the least solution of a finite set of recursive equations in an appropriately chosen algebra of finite and countable structures (eg. VR-equational structures).
>
> **Logical:** structures are described by interpreting them, using a finite collection of formulae, in a fixed structure (eg. tree-interpretable structures). A different approach consists in (recursively) axiomatising the isomorphism class of the structure to be represented.
>
> **Transformational:** structures are defined by sequences of prescribed transformations, such as graph-unraveling, or Muchnik's iterations applied

to certain fixed initial structures (which are already known to have a decidable theory). Transformations can also be transductions, logical interpretations, etc. [23]

The last two approaches overlap somewhat. Also, the algebraic approach can be viewed *generatively*: convert the equational system into an appropriate *deterministic grammar* generating the solution of the original equations [44]. The grammar is thus the finite presentation of the graph. One may also say that internal presentations and generating grammars provide descriptions of the *local structure* from which the whole arises, as opposed to descriptions based on *global symmetries* typical of algebraic specifications.

Prerequisites and notation

We assume rudimentary knowledge of finite automata on finite and infinite words and trees, their languages and their correspondence to monadic second-order logic (MSO) [133, 79]. Undefined notions from logic and algebra (congruence on structures, definability, isomorphism) can be found in any standard textbook. We mainly consider the following logics $\mathcal{L}$: first-order (FO), monadic second order (MSO), and weak monadic second-order (wMSO) which has the same syntax as MSO, but the intended interpretation of the set variables is that they range over *finite* subsets of the domain of the structure under consideration.

We mention the following to fix notation: infinite words are called ω-words and infinite trees are called ω-trees (to distinguish them from finite ones); relations computable by automata will be called *regular*; the domain of a *structure* $\mathfrak{B}$ is usually written B and its relations are written $R^{\mathfrak{B}}$. An MSO-formula $\phi(X_1, \ldots, X_j, x_1, \ldots, x_k)$ interpreted in $\mathfrak{B}$ *defines* the set $\phi^{\mathfrak{B}} := \{(B_1, \ldots, B_j, b_1, \ldots, b_k) \mid B_i \subset B, b_i \in B, \mathfrak{B} \models \phi(B_1, \ldots, B_j, b_1, \ldots, b_k)\}$. A wMSO-formula is similar except that the B_i range over finite subsets of B. The *full binary tree* $\mathfrak{T}_2$ is defined as the structure

$$\left(\{0, 1\}^*, \mathrm{suc}_0, \mathrm{suc}_1\right)$$

where the successor relation suc_i consists of all pairs (x, xi). Tree automata operate on Σ-*labelled trees* $T : \{0, 1\}^* \to \Sigma$. Such a tree is identified with the structure

$$\left(\{0, 1\}^*, \mathrm{suc}_0, \mathrm{suc}_1, \{T^{-1}(\sigma)\}_{\sigma \in \Sigma}\right).$$

Rabin proved the decidability of the MSO-theory of $\mathfrak{T}_2$ and the following fundamental correspondence between MSO and tree automata (see [132] for an overview):

For every monadic second-order formula $\varphi(\overline{X})$ in the signature of $\mathfrak{T}_2$ there is a tree automaton $\mathcal{A}$ (and vice versa) such that

$$L(\mathcal{A}) = \{T_{\overline{X}} \mid \mathfrak{T}_2 \models \varphi(\overline{X})\} \qquad (1.1)$$

where $T_{\overline{X}}$ denotes the tree with labels for each X_i.

Similar definitions and results hold for r-ary trees, in which case the domain is $[r]^*$ where $[r] := \{0, \dots, r - 1\}$, and finite trees.

In section 1.2.2 and elsewhere we do not distinguish between a term and its natural representation as a tree. Thus we may speak of infinite terms. We consider countable, vertex- and edge-labelled graphs possibly having distinguished vertices (called sources), and no parallel edges of the same label. A graph is *deterministic* if each of its vertices is the source of at most one edge of each edge label.

Interpretations

Interpretations allow one to define an isomorphic copy of one structure in another. Fix a logic $\mathcal{L}$. A d-dimensional $\mathcal{L}$-*interpretation* $\mathcal{I}$ of structure $\mathfrak{B} = (B; (R_i^{\mathfrak{B}})_i)$ in structure $\mathfrak{A}$, denoted $\mathfrak{B} \leq_{\mathcal{L}}^{\mathcal{I}} \mathfrak{A}$, consists of the following $\mathcal{L}$-formulas in the signature of $\mathfrak{A}$,

– a domain formula $\Delta(\overline{x})$,
– a relation formula $\Phi_{R_i}(\overline{x}_1, \dots, \overline{x}_{r_i})$ for each relation symbol R_i, and
– an equality formula $\epsilon(\overline{x}_1, \overline{x}_2)$,

where each $\Phi_{R_i}^{\mathfrak{A}}$ is a relation on $\Delta^{\mathfrak{A}}$, each of the tuples $\overline{x}_i, \overline{x}$ contain the same number of variables, d, and $\epsilon^{\mathfrak{A}}$ is a congruence on the structure $(\Delta^{\mathfrak{A}}, (\Phi_{R_i}^{\mathfrak{A}})_i)$, so that $\mathfrak{B}$ is isomorphic to

$$(\Delta^{\mathfrak{A}}, (\Phi_{R_i}^{\mathfrak{A}})_i) / \epsilon^{\mathfrak{A}} .$$

If $\mathcal{L}$ is FO then the free $\overline{x}$ are FO and we speak of a *FO interpretation*. If $\mathcal{L}$ is MSO (wMSO) but the free variables are FO, then we speak of a *(weak) monadic second-order interpretation*.

We associate with $\mathcal{I}$ a transformation of formulas $\psi \mapsto \psi^{\mathcal{I}}$. For illustration we define it in the first-order case: the variable x_i is replaced by the d-tuple $\overline{y}_i$, $(\psi \vee \phi)^{\mathcal{I}}$ by $\psi^{\mathcal{I}} \vee \phi^{\mathcal{I}}$, $(\neg \psi)^{\mathcal{I}}$ by $\neg \psi^{\mathcal{I}}$, $(\exists x_i \psi)^{\mathcal{I}}$ by $\exists \overline{y}_i \Delta(\overline{y}_i) \wedge \psi^{\mathcal{I}}$, and $(x_i = x_j)^{\mathcal{I}}$ is replaced by $\epsilon(\overline{y}_i, \overline{y}_j)$. Thus one can translate $\mathcal{L}$ formulas from the signature of $\mathfrak{B}$ into the signature of $\mathfrak{A}$.

Proposition 1.1.1 *If $\mathfrak{B} \leq_{\mathcal{L}}^{\mathcal{I}} \mathfrak{A}$, say the isomorphism is f, then for every formula $\psi(x_1, \dots, x_k)$ in the signature of $\mathfrak{B}$ and all k-tuples $\overline{b}$ of elements of*

$\mathfrak{B}$ *it holds that*

$$\mathfrak{B} \models \psi(b_1, \ldots, b_k) \iff \mathfrak{A} \models \psi^{\mathcal{I}}(f(b_1), \ldots, f(b_k))$$

In particular, if $\mathfrak{A}$ has decidable $\mathcal{L}$-theory, then so does $\mathfrak{B}$.

Set interpretations

When $\mathcal{L}$ is MSO (wMSO) and the free variables are MSO (wMSO) the interpretation is called a *(finite) set interpretation*. In this last case, we use the notation $\mathfrak{B} \leq_{\text{set}}^{\mathcal{I}} \mathfrak{A}$ or $\mathfrak{B} \leq_{\text{fset}}^{\mathcal{I}} \mathfrak{A}$. We will only consider (finite) set interpretations of dimension 1.

If finiteness of sets is MSO-definable in some structure $\mathfrak{A}$ (as for linear orders or for finitely branching trees) then every structure $\mathfrak{B}$ having a finite-set interpretation in $\mathfrak{A}$ can also be set interpreted in $\mathfrak{A}$.

Example 1.1.2 An interpretation $(\mathbb{N}, +) \leq_{\text{fset}}^{\mathcal{I}} (\mathbb{N}, 0, \text{suc})$ based on the binary representation is given by $\mathcal{I} = (\varphi(X), \varphi_+(X, Y, Z), \varphi_=(X, Y))$ with $\varphi(X)$ always true, $\varphi_=$ the identity, and $\varphi_+(X, Y, Z)$ is

$$\exists C \, \forall n \, [(Zn \leftrightarrow Xn \oplus Yn \oplus Cn) \wedge (C(\text{suc}n) \leftrightarrow \mu(Xn, Yn, Cn)) \wedge \neg C0]$$

where C stands for carry, $\oplus$ is exclusive or, and $\mu(x_0, x_1, x_2)$ is the majority function, in this case definable as $\bigvee_{i \neq j} x_i \wedge x_j$.

To every (finite) subset interpretation $\mathcal{I}$ we associate, as usual, a transformation of formulas $\psi \mapsto \psi^{\mathcal{I}}$, in this case mapping first-order formulas to (weak) monadic second-order formulas.

Proposition 1.1.3 *Let $\mathfrak{B} \leq_{\text{(f)set}}^{\mathcal{I}} \mathfrak{A}$ be a (finite) subset interpretation with isomorphism f. Then to every first-order formula $\psi(x_1, \ldots, x_k)$ in the signature of $\mathfrak{B}$ one can effectively associate a (weak) monadic second-order formula $\psi^{\mathcal{I}}(X_1, \ldots, X_k)$ in the signature of $\mathfrak{A}$ such that for all k-tuples $\bar{b}$ of elements of $\mathfrak{B}$ it holds that*

$$\mathfrak{B} \models \psi(b_1, \ldots, b_k) \iff \mathfrak{A} \models \psi^{\mathcal{I}}(f(b_1), \ldots, f(b_k)).$$

Consequently, if the (weak) monadic-second order theory of $\mathfrak{A}$ is decidable then so is the first-order theory of $\mathfrak{B}$.

For more on subset interpretations we refer to [23].

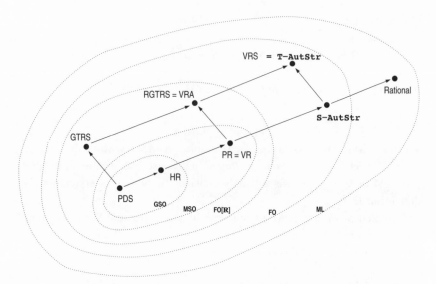

Figure 1.1 Relationship of graph classes and logical decidability boundaries.

1.2 A hierarchy of finitely presentable structures

This section provides an overview of some of the prominent classes of graphs and their various finite presentations.

These developments are the product of over two decades of research in diverse fields. We begin our exposition with the seminal work of Muller and Schupp on context-free graphs, we mention prefix-recognisable structures, survey hyperedge-replacement and vertex-replacement grammars and their corresponding algebraic frameworks leading up to equational graphs in algebras with asynchronous or synchronous product operation. These latter structures are better known in the literature by their automatic presentations, and constitute the topic of the rest of this survey.

As a unifying approach we discuss how graphs belonging to individual classes can be characterised as least fixed-point solutions of finite systems of equations in a corresponding algebra of graphs. We illustrate on examples how to go from graph grammars through equational presentations and interpretations to internal presentations and vice versa.

We briefly summarise key results on Caucal's pushdown hierarchy and more recent developments on simply-typed recursion schemes and collapsible pushdown automata.

Figure 1.1 provides a summary of some of the graph classes discussed in this section together with the boundaries of decidability for relevant logics.

Rational graphs and automatic graphs featured on this diagram are described in detail in Section 1.3.

1.2.1 From context-free graphs to prefix-recognisable structures

Context-free graphs were introduced in the seminal papers [110, 111, 112] of Muller and Schupp. There are several equivalent definitions. The objects of study are countable directed edge-labelled, finitely branching graphs. An *end* is a maximal connected[4] component of the induced subgraph obtained by removing, for some n, the n-neighbourhood of a fixed vertex v_0. A vertex of an end is on the *boundary* if it is connected to a vertex in the removed neighbourhood. Two ends are end-isomorphic if there is a graph isomorphism (preserving labels as well) between them that is also a bijection of their boundaries. A graph is *context-free* if it is connected and has only *finitely many ends* up to end-isomorphism. This notion is independent of the v_0 chosen.

A graph is context-free if and only if it is isomorphic to the connected component of the configuration graph of a pushdown automaton (without ϵ-transitions) induced by the set of configurations that are reachable from the initial configuration [112].

A *context-free group* is a finitely generated group G such that, for some set S of semigroup generators of G, the set of words $w \in S^*$ representing the identity element of G forms a context-free language. This is independent of the choice of S. Moreover, a group is context-free if and only if its Cayley graph for some (and hence all) sets S of semigroup generators is a context-free graph. Finally, a finitely generated group is context-free if and only if it is *virtually free*, that is, if it has a free subgroup of finite index [111].[5]

Muller and Schupp have further shown that context-free graphs have a decidable MSO-theory. Indeed, every context-free graph can be MSO-interpreted in the full binary tree.

Example 1.2.1 Consider the group G given by the finite presentation $\langle a, b, c \mid ab, cc, acac, bcbc \rangle$. The Cayley graph $\Gamma(G, S)$ of G with respect

---

[4] connectedness is taken with respect to the underlying undirected graph.

[5] Originally [111] proved this under the assumption of *accessibility*, a notion related to group decompositions introduced by Wall who conjectured that all finitely generated groups would have this property. Muller and Schupp conjectured every context-free group to be accessible, but it was not until Dunwoody [64] proved that all finitely presentable groups are accessible that this auxiliary condition could be dropped from the characterisation of [111]. Unfortunately, many sources forget to note this fact. Later Dunwoody also gave a counterexample refuting Wall's conjecture.

to the set of semigroup generators $S = \{a, b, c\}$ is depicted below.

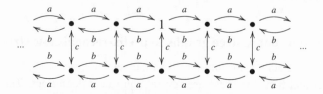

Notice that $\Gamma(G, S)$ has two ends, for any n-neighbourhood of the identity with $n > 1$. These are

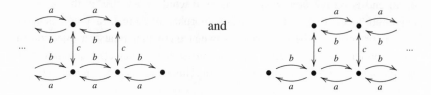

A word $w \in \{a, b, c\}^*$ represents the identity of G if, and only if, w has an even number of c's and the number of a's equals the number of b's. We present a pushdown automaton $\mathcal{A}$ which recognises this set of words and, moreover, has a configuration graph that is isomorphic to $\Gamma(G, S)$. The states of $\mathcal{A}$ are $Q = \{1, c\}$ with $q_0 = 1$ as the initial state, the stack alphabet is $\Gamma = \{a, b\}$, the input alphabet is $\{a, b, c\}$ and $\mathcal{A}$ has the following transitions:

$$
\begin{array}{llll}
\text{internal:} & 1\theta & \xrightarrow{c} & c\theta \\
\text{internal:} & c\theta & \xrightarrow{c} & 1\theta \\
\text{push:} & q\sigma\theta & \xrightarrow{\sigma} & q\sigma\sigma\theta & \text{for } q = 1, c \text{ and } \sigma = a, b \\
\text{push:} & q\perp & \xrightarrow{\sigma} & q\sigma\perp & \text{for } q = 1, c \text{ and } \sigma = a, b \\
\text{pop:} & q\sigma\theta & \xrightarrow{\bar{\sigma}} & q\theta & \text{for } q = 1, c \text{ and } \{\sigma, \bar{\sigma}\} = \{a, b\}
\end{array}
$$

Here θ is the stack content written with its top element on the left and always ending in the special symbol $\perp$ marking the bottom of the stack.

In every deterministic edge-labelled connected graph and for any ordering of the edge labels one obtains a spanning tree by taking the shortest path with the lexicographically least labeling leading to each node from a fixed source. Take such a spanning tree T for the example graph $\Gamma(G, S)$ with root 1_G. Observe that T is regular, having only finitely many subtrees (ends) up to isomorphism. The ordering $a < b < c$ induces the spanning tree depicted below. The Cayley graph $\Gamma(G, S)$ is MSO-interpretable in this regular spanning tree by defining the missing edges using the relators from the presentation of the group.

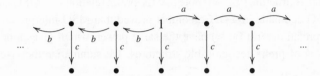

In particular $\Gamma(G, S)$ is MSO-interpretable in the full binary tree, and hence has decidable MSO.

A mild generalisation of pushdown transitions, *prefix-rewriting* rules, take the form $uz \mapsto vz$ where u and v are fixed words and z is a variable ranging over words. As in the previous example, pushdown transitions are naturally perceived as prefix-rewriting rules affecting the state and the top stack symbols. Conversely, Caucal [40] has shown that connected components of configuration graphs of prefix-rewriting systems given by finitely many prefix-rewriting rules are effectively isomorphic to connected components of pushdown graphs. Later, Caucal introduced *prefix-recognisable graphs* as a generalisation of context-free graphs and showed that these are MSO-interpretable in the full binary tree and hence have a decidable MSO-theory [42].

Definition 1.2.2 (Prefix-recognisable relations) Let Σ be a finite alphabet. The set $\mathsf{PR}(\Sigma)$ of prefix-recognisable relations over Σ^* is the smallest set of relations such that

- every regular language $L \subseteq \Sigma^*$ is a prefix-recognisable unary relation;
- if $R, S \in \mathsf{PR}$ (arities r and s) and L is regular then $L \cdot (R \times S) = \{(uv_1, \ldots, uv_r, uw_1, \ldots, uw_s) \mid u \in L, \bar{v} \in R, \bar{w} \in S\} \in \mathsf{PR}$;
- if $R \in \mathsf{PR}$ of arity $m > 1$ and $\{i_1, \ldots, i_m\} = \{1, \ldots, m\}$, then $R^{(\bar{i})} = \{(u_{i_1}, \ldots, u_{i_m}) \mid (u_1, \ldots, u_m) \in R\} \in \mathsf{PR}$;
- if $R, S \in \mathsf{PR}$ are of the same arity, then $R \cup S \in \mathsf{PR}$.

Example 1.2.3 Consider the lexicographic ordering $<_{\mathrm{lex}}$ on an ordered alphabet Σ. It is prefix-recognisable being the union of

$$\Sigma^* \cdot (\{\varepsilon\} \times \Sigma^+) \quad \text{and} \quad \Sigma^* \cdot (a\Sigma^* \times b\Sigma^*) \quad \text{for all } a < b \in \Sigma.$$

Following [22] we say that a structure $\mathfrak{A} = (A, \{R_i\}_i)$ is *prefix-recognizable* if A is a regular set of words over some finite alphabet Σ and each of the relations R_i is in $\mathsf{PR}(\Sigma)$. Prefix-recognisable structures can be characterized in terms of interpretations. On the basis of tree automata, it is relatively straightforward to show that the prefix-recognisable structures coincide with the structures that are MSO-interpretable in the binary tree $\mathfrak{T}_2$ [97, 42, 22]. This result has been strengthened by Colcombet [51] to first-order interpretability in the expanded structure $(\mathfrak{T}_2, \prec)$ (note that the prefix relation $\prec$ is MSO-definable but not FO definable in $\mathfrak{T}_2$). Colcombet proved that MSO-interpretations and FO-interpretations in $(\mathfrak{T}_2, \prec)$ have the same power, which gives a new characterisation of prefix-recognisable structures. We summarize these results as follows.

Theorem 1.2.4   *For every structure $\mathfrak{A}$, the following are equivalent.*

(1) $\mathfrak{A}$ is isomorphic to a prefix-recognisable structure;
(2) $\mathfrak{A}$ is MSO-interpretable in the full binary tree $\mathfrak{T}_2$;
(3) $\mathfrak{A}$ is FO-interpretable in $(\mathfrak{T}_2, \prec)$.

In particular, every prefix-recognisable structure has a decidable MSO-theory.

Below we discuss further characterisations of prefix-recognisable structures in terms of vertex-replacement grammars, or as least solutions of VR-equational systems.

1.2.2   Graph grammars and graph algebras

In this section we consider vertex- and edge-labelled graphs. In formal language theory grammars generate sets of finite words. Similarly, context-free graph grammars produce sets of finite graphs – start from an initial nonterminal and rewrite nonterminal vertices and edges according to the derivation rules. Just as for languages, the set of valid derivation trees, or parse trees, forms a regular set of trees labelled by derivation rules of the graph grammar. Conversely, consider a collection Θ of graph operations – such as disjoint union, recolourings, etc. – as primitives. Every closed Θ-term t evaluates to a finite graph $[\![t]\!]$, and similarly every Θ-term $t(\overline{x})$ evaluates to a finite graph $[\![t(\overline{x})]\!]$ with non-terminal (hyper)-edges and/or vertices. Formally, evaluation is the unique homomorphism from the initial algebra of Θ-terms to the Θ-algebra of finite graphs with non-terminals. Each regular tree language L of closed terms thus represents a family of finite graphs $\{[\![t]\!] \mid t \in L\}$. For a concise treatment of graph grammars and finite graphs we refer to the surveys [69, 59] and the book [53].

Our focus here is on individual countable graphs generated by *deterministic* grammars via 'complete rewriting'. A suitable framework for formalising complete rewriting, in the context of term rewriting, is convergence in complete partial orders (cpo's). Since no classical order- or metric-theoretic notion of limit seems to exist for graphs, we use the more general categorical notion of colimit [11]. We outline this framework in which an infinite term (over the graph operations Θ) yields a countable graph; details may be found in [55, 11, 53].

In the category $\mathbb{G}$ of graphs and their homomorphisms every diagram of the form

$$G_0 \xrightarrow{f_0} G_1 \xrightarrow{f_1} G_2 \xrightarrow{f_2} \cdots \xrightarrow{f_{n-1}} G_n \xrightarrow{f_n} G_{n+1} \xrightarrow{f_{n+1}} \cdots$$

has a colimit G, i.e. a kind of least common extension G of the G_ns with homomorphisms $g_n : G_n \to G$ such that $g_n = g_{n+1} f_n$ for all n.[6] We assume that the graph operations in Θ determine endofunctors of $\mathbb{G}$ that are cocontinuous i.e. colimit preserving.

On the other side, take the cpo of finite and infinite terms over the signature $\Theta \cup \{\bot\}$, with the empty term $\bot$ and the extension ordering $s \sqsubseteq t$. We may turn it into a category $\mathbb{T}_\Theta$ with each relation $s \sqsubseteq t$ inducing a unique arrow $s \to t$. Moreover, in this category, colimits (of diagrams as above) exist and an infinite term t is the colimit of approximations $t_0 \to t_1 \to \cdots$ (think that t_i is the restriction of t to the first i levels). The evaluation mapping $[\![\cdot]\!]$ has a unique cocontinuous extension, also denoted $[\![\cdot]\!]$, mapping infinite terms to colimits of graphs.

This completes the basic description. Now consider a grammar $\mathcal{G}$ whose derivation rules $\langle X_i \mapsto t_i(\overline{X}) \rangle$ can be expressed by Θ-terms. These terms determine cocontinuous endofunctors in the category of terms $\mathbb{T}_\Theta$. By the Knaster-Tarski theorem the functors have a least fixed-point $\overline{G}$, which by Kleene's Theorem is attained as the colimit of the chain $\langle \gamma^n(\emptyset) \rangle_n$ with the natural homomorphisms. The graph *generated* by the grammar from the corresponding non-terminal X_i is defined to be the component G_i of the colimit $\overline{G}$.

Equivalently, given the system of equations $\mathcal{E}_\mathcal{G} = \langle X_i = t_i(\overline{X}) \rangle$ one can construct a syntactic (uninterpreted) solution of $\mathcal{E}_\mathcal{G}$ by 'unraveling' these equations from the initial non-terminal X_0 of the grammar. This results in a possibly infinite regular term $t_\mathcal{G}$, which is precisely the least fixed-point solution for X_0 in $\mathbb{T}_\Theta$. By cocontinuity of the evaluation mapping $[\![t_\mathcal{G}]\!]$ is isomorphic to the least fixed-point solution of $\mathcal{E}_\mathcal{G}$ in $\mathbb{G}$, that is to the graph generated by $\mathcal{G}$.

In what follows we focus on different sets of graph operations Θ (namely, HR, VR and some extensions). It has been observed that for suitable choices of operations, most notably avoiding products, the evaluation mapping can be realised as a monadic second-order interpretation or transduction [11, 60]. Consequently every interpretation $[\![t]\!] \leq^{\mathcal{I}}_{\mathrm{MSO}} t$ naturally translates to an internal

---

[6] There are examples of ascending chains $G_0 \xrightarrow{f_0} G_1 \xrightarrow{f_1} \cdots$ and $G_0 \xrightarrow{g_0} G_1 \xrightarrow{g_1} \cdots$ with identical graphs but different embeddings yielding different colimits, whence there is no apparent canonical way of defining a limit knowing only that each G_n is embeddable into G_{n+1}.

presentation of $[\![t]\!]$ using tree automata. Moreover, for a regular term t the MSO-theory of $[\![t]\!]$ is decidable by Rabin's Theorem.

Finally we mention that all this smoothly extends to solutions of infinite sets of equations [33]. Although unraveling might not result in a regular solution term, as long as it has a decidable MSO-theory so does the solution graph.

Equational graphs and hyperedge-replacement grammars

Hyperedge-replacement (HR) grammars are a very natural generalisation of context-free grammars from formal language theory. Every HR-grammar defines a 'language' of finite graphs just as context-free grammars define languages of finite words. The class of graph languages defined by HR-grammars possesses many structural properties akin to those well-known for context-free languages. The interested reader is referred to the monograph [80].

An HR-grammar is given as a finite collection of rules that allow the replacement of any hyperedge of a hypergraph bearing a non-terminal label by the right hand side of a matching rule, which is a given finite hypergraph with a number of distinguished vertices equal to the arity of the hyperedge to be replaced. A copy of the right-hand side of a matching rule is then glued to the original hypergraph precisely at these distinguished vertices and corresponding end vertices of the hyperedge being replaced. Derivation begins with a distinguished non-terminal.

As outlined at the start of section 1.2.2, each *deterministic* HR-grammar determines a unique countable graph constructed from the initial graph by complete rewriting in the course of which every non-terminal hyperedge is eventually replaced by the right-hand side of the unique matching rule. A countable graph is HR-*equational*, or simply *equational*, if it is generated by a deterministic HR grammar [55]. The class of equational graphs will be denoted by HR. Equational graphs constitute a proper extension of the class of context-free graphs [41].

Proposition 1.2.5 *A connected graph is context-free if, and only if, it is equational and of finite degree.*

Example 1.2.6 To generate the context-free graph of Example 1.2.1 with a deterministic HR grammar we take as our initial graph the 1-neighbourhood of the root node (labelled with 1 above) and attach to it non-terminal hyperedges labelled with X and with Y, respectively, whose vertices enumerate the boundaries of either ends. Similarly, the 1-neighbourhood of the boundary of each end, that is the vertices of the corresponding non-terminal hyperedge, constitutes the right-hand side of the matching rule. Again, non-terminal hyperedges are attached to mark the new boundary. The initial graph and the rule for the non-terminal X obtained this way are pictured below.

Notice how the linearity of the generated graph is reflected in the linearity of the replacement rules each having only a single non-terminal hyperedge on the right. In the next example a non-linear rule is used to generate a tree, which is not context-free.

Example 1.2.7 The complete bipartite graph $K_{1,\omega}$ and the full ω-branching tree $\mathfrak{T}_\omega$ (in the signature of graphs) are not context-free, but can be generated by the following rules from the initial graph $\bullet \overset{X}{\dashrightarrow} \bullet$.

The HR-algebra of finite and countable graphs corresponding to hyperedge-replacement grammars is a many-sorted algebra defined as follows. For each n there is a separate sort $\mathbb{G}_n$ of graphs with n *sources*. These are distinguished vertices, though not necessarily distinct, named $v_1, \ldots, v_n$. There are constants of each sort $\mathbb{G}_n$: these are hypergraphs having at most one hyperedge, exactly n vertices, each vertex a distinct source. The HR-algebra is built on the following operations: *disjoint union* $\oplus$, *renaming of sources* $\mathrm{rename}_{c \mapsto c'}$, and *fusion of sources* $\mathrm{fuse}_\approx$ according to an equivalence $\approx$ on source names. By convention $\oplus$ is understood to automatically shift the source names of its second argument by the maximum of the source names used in the first to avoid naming conflict. Also fuse assigns the least source name of a class to each fused node while dropping the others.

It is intuitively clear how a hyperedge-replacement step can be expressed using disjoint union with the right-hand side of the rule followed by a fusion and renaming of sources. Formally, one transforms an HR-grammar $\mathcal{G}$ into a system of finitely many equations $X_i = t_i(\overline{X})$ where variables play the role of non-terminals of the grammar and the terms t_i are chosen such that, when variables are interpreted as individual hyperedges, $[\![t_i(\overline{X})]\!]$ is the right hand side of the matching rule for a hyperedge labelled X_i.

Example 1.2.8   The equation corresponding to the single rule of the HR grammar of Example 1.2.7 generating $\mathfrak{T}_\omega$ is

$$X = \mathrm{rename}_{0\mapsto 0, 1\mapsto 1}(\; \mathrm{fuse}_{\{0,2\},\{1,4\}}(\; \overset{0}{\bullet} \to \overset{1}{\bullet} \oplus X \oplus X\;)).$$

Note that the source names of the first and second occurrences of X are shifted by 2 and by 4, respectively, while forming their disjoint union. Thus, after fusion we obtain precisely the right hand side of the HR-rule generating $\mathfrak{T}_\omega$, however, with additional source names. The renaming operation in this term has the effect of forgetting the source names 2 and above. So the least solution of this equation is indeed $\mathfrak{T}_\omega$ with its root labelled 0 and one of its children with 1.

The generating power of HR-grammars is limited by the fact that edges can only be 'created' via fusion of sources (after having taken the disjoint union of two graphs). Because there are only a fixed number of source names available in a finite HR-equational system there is a bound on the size of complete bipartite subgraphs $K_{n,n}$ that can be created [12], cf. Theorem 1.2.12. The infinite bipartite graph $K_{\omega,\omega}$ is thus an example of a prefix-recognisable graph which is not HR-equational.

It is a key observation that in case of HR-terms the evaluation mapping $t \mapsto [\![t]\!]$ is expressible as an MSO-interpretation. In fact, since edges cannot be created by any of the HR operations, the vertex-edge-adjacency graph of $[\![t]\!]$ is MSO-interpretable in the tree representation of t, whether t is finite or infinite.

Theorem 1.2.9   *For a countable graph G the following are equivalent.*

(1) G is generated by a deterministic HR grammar;
(2) G is HR-equational, i.e. the evaluation of a regular HR-term, i.e. the least solution of a finite system of HR-equations;
(3) The two-sorted incidence graph $\hat{G}$ of G is monadic second-order interpretable in the full binary tree, i.e. $\hat{G} \leq_{\mathrm{MSO}} \mathfrak{T}_2$.

For a detailed presentation of these and other algebraic frameworks and their connections to the generative approach based on graph grammars we advise consulting [55, 12, 21]. In [54] Courcelle considered an extension of monadic second-order logic, denoted CMSO_2, in which one can quantify over sets of edges as well as over sets of vertices and, additionally, make use of modulo counting quantifiers. Notice that the last item of the previous theorem implies that the CMSO_2-theory of equational graphs is interpretable in S2S and is thus decidable. Further, Courcelle proved that $CMSO_2$ is able to axiomatise each and every equational graph up to isomorphism.

Theorem 1.2.10 *Each* HR-*equational graph is axiomatisable in* $CMSO_2$. *Consequently the isomorphism problem of equational graphs is decidable.*

Sénizergues considered HR-equational graphs of finite out-degree and proved that they are, up to isomorphism, identical with the ε-closures of configuration graphs of normalised[7] pushdown automata restricted to the set of reachable configurations. Further, he proved that bisimulation equivalence of HR-equational graphs of finite out-degree is decidable [128]. This last result is an improvement on the decidability of bisimulation equivalence for deterministic context-free processes, which is a consequence of the celebrated result of Sénizergues establishing decidability of the DPDA language equivalence problem.

Vertex-replacement grammars

Vertex replacement systems are a finite collection of graph rewriting rules that allow one to substitute given finite graphs in place of single vertices while keeping all the connections. This form of graph rewriting emerged as the most robust and manageable from among a host of different notions within a very general framework [55, 69, 59, 58]. The corresponding VR-algebra of graphs is built on the following operations: constant graphs of a single c-coloured vertex $\overset{a}{\bullet}$, *disjoint union* $\oplus$, *recolouring of vertices* $\mathrm{recol}_{c\mapsto c'}$ and *introduction of a-coloured edges* $\mathrm{edge}_{c\overset{a}{\to}d}$ from every c-coloured vertex to every d-coloured vertex.

The evaluation of VR-terms, whether finite or infinite, is realisable as a monadic second-order interpretation. More precisely, as VR-equational graphs are interpretations of regular terms obtained by unfolding a finite system of VR equations, they can be MSO-interpreted in a regular tree, hence also in the full binary tree $\mathfrak{T}_2$, and thus are prefix-recognisable. These and other characterisations, together with our previous discussion of prefix-recognisable structures are summarised in the next theorem.

Theorem 1.2.11 *For a countable graph G the following are equivalent.*

(1) G is isomorphic to a prefix-recognisable structure;

(2) G is generated by a deterministic VR *grammar;*

---

[7] Here a PDA is said to be normalised, if in addition to being in a familiar normal-form its ε-transitions may not push anything on the stack. Hence the finiteness bound on the out-degree of configurations. For precise definitions see [128].

(3) *G is* VR-*equational, i.e. the evaluation of a regular* VR-*term, i.e. the least solution of a finite system of equations of the form* $X_i = t_i(\overline{X})$ *with finite* VR-*terms* $t_i(\overline{X})$;

(4) $G \leq_{\mathrm{MSO}} \mathfrak{T}_2$;

(5) $G = h^{-1}(\mathfrak{T}_2)|_C$, *i.e. the vertices of G are obtained by restricting the nodes of* $\mathfrak{T}_2$ *to a regular set C, and its edges are obtained by taking the inverse of a rational substitution h to* $\mathfrak{T}_2$;

(6) *G is isomorphic to the* ∈-*closure of the configuration graph of a pushdown automaton.*

Further, the HR-equational graphs can be characterised as the class of VR-equational graphs of finite tree width [11].

Theorem 1.2.12 VR-*equational graphs of finite tree width are* HR-*equational.*

Example 1.2.13 The complete bipartite graph $K_{\omega,\omega}$ is a prominent example of a VR-equational graph that is not HR-equational. A VR grammar and the corresponding system of VR equations generating $K_{\omega,\omega}$ are given below.

$$X = \mathrm{edge}_{a \leftrightarrow b}(A \oplus \mathrm{recol}_{a \mapsto b}(A))$$

$$A = \overset{a}{\bullet} \oplus A$$

The expressive power of this formalism (for describing families of finite graphs) is not increased by extending the VR operations by graph transformations that are definable using quantifier-free formulas (of which $\mathrm{recol}_{c \mapsto c'}$ and $\mathrm{edge}_{c \overset{a}{\to} d}$ are particular examples), nor by the *fusion* operations fuse_c identifying all nodes bearing a certain colour c [60]. Care has to be taken when defining countable graphs as evaluations of infinite terms, for it is unclear how to deal with infinite terms built with non-monotonic operations. Nonetheless, infinite terms built with operations definable by *positive* quantifier-free formulas can be evaluated unambiguously [11].

In this setting Theorem 1.2.11 can be generalised to infinite systems of equations (whose unfoldings are typically non-regular terms) using infinite deterministic automata [33], leading us to the following families of transition graphs.

1.2.3 Higher-order data structures

Tree-constructible graphs and Caucal's pushdown hierarchy

Courcelle introduced MSO-*compatible transductions* in the investigation of structures with decidable monadic theories. Let C and C' be classes of structures on signatures σ and σ', respectively. Following [57] we say that a functional transduction $T : C \to C'$ is MSO-compatible if there is an algorithm mapping each monadic formula φ of signature σ' to a monadic formula φ^T in the signature σ such that

$$\mathfrak{A} \models \varphi^T \quad \Longleftrightarrow \quad T(\mathfrak{A}) \models \varphi .$$

MSO-interpretations are the most natural examples of MSO-compatible transductions. Slightly more generally, the MSO-*definable transductions* of Courcelle are MSO-compatible. Recall that these are given by a k-copying operation (for some k) followed by an MSO-interpretation and in particular the resulting structure may have k times the cardinality of the original one.

The more difficult result that the *unfolding* operation, mapping graphs $(\mathfrak{G}, v)$ to trees $\mathfrak{T}_{(\mathfrak{G},v)}$, is also MSO-compatible appeared in [61] (see also [57] for an exposition and a treatment of the simpler case of deterministic graphs). We note that this result also follows from Muchnik's Theorem [126, 138, 17] and that it generalises Rabin's theorem.

A rich class of graphs, each with decidable monadic theory, can now be constructed. Caucal [43] proposed the hierarchies of graphs and trees obtained by alternately applying unfoldings and MSO-interpretations starting with finite graphs:

Definition 1.2.14

$\mathsf{Graphs}_0 = \{\text{finite edge- and vertex-labelled graphs}\}$
$\mathsf{Trees}_{n+1} = \{\mathfrak{T}_{\mathfrak{G},v} \mid (\mathfrak{G}, v) \in \mathsf{Graphs}_n\}$
$\mathsf{Graphs}_{n+1} = \{\mathcal{I}(\mathfrak{T}) \mid \mathfrak{T} \in \mathsf{Trees}_{n+1}, \mathcal{I} \text{ is an MSO interpretation}\}$

By the results above, we have

Theorem 1.2.15 *For every $n \in \mathbb{N}$ every graph G from* Graphs_n *has a decidable* MSO-*theory.*

Fratani [72, 73] provided an alternative proof of the above theorem, among a host of other results on higher-order pushdown graphs, using a different kind of MSO-compatible operation. Indeed, she established that if a homomorphism of words maps the branches of a tree T to those of T' surjectively while also preserving the node-labeling then definability and decidability results for MSO over T' can be transferred to T.

The Caucal hierarchy is very robust. Various weakenings and strengthenings of the definition yield exactly the same classes [37]. In fact, in place of MSO-interpretations, Caucal originally used inverse rational mappings in the style of item (5) of Theorem 1.2.11. Recently Colcombet [51] proved that every graph of Graphs_{n+1} can in fact be obtained via a first-order interpretation in some tree belonging to Trees_{n+1}. The next theorem provides internal presentations of graphs of each level as a generalisation of Theorem 1.2.11 item (6) thereby justifying the name pushdown hierarchy.

Theorem 1.2.16 ([37])   *For every n a graph G is in Graphs_n if, and only if, it is isomorphic to the ϵ-closure of the configuration graph of a higher-order pushdown automaton at level n.*

The strictness of the hierarchy was also shown in [37]. The level-zero graphs are the finite graphs, trees at level one are the regular trees, and as we have seen in Theorem 1.2.11 the level-one graphs are the prefix-recognisable ones. The deterministic level-two trees are known as algebraic trees. From the second level onwards we have no clear structural understanding of the kind of graphs that inhabit the individual levels. We recommend [134] for an exposition.

Term-trees defined by recursion schemes
Caucal also gave a kind of algebraic characterisation of term-trees at level n as fixed points of *safe* higher-order recursion schemes.

Theorem 1.2.17 ([43])   *For every n, the class of term-trees Trees_n coincides with that of term-trees generated by safe higher-order recursion schemes of level at most n.*

The notion of higher-order schemes is a classical one [62, 56]. Safety is a technical restriction (implicit in [62]) ensuring that no renaming of variables (α-conversion) is needed during the generative substitutive reduction (β-reduction) process constructing the solution-term [1, 117]. Safe schemes are intimately related to the pushdown hierarchy. This connection is well explained in [1] showing that while on the one hand order-n schemes can define the behaviour and hence (the unfolding of) the configuration graphs of level-n deterministic pushdown automata, on the other hand, deterministic pushdown automata of level n can evaluate safe order-n schemes. Safety is hereto essential.

In order to evaluate arbitrary schemes [81] introduced *higher-order collapsible pushdown automata* (CPDA), a kind of generalisation of panic automata [92], and gave in essence the following characterisation in the spirit of Theorem 1.2.16.

Theorem 1.2.18   *The term-trees defined by order-n recursion schemes are up to isomorphism identical with the unfoldings of ϵ-closures of configuration graphs of level-n collapsible higher-order pushdown automata.*

As shown in [117, 81], it is not necessary to assume safety for establishing decidability of the MSO-theories of term-trees that are solutions of higher-order schemes.

Theorem 1.2.19 *The* MSO-*theory of a term-tree defined by an arbitrary higher-order recursion scheme is decidable.*

Consequently, configuration graphs of higher-order collapsible pushdown automata can be model-checked against modal μ-calculus formulas. However, there is a second-order CPDA whose configuration graph interprets the infinite grid and whose MSO-theory is thus undecidable [81]. This shows that higher-order CPDA configuration graphs constitute a proper extension of Caucal's pushdown hierarchy.

1.2.4 Introducing products

There is a connection between the internal presentations of graphs seen so far and the graph operations used in the corresponding equational framework. Pushdown stacks are naturally represented as strings. The set of strings over some alphabet can in turn be modelled as an algebra of terms built with unary functions, one for each letter of the alphabet. Strings thus correspond to terms and letters to unary functions. In functional programming terminology the abstract data type of, say, binary strings has the recursive type definition

$$T = \perp \oplus 0(T) \oplus 1(T) \tag{1.2}$$

Here the letters 0 and 1 are seen as type constructors and the empty string $\perp$ is a constant type constructor. The set of finite strings is the least fixed-point solution of this equation.

Automata operating on terms of type T can be viewed as functions mapping terms to states. Moreover these functions are defined according to structural recursion. Analogously, recursion schemes (fix-point equations) in an algebra of graph operations transform automata-based internal presentations of a graph into equational specifications. We can use the recursion scheme associated to the type definition (1.2) to define any PR-graph by a VR equation extending the type definition. For instance, the graph of the lexicographic order from Example 1.2.3 satisfies the following equation

$$L = \text{edge}_{0 \to 1, \varepsilon \to 0, \varepsilon \to 1}(\bullet^\varepsilon \oplus \text{recol}_{0,1,\varepsilon \mapsto 0}(L) \oplus \text{recol}_{0,1,\varepsilon \mapsto 1}(L)).$$

We briefly explain how to go from automata presenting a PR-graph to a VR-equation. For a language $V \subset \{0, 1\}^*$ recognised by an automaton with transition table $\Delta \subset Q \times \Sigma \times Q$ and final states F the following VR-equation colours each word $w \in \{0, 1\}^*$ by those states q such that the automaton starting from q accepts w. (N.B. in accordance with (1.2) the simulation proceeds right-to-left.)

$$X = \bullet^F \oplus \texttt{recol}_{\{q' \mapsto q : \Delta(q,0,q')\}}(X) \oplus \texttt{recol}_{\{q' \mapsto q : \Delta(q,1,q')\}}(X)$$

In general, every PR-graph $\bigcup_i U_i \cdot (V_i \times W_i)$ is the recolouring of a graph satisfying a VR-equation of the form

$$X = \vartheta(\vartheta_\varepsilon(\bullet) \oplus \vartheta_0(X) \oplus \vartheta_1(X)) . \tag{1.3}$$

Here, the states of the automata recognising V_i or W_i are encoded as vertex colours (just as above) and ϑ_ε colours $\bullet$ by the final states of the V_i's and W_i's. Edge colours are used to represent states of automata for each U_i. For every $v \in V_i$ and $w \in W_i$, and z accepted by the automaton for U_i from state q there is a q-coloured edge (zv, zw). To this end, ϑ_0 and ϑ_1 recolour the vertices and edges, and ϑ adds an edge between all $x \in V_i$ and $y \in W_i$ coloured by the final states of U_i.

In passing we mention that higher-order stacks can also be represented as strings: either as well-bracketed sequences of stack symbols, or as strings of stack operations yielding the particular stack configuration. The former comes at the cost of losing regularity of the domain and has no apparent algebraic counterpart. The latter gives rise to a unary algebra of higher-order stacks that is not, except for level 1 pushdown stacks, freely generated by the stack operations. Thus there is no unique term representing a general stack. The work of Fratani, Carayol and others [72, 73, 33, 32] has shown that both of these deficiencies can be turned into features.

We now turn to graphs internally presented by finite trees. A type definition for $\{0, 1\}$-labelled binary branching trees is

$$\mathcal{T} = \bot \oplus 0(\mathcal{T} \otimes \mathcal{T}) \oplus 1(\mathcal{T} \otimes \mathcal{T}) \tag{1.4}$$

where $\otimes$ denotes direct product. Later we will compare this with another type definition (1.6). Colcombet observed that this schema can be used to define graphs with internal presentations involving tree automata operating on finite trees. He proposed extensions of the VR-algebraic framework by the *asynchronous product* $\otimes_A$ [48] and by the *synchronous product* $\otimes_S$ [50, 49] which we shall denote here by VRA and VRS, respectively.

Definition 1.2.20 (Synchronous and asynchronous product) The products are defined for vertex and edge-coloured graphs $\mathcal{G}$ and $\mathcal{H}$ as follows. In the synchronous product there is a d-coloured edge from (g, h) to (g', h') if, and only if, both (g, g') and (h, h') are connected by a d-edge in $\mathcal{G}$ and $\mathcal{H}$, respectively. The edge relation E_d of the asynchronous product $\mathcal{G} \otimes_A \mathcal{H}$ is defined as the union of $\{((g, h), (g', h)) \mid E_d^{\mathcal{G}}(g, g'), h \in H\}$ and $\{((g, h), (g, h')) \mid E_d^{\mathcal{H}}(h, h'), g \in G\}$. The definition of vertex colours requires a little care. In both cases a vertex (g, h) of the product has colour $\delta(c, c')$ whenever g has colour c and h has colour c'. Here the function $\delta : C^2 \to C$ is a parameter of the product operation. However, it is really only relevant that δ acts as a pairing function on some sufficiently large subsets of the colours. For instance, Colcombet identifies C with $\{0, 1, \ldots, N - 1\}$ and defines δ as addition modulo N [48].

As before, VRA-equational and VRS-equational graphs are defined as least fixed-point solutions of a finite system of equations in the respective algebra. Both product operations are cocontinuous with respect to graph embeddings. Therefore the evaluation mapping of both VRA and VRS terms uniquely extends from finite terms to infinite terms. Hence, just as for HR- and VR-equational graphs, the solution of a system of VRA or VRS equations is the evaluation of the regular term obtained by unraveling the system of equations.

Example 1.2.21   The infinite two-dimensional grid $(\mathbb{N} \times \mathbb{N}, \mathrm{Up}, \mathrm{Right})$ is easily constructed as the asynchronous product of the VR-equational, even context-free, graphs $(\mathbb{N}, \mathrm{Up})$ and $(\mathbb{N}, \mathrm{Right})$:

$$
\begin{aligned}
G &= \otimes_A(N_u, N_r) \\
N_u &= \mathrm{edge}_{a\overset{\mathrm{Up}}{\to}b}\left(\bullet \oplus \mathrm{recol}_{a\mapsto b, b\mapsto c}(N_u)\right) \\
N_r &= \mathrm{edge}_{a\overset{\mathrm{Right}}{\to}b}\left(\bullet \oplus \mathrm{recol}_{a\mapsto b, b\mapsto c}(N_r)\right)
\end{aligned}
$$

The unfolding of this system of equations is, schematically, an infinite term consisting of two periodic branches joined at the root. Elements of the grid G, by definition of asynchronous product, are represented as pairs of nodes of this term-tree with one node on either branch, corresponding to the respective co-ordinates. The example of the grid, whose MSO theory is undecidable, shows that the evaluation mapping of VRA terms (also of VRS terms) can not be realised by an MSO-interpretation.

For any VRA or VRS-term t, vertices of $[\![t]\!]$ can be identified with maximal *subsets* of nodes of t belonging to sub-terms joined by a product operator. It is thus easily expressible in MSO whether a set X of nodes (finite or infinite[8]) is actually well-formed in this sense, i.e. whether it represents an element of $[\![t]\!]$.

---

[8] In least fixed-point semantics only finite sets are considered, whereas in greatest fixed-point semantics both finite and infinite sets can represent elements of the solution, provided that there is an infinite nesting of product operators in t.

VR with asynchronous product and ground term rewriting

Ground term rewrite systems (GTRSs) are a natural generalisation of prefix-rewriting to trees. They are term rewrite systems given by rewriting rules in which no variables occur. Tree automata are a special case of GTRSs (see [52]).

Example 1.2.22 The rewrite rule $a \to f(a)$ confined to terms of the form $d(f^n(a), f^m(a))$ is a GTRS whose configuration graph is isomorphic to the infinite square grid.

We have noted that prefix-recognisable graphs are identical to ε-closures of pushdown graphs. This correspondence is achieved by generalising the simple prefix-rewriting rules of pushdown systems of the form $v \to w$ where v and w are strings to replacement rules $V \to W$ for given regular languages V, W. The latter rule allows one to rewrite any prefix $v \in V$ of a given string by any word from W. Regular Ground Term Rewrite Systems (RGTRS) generalise GTRS in the exact same manner: simple ground rewrite rules $s \to t$ with ground terms s, t are replaced by 'rule schemes' $S \to T$ with regular sets of terms on both left and right-hand side.

Löding [99, 100] and Colcombet [48] studied transition graphs of GTRSs and RGTRSs from a model-checking point of view. In Löding's work vertices of the transition graph are those terms reachable from an initial term, whereas Colcombet considers all terms of a given type as vertices.

The VR-equations defining PR graphs (1.3) easily generalise to VRA-equations defining graphs of RGTRSs using the recursion scheme (1.4):

$$X = \vartheta(\vartheta_\varepsilon(\bullet) \oplus \vartheta_0(X \otimes_A X) \oplus \vartheta_1(X \otimes_A X)) \qquad (1.5)$$

For each rule $S_i \to T_i$ of the RGTRS we simulate (frontier to root) tree automata recognising S_i and T_i. Vertices of X represent terms, so we call these vertex-terms. A vertex-term is coloured by those states q occurring at the root of the term after being processed by the automata. The simulation is initialised as follows: ϑ_ε labels $\bullet$ by initial states, and ϑ adds edges between all vertex-terms coloured by accepting states of automata for S_i and T_i. Updates occur in ϑ_js according to the transition rules, similarly to (1.3). To this end assume that two vertex-terms v', v'' are coloured by states q' and q'' respectively. After taking the product the paired vertex-term $j(v', v'')$ is initialised with colour (q', q'') (cf. Def. 1.2.20). This pair is then recoloured to q by ϑ_j whenever (q, j, q', q'') is a transition.

Notice how naturally the asynchronous product captures closure of RGTR rewriting under contexts: if there was an edge between v and v' then there is an edge between $j(v, v'')$ and $j(v', v'')$, and, symmetrically, between $j(v'', v)$ and $j(v'', v')$. One obtains along these lines the following generalisations of Theorem 1.2.11 (cf. examples 1.2.22 and 1.2.21).

Theorem 1.2.23 (Colcombet [48])

(i) *A countable graph is* **VRA**-*equational if, and only if, it is (after removal of certain colours) isomorphic to an RGTRS graph*[9].

(ii) *Each* **VRA**-*equational graph is finite-subset interpretable in a regular term-tree, hence also in the full binary tree.*

Theorem 1.2.12 also extends to **VRA**-equational graphs [48, 100].

Theorem 1.2.24 **VRA**-*equational graphs of finite tree-width are HR-equational.*

An immediate consequence of Theorem 1.2.23 is that the FO-theory of every **VRA**-equational structure is decidable via interpretation in S2S. In fact, for any **VRA**-equational graph $G = (V, \{E_a\}_a)$ the subset interpretation, hence also first-order decidability, extends to G with additional reachability predicates $R_C = \{(v, w) \mid w$ can be reached from v using edges of colours from $C \}$ for arbitrary subsets C of edge colours [48].

Theorem 1.2.25 **VRA**-*equational graphs have a decidable first-order theory with reachability.*

This result cannot be improved much further. Examples of [139] show that 'regular reachability', i.e. the problem whether there exists a path in a given **VRA**-equational graph between two given nodes and such that the labeling of the path belongs to a given regular language over the set of colours, is undecidable. In [100] Löding identified a maximal fragment of CTL that is decidable on every **GTRS** graph (with vertices restricted to terms reachable from an initial one) that can express, besides reachability, recurring reachability.

VR with synchronous product and tree-automatic structures

We have remarked that in the subset interpretation of **VRA** terms the subsets are used in a special form. Indeed, in the evaluating interpretation they merely serve the purpose of outlining the shape of a finite term. General finite-subset interpretations are more powerful and are capable of expressing the evaluation of **VRS** terms. In fact, these two formalism are equally expressive.

---

[9] Here RGTRS graphs are taken in the sense of [48] as being restricted to the set of terms of a given type.

This is best explained by *tree-automatic presentations*. These are internal presentations of VRS-structures which will be formally introduced in the next section. For now it suffices to use the characterisation (Theorem 1.3.18) that tree-automatic graphs are those that are wMSO-interpretable in a regular tree (reflected in the equivalence of (1) and (2) below).

Theorem 1.2.26 (Colcombet [50])
For every countable graph G the following are equivalent

(1) G is isomorphic to a tree-automatic graph.
(2) G is interpretable in a regular tree (wlog. the full binary tree) via a finite-subset interpretation.
(3) G is the restriction of a VRS-equational vertex-labelled graph G' to its set of vertices of a given colour;

We have noted that the evaluation mapping of VRS-terms can be naturally defined as a finite subset interpretation – this justifies (3) → (2). Continuing our discussion of translations from automata-based internal presentations into equational specifications using graph products we illustrate the remaining translation (2) → (3) from finite-tree automatic to VRS-equational presentations on graphs as we did for PR and RGTRS. That is, we build the terms of the presentation from the bottom up while also simulating the automata constituting the tree-automatic presentation by VRS-operations.

Start with a graph (V, E) that is definable via finite-subset interpretation in the full binary tree. By the fundamental correspondence that wMSO-definable relations in a regular tree are exactly those that are recognised by tree automata operating on finite trees, we see that V may be taken to be a regular set of finite Σ-labelled binary trees, and E is recognised by an automaton $\mathcal{A}$ accepting pairs of such trees.

The tree automaton $\mathcal{A}$ has transition rules (here we read them from left-to-right, i.e in top-down fashion, but that is a matter of choice and the simulation will actually proceed from bottom up) of the form

$$r : (q, \langle a, b \rangle, q_0, q_1) \qquad \text{with } a, b \in \{0, 1, \square\}$$

where the symbol $\square$ is necessary for padding either component of a pair of trees so that they have the same shape. It indicates the fact that no node is defined in the current position, i.e. that the automaton finds itself below a leaf of the respective tree (while still reading the other). We may assume that the transition rules enforce a proper usage of the padding symbols.

We introduce edge relations E_q and E_r for each state q and each rule r of the automaton. The simulation of transitions of the synchronous automaton on *pairs of labelled trees* necessitates a more sophisticated recursion scheme associated to the following type definition of $\{0, 1\}$-labelled binary branching trees.

$$\mathcal{T} = \bot \oplus (\{0, 1\} \otimes \mathcal{T} \otimes \mathcal{T}) \tag{1.6}$$

There is a natural identification of terms of this type and of those of the more natural type definition (1.4). As far as unary predicates are concerned the current type definition does not provide any advantage. However, compared with (1.4) the current type definition has a more powerful associated recursion scheme allowing for defining non-trivial *binary* relations between terms with different root labels. This will allow us to specify tree-automatic graphs via VRS-equations of the following form analogous to (1.6)

$$X = \vartheta \left(\bullet^\bot \oplus (\vartheta_0 \otimes_S \vartheta_1(X) \otimes_S \vartheta_2(X)) \right) \tag{1.7}$$

Here too, as in (1.3) and in (1.5) the ϑ's are VR-expressions facilitating the simulation of the automaton. The expression ϑ_0 specifies the graph with vertex set $\{0, 1\}$ and having an r-labelled edge from a to b for each rule r such that $r = (\cdot, \langle a, b \rangle, \cdot, \cdot)$ and with VR operations (here equivalently expressed as positive quantifier-free definable operations) responsible for updating the edge relations to simulate the transitions of $\mathcal{A}$. This is done in two phases.

- First, in preparation, state-labelled edges are used to 'enable' compatible rule-labelled edges in either copy of the graph: for each rule $r = (\cdot, \langle \cdot, \cdot \rangle, q_1, q_2)$ and $i \in \{1, 2\}$ the expression ϑ_i adds an E_r-edge from x to y for every E_{q_i}-edge from x to y in the graph.
- Then, after the synchronous product of rule-labelled edges has been taken, edges labelled by rules are renamed to their resulting states: ϑ adds for each state q an E_q-edge from x to y for every E_r-edge from x to y such that $r = (q, \langle \cdot, \cdot \rangle, \cdot, \cdot)$. In addition, ϑ deals with the case when either x or y is the singleton tree $\bot$. For this we may assume that all necessary information is coded in vertex labels implemented as reflexive edges and maintained along with the rest of the edge labels as explained here.

Finally, to obtain the graph G' as required in item (3) of Theorem 1.2.26 we also use vertex colours to keep track of the states of the tree automaton recognising V. The generalisation of this construction to arbitrary relational structures is straightforward.

1.3 Automatic Structures

1.3.1 Fundamentals

This section concerns structures with internal presentations consisting of automata operating synchronously on their inputs. The starting point of this investigation is the robust nature of finite automata. In particular, synchronous automata are effectively closed under certain operations that can be viewed in logical terms, i.e. Boolean operations, projection, cylindrification and permutation of arguments. Thus a structure whose domain and atomic operations are computable by such automata has decidable first-order theory (Definition 1.3.2 and Theorem 1.3.4).

Example 1.3.1   (i)   The domain and relations of the following structure are regular.

$$S_\Sigma = (\Sigma^*, \{\mathtt{suc}_a\}_{a \in \Sigma}, \prec_{\mathtt{prefix}}, \mathtt{el})$$

where Σ^* is the set of finite words over alphabet Σ, the binary relation $\mathtt{suc}_a$ is the successor relation (x, xa) for $x \in \Sigma^*$, the binary relation $\prec_{\mathtt{prefix}}$ is the prefix relation and the binary relation $\mathtt{el}$ is the equal-length relation.

(ii)   The following structure can be coded (eg. in base k least significant digit first) so that the domain and atomic operations are regular.

$$\mathcal{N}_k = (\mathbb{N}, +, |_k)$$

where $+$ is the usual addition on natural numbers and $x \mid_k y$ holds precisely when x is a power of k and x divides y.

Actually the link between synchronous automata and logic goes both ways. It was first expressed in terms of weak monadic second-order logic: a set of tuples $(A_1, \ldots, A_n)$ of finite sets of natural numbers is weak monadic second-order definable in $(\mathbb{N}, S)$ if and only if the corresponding n-ary relation of characteristic strings (a subset of $(\{0, 1\}^*)^n$) is synchronous rational. This was proved by [27] and [68], and is implicit in [135].

A first-order characterisation was provided by [65]: a relation $R \subset (\Sigma^*)^n$ is synchronous rational if and only if R is first-order definable in S_Σ for $|\Sigma| \geq 2$. Similarly, the Büchi-Bruyère Theorem states that a relation $R \subset \mathbb{N}^n$ (coded in base $k \geq 2$ least significant digit first) is synchronous rational if and only if it is first-order definable in $\mathcal{N}_k$ (proofs of which can be found in [104] and [137]).

These results were generalised to full MSO on the line $(\mathbb{N}, S)$ and weak MSO and full MSO on the tree $(\{0, 1\}^*, \mathtt{suc}_0, \mathtt{suc}_1)$ and form the basis of the logical characterisation of automatic structures (Section 1.3.4). However, we start with the more common internal definition.

Recall that the four basic types of automata operate on finite or infinite words or trees. So, let $\square$ be one of word, ω-word, tree, ω-tree.

We consider a structure $\mathfrak{B} = (B, \{R_i\})$ comprising relations R_i over the domain $\text{dom}(\mathfrak{B}) = B$. Thus constants and operations are implicitly replaced by their graphs.

Definition 1.3.2 (Automatic presentation)   A $\square$-automatic presentation of $\mathfrak{B}$ consists of a tuple $\mathfrak{d} = (\mathcal{A}, \mathcal{A}_{\approx}, \{\mathcal{A}_i\})$ of finite synchronous $\square$-automata and a *naming function* $f : \mathcal{L}(\mathcal{A}) \to B$ such that

– Each $\mathcal{L}(\mathcal{A}_i)$ is a relation on the set $\mathcal{L}(\mathcal{A})$.
– $\mathcal{L}(\mathcal{A}_{\approx})$ is a congruence relation on the structure $(\mathcal{L}(\mathcal{A}), \{\mathcal{L}(\mathcal{A}_i)\}_i)$.
– The quotient structure is isomorphic to $\mathfrak{B}$ via f.

Moreover, the quotient structure is called an *automatic copy* of $\mathfrak{B}$. We say that the presentation is *injective* whenever f is, in which case $\mathcal{A}_{\approx}$ can be omitted.

Definition 1.3.3 (Automatic structure[10])   A structure $\mathfrak{B}$ is $\square$-*automatic* if it has an $\square$-automatic presentation. If $\mathfrak{B}$ is $\square$-automatic for some $\square$ then $\mathfrak{B}$ is simply called *automatic*. The classes of automatic structures are respectively denoted by S-AutStr, ωS-AutStr, T-AutStr and ωT-AutStr.

The following theorem motivates the study of automatic structures and so may be called the *Fundamental Theorem* of automatic structures/presentations.

Theorem 1.3.4 (Definability)   *There is an algorithm that given a $\square$-automatic presentation $(\mathfrak{d}, f)$ of a structure $\mathfrak{A}$ and a FO-formula $\varphi(\bar{x})$ in the signature of $\mathfrak{A}$ defining a k-ary relation R over $\mathfrak{A}$, effectively constructs a synchronous $\square$-automaton recognising $f^{-1}(R)$.*

Immediate corollaries are

 (i) *Decidability*: The FO-theory of every automatic structure is decidable.
(ii) *Interpretations*: The class of $\square$-automatic structures is closed under FO-interpretations.

We point out that the Fundamental Theorem implies that every relation first-order definable from $\square$-regular relations is itself $\square$-regular.

Remark 1.3.5   One may allow finitely many parameters $\varphi(\bar{a}, \bar{x})$ under the following conditions. For finite-word and finite-tree presentations any parameters

---

[10] Some authors write *automatically presentable*.

can be used. However, for ω-tree (and ω-word) presentations a parameter a can be used if $f^{-1}(a)$ contains a regular ω-tree (ultimately periodic ω-word).

Consequently □-automatic structures (on a given signature) are closed with respect to operations such as disjoint union, ordered sum and direct product – each a special case of generalised products treated in [20, 23]. However AutStr and ωS-AutStr are not closed under weak direct-power. For instance, $(\mathbb{N}, +)$ is in S-AutStr but its weak direct-power is isomorphic to $(\mathbb{N}, \times)$, which is not in S-AutStr (see [20]). On the other hand, it is straightforward to see that T-AutStr and ωT-AutStr are closed under weak direct-power.

1.3.2 Examples

Obviously every finite structure is automatic. Here are a some examples of structures with automatic presentations.

Example 1.3.6 (Ordinals)   (i) $(\omega, <) \in$ S-AutStr: The simplest automatic copy is the unary one: $(0^*, \{(0^k, 0^l) \mid k < l\})$.

(ii) Every ordinal below ω^ω is in S-AutStr: An automatic copy of ω^k is $((0^*1)^k, <_{\text{lex}})$ where $<_{\text{lex}}$ denotes the lexicographic order[11] which is clearly regular. In this presentation the naming function is

$$0^{n_{k-1}}1 \ldots 0^{n_0}1 \mapsto n_{k-1}\omega^{k-1} + \cdots + n_1\omega^1 + n_0.$$

(iii) Every ordinal below ω^{ω^ω} is in T-AutStr: recall that the ordinal ω^α has a representation as the set of functions $f : \alpha \to \omega$ with f equal to 0 in all but finitely many places. These functions are ordered as follows: $f < g$ if the largest β with $f(\beta) \neq g(\beta)$ has that $f(\beta) < g(\beta)$. Then for fixed k, a function $f : \omega^k \to \omega$ is coded by the tree T_f with domain a finite subset of $0^*1^*2^* \cdots k^*$ so that for every β, expressed in Cantor-normal-form as $\omega^{k-1}c_0 + \omega^{k-2}c_1 \cdots + \omega^0 c_{k-1}$, $0 \leq c_i < \omega$, we have $T_f(0^{c_0}1^{c_1} \cdots (k-1)^{c_{k-1}}k^{f(\beta)}) = 1$.

Example 1.3.7 (Orderings)   (i) $(\mathbb{Q}, <) \in$ S-AutStr: The countable linear order $(\{0, 1\}^*1, <_{\text{lex}})$ is dense without endpoints.

(ii) $(\mathbb{R}, <) \in \omega$S-AutStr.

---

[11] Given an ordering on the symbols of the alphabet a word u is lexicographically smaller than w if either u is a proper prefix of w or if in the first position where u and w differ there is a smaller symbol in u than in w.

Example 1.3.8 (Groups)   (i) Every finitely-generated group with an Abelian group of finite index is in S-AutStr. And these are the only finitely generated word-automatic groups [116].

(ii)   The direct sum of countably many copies of $\mathbb{Z}/m\mathbb{Z}$ is in S-AutStr.

(iii)   The subgroup $\mathbb{Z}[1/k]$ of rationals of the form $\{zk^{-i} \mid z \in \mathbb{Z}, i \in \mathbb{N}\}$ for fixed $k \in \mathbb{N}$ is in S-AutStr.

(iv)   The Prüfer p-group $\mathbb{Z}(p^{\infty}) = \mathbb{Z}[1/p]/\mathbb{Z}$ (prime p) is in S-AutStr [114].

(v)   Real addition $(\mathbb{R}, +)$ is in ωS-AutStr.

However, the additive group of the rationals $(\mathbb{Q}, +)$ is not automatic [136]. In fact, Tsankov shows that no torsion free Abelian group that is p-divisible for infinitely many primes p is automatic.

Example 1.3.9 (Arithmetics)   (i) $(\mathbb{N}, +)$ is in S-AutStr: For every natural $k > 1$, the base k least-significant-digit-first presentation of naturals (with or without leading zeros) constitutes a naming function of an automatic presentation. A finite automaton can perform the schoolbook addition method while keeping track of the carry in its state. Such a presentation is injective when leading zeros are suppressed.

(ii) $(\mathbb{N}, \cdot)$ is in T-AutStr: The presentation is based on the unique factorisation of every natural number n into prime powers $2^{n_2} 3^{n_3} \cdots p^{n_p}$. Each n_k is written, say in binary notation, on a single branch of a tree with domain 0^*1^*. Multiplication is reduced to the addition of corresponding exponents. This construction can naturally be generalised to give tree-automatic presentations of weak direct powers of word-automatic structures [20, 25].

Example 1.3.10 (Equivalence relations)   The following have finite-word automatic presentations.

(i)   There is one class of size n for every $n \in \mathbb{N}$.

(ii)   There are $d(n)$ classes of size $n \in \mathbb{N}$ where $d(n)$ is the number of divisors of n. (This is the direct product of the previous equivalence relation with itself).

Example 1.3.11 (Free algebras)   (i) The free algebra with n unary operations and at most ω many constants is in S-AutStr.

(ii)   The free monoid generated by a single constant is in S-AutStr. However, no non-unary free or even free-associative algebra on two or more constants is in S-AutStr.

(iii) The free algebra generated by countably many constants and any finite number of operations is in T-AutStr.[12] For instance suppose there is one binary operation F. The domain of the presentation consists of all $\{F, c, \bot\}$-labelled binary trees. The operation (representing F) takes trees S and T as input and returns the tree with domain the prefix-closure of $(\mathrm{dom}(S) \cup \mathrm{dom}(T))\{0, 1\}$ and taking the following values: the root position is labelled F; position $\alpha 0$ is labelled by the label of S at position α; position $\alpha 1$ by the label of T at position α (if either of these latter positions does not exist, the label is $\bot$). It is not known whether finitely generated (non-unary) term algebras are in T-AutStr.

Example 1.3.12 (Boolean Algebras)   The signature we work in consists of the symbols for boolean operations $\cap$, $\cup$, $\cdot^c$ and constants $\bot$, $\top$.

  (i) Every finite power of the algebra of finite and co-finite subsets of $\mathbb{N}$ is in S-AutStr.
 (ii) The countable atomless Boolean algebra is in T-AutStr: It is isomorphic to the algebra of sets consisting of the clopen sets in Cantor space. Each clopen set has a natural representation as a finite tree.
(iii) The algebra of all subsets of $\mathbb{N}$ is in ωS-AutStr.
 (iv) The algebra of all subsets of $\mathbb{N}$ factored by the congruence of having finite symmetric difference is in ωS-AutStr. It is unknown whether this structure can be injectively presented in ωS-AutStr.
  (v) The interval algebra of the real interval $[0, 1)$ is in ωT-AutStr.
 (vi) The algebra of all subsets of $\{0, 1\}^*$ with a distinguished set $\mathcal{F}$ consisting of those $X \subset \{0, 1\}^*$ such that for every path $\pi \in \{0, 1\}^\omega$ only finitely many prefixes of π are in X.

Example 1.3.13 (Graphs)   (i) The infinite upright grid is in S-AutStr: Here the structure is $(\mathbb{N} \times \mathbb{N}, \mathrm{Up}, \mathrm{Right})$ with the functions $\mathrm{Right} : (n, m) \mapsto (n + 1, m)$ and $\mathrm{Up} : (n, m) \mapsto (n, m + 1)$. It can be automatically presented on the domain $a^* b^*$ with relations

$$R = \binom{a}{a}^* \binom{b}{a} \binom{b}{b}^* \binom{\square}{b}$$

and U defined by a similar regular expression.
 (ii) The transition graphs of pushdown automata are in S-AutStr:[13] Given a pushdown automaton $\mathcal{A}$ with states Q, stack alphabet Γ, input alphabet

---

[12] Communicated by Damian Niwinski.
[13] For *visibly pushdown automata* the same representation of configurations also allows for the trace equivalence relation to be recognised by a finite automaton. In [10] this presentation was utilised to obtain a decidability result.

Σ and transition relation Δ we can construct an automatic presentation of the transition graph of its configurations as follows. We take $Q\Gamma^*$ to be the domain of the presentation in which $q\gamma$ represents the configuration of state q and stack $\gamma \in \Gamma^*$. For each $a \in \Sigma$ there is an a-transition from $q\gamma$ to $q'\gamma'$ if, and only if, $\gamma = z\alpha$, $\gamma' = w\alpha$ and $(q, z, q', w) \in \Delta$ for some $z \in \Gamma$ and $w \in \Gamma^*$. Since Δ is finite, this relation is obviously regular for each a. Notice that in these presentations the transition relations are not only regular but in fact defined by prefix-rewriting rules (cf. Section 1.2.1 on context-free graphs).

(iii) The transition graphs of Turing machines are in $\mathtt{S\text{-}AutStr}$ [87]. We can give an automatic presentation of each TM $\mathcal{M}$ similar to those of pushdown automata. Configurations are encoded as strings $\alpha q\beta \in \Gamma^* Q\Gamma^*$ where α and β are the tape contents to the left, respectively, to the right of the head of $\mathcal{M}$, and q is the current state. Observe that, as opposed to presentations of pushdown graphs, the state is now positioned not at the left of the string but at the location of the head. Consequently, rewriting is not confined to prefixes, but rather occurs around the state symbol: transitions are of the form $\alpha u q w \beta \mapsto \alpha u' q' w' \beta$ for adequate u, w, u', w' and q, q' as determined by the transition function of $\mathcal{M}$. The fact that TM graphs are presentable using *infix rewriting* has the profound consequence that reachability questions in infix-rewriting systems are generally undecidable, as opposed to graphs of *prefix-rewriting* systems, whose monadic second-order theory is decidable (cf. Theorem 1.2.4).

Example 1.3.14 (Automata-theoretic structures) The following structures turn out to be universal for their respective classes (see Theorem 1.3.17).

(i) Let

$$\mathcal{S}_\Sigma = (\Sigma^*, \{\mathrm{suc}_a\}_{a\in\Sigma}, \prec_{\mathrm{prefix}}, \mathtt{el})$$

and

$$\mathcal{S}_\Sigma^\omega = (\Sigma^{\leq\omega}, \{\mathrm{suc}_a\}_{a\in\Sigma}, \prec_{\mathrm{prefix}}, \mathtt{el})$$

be the structures defined on finite, respectively on finite and ω-words, comprising the successor relations $\mathrm{suc}_a = \{(w, wa) \mid w \in \Sigma^*\}$; the prefix relation $u \prec_{\mathrm{prefix}} w$ (where u is finite and w is finite or infinite); and the equal-length relation: $u \mathbin{\mathtt{el}} w$ if, and only if, $|u| = |w|$. Clearly $\mathcal{S}_\Sigma \in \mathtt{S\text{-}AutStr}$ and $\mathcal{S}_\Sigma^\omega \in \omega\mathtt{S\text{-}AutStr}$. Note that if Σ is unary, then $\mathcal{S}_\Sigma$ reduces to $(\mathbb{N}, +1, <, =)$.

(ii)   The structure $\mathcal{T}_\Sigma \in \texttt{T-AutStr}$ has domain consisting of all finite binary
       Σ-labelled trees and has operations

$$(\preceq_{\text{ext}}, \equiv_{\text{dom}}, (\texttt{suc}_a^d)_{d\in\{l,r\},a\in\Sigma}, (\epsilon_a)_{a\in\Sigma})$$

where $T \preceq_{\text{ext}} S$ if $\text{dom}(T) \subset \text{dom}(S)$ and $S(\alpha) = T(\alpha)$ for $\alpha \in \text{dom}(T)$;
$T \equiv_{\text{dom}} S$ if $\text{dom}(T) = \text{dom}(S)$; $\texttt{suc}_a^d(T) = S$ if S is formed from T by
extending its leaves in direction d and labeling each new such node by a;
and ϵ_a is the tree with a single node labelled a.

   Similarly the structure $\mathcal{T}_\Sigma^\omega \in \omega\texttt{T-AutStr}$ has domain consisting of
all finite and infinite trees and operations

$$(\preceq_{\text{ext}}, \equiv_{\text{dom}}, (\texttt{suc}_a^d)_{d\in\{l,r\},a\in\Sigma}, (\epsilon_a)_{a\in\Sigma}).$$

that are restricted to finite trees, except that $T \preceq_{\text{ext}} S$ is defined as above
but allows S to be an infinite tree.

1.3.3 Injectivity

Recall that an automatic presentation is injective if the naming function is
injective. The problem of injectivity is this:

Does every $\square$-automatic structure have an injective $\square$-automatic presentation?

   An injective presentation has the advantage that it is easier to express certain
cardinality-properties of sets of elements (Theorem 1.4.6). We consider the four
cases.

Finite words

From a finite-word automatic presentation of $\mathfrak{A}$ one defines an injective presentation of $\mathfrak{A}$ by restricting to a regular set D of unique representatives. These
can be chosen using a regular well-ordering of the set of all finite words. For
instance, define $D \subset \mathcal{L}(\mathcal{A})$ to be the length-lexicographically least words from
each $\mathcal{L}(\mathcal{A}_\approx)$ equivalence class.

Finite trees

Except in the finite word case, there is no regular well ordering of the set
of all finite trees [39]. However one can still convert a finite-tree automatic
presentation into an injective one [47]. The idea is to associate with each tree t
a new tree $\hat{t}$ of the following form: the domain is the intersection of the prefix-
closures of the domains of all trees that are $\mathcal{L}(\mathcal{A}_\approx)$-equivalent to t; a node is
labelled σ if t had label σ in that position; a leaf x is additionally labelled by

those states q from which the automaton $\mathcal{A}_\approx$ accepts the pair consisting of the subtree of t rooted at x and the tree with empty domain.[14] Using transitivity and symmetry of $\mathcal{L}(\mathcal{A}_\approx)$, if $\hat{t} = \hat{s}$ then t is $\mathcal{L}(\mathcal{A}_\approx)$-equivalent to s. Moreover each equivalence class is associated with finitely many new trees, and so a representative may be chosen using any fixed regular linear ordering of the set of all finite trees.

ω-words

There is a structure in ωS-AutStr that does not have an injective ω-word automatic presentation [82]. The proof actually shows that the structure has no injective presentation in which the domain and atomic relations are Borel.

However, every *countable* structure in ωS-AutStr does have an injective ω-word automatic presentation [85] (and consequently is also in S-AutStr). This follows from the more general result that every ω-word regular equivalence relation with countable index has a regular set of representatives [85].

ω-trees

It has not yet been settled whether injective presentations suffice, even for the countable structures.

1.3.4 Alternative characterisations

Automatic structures were defined internally. We now present equivalent characterisations: logical (FO and MSO) and equational.

First-order characterisations

In order to capture regularity in the binary representation of $\mathbb{N}$ using first-order logic Büchi suggested the expansion $(\mathbb{N}, +, \{2^n \mid n \in \mathbb{N}\})$ of Presburger arithmetic, which is, however, insufficient (see [26]). Boffa and Bruyère considered expressively complete expansions of $(\mathbb{N}, +)$ by relations of the form $x \mid_k y$ (defined to hold precisely when x is a power of k and x divides y).

Theorem 1.3.15 (Büchi-Bruyère, cf. [26]) *A relation $R \subseteq \mathbb{N}^r$ is regular in the least-significant-digit-first base k presentation of $\mathbb{N}$ if, and only if, R is first-order definable in the structure $\mathcal{N}_k = (\mathbb{N}, +, \mid_k)$.*

Closer to automata, the structures $\mathcal{S}_\Sigma$ on words (see example 1.3.14) allow one to define every regular relation on alphabet Σ.

---

[14] The construction given in [47] is slightly more general and allows one to effectively factor finite-subset interpretations in any tree.

Theorem 1.3.16 ([65])   *Let Σ be a finite, non-unary alphabet. A relation over Σ^* is regular if, and only if, it is first-order definable in $\mathcal{S}_\Sigma$.*

The proofs of these theorems are by now standard. From left to write one writes a formula $\phi_{\mathcal{A}}(x)$ that expresses the existence of a successful run in automaton $\mathcal{A}$ on input x. For the other direction the atomic operations of the structures are regular forms the base case for structural induction on the formula. Both theorems transfer to automatic structures by replacing definability with interpretability [24, 25].

Theorem 1.3.17 (First-order characterisation of S-AutStr)   *The following conditions are equivalent.*

– $\mathfrak{A} \in$ S-AutStr.
– $\mathfrak{A}$ *is first-order interpretable in $\mathcal{S}_\Sigma$ (for some/all Σ with $|\Sigma| \geq 2$).*
– $\mathfrak{A}$ *is first-order interpretable in $\mathcal{N}_k$ (for some/all $k \geq 2$).*

These structures have been called *universal* or *complete* (with respect to FO-interpretations) for the class of finite-word automatic structures. There are similar universal structures for the other classes of automatic structures. These are the structures $\mathcal{S}_\Sigma^\omega$, $\mathcal{T}_\Sigma$ and $\mathcal{T}_\Sigma^\omega$ from Example 1.3.14 [20, 14].

Finite set interpretations

The four notions of automatic presentation have straightforward reformulations in terms of subset interpretations either in the line $\Delta_1 = (\mathbb{N}, \mathtt{suc})$ or in the tree $\Delta_2 = (\{0, 1\}^*, \mathtt{suc}_0, \mathtt{suc}_1)$.

Theorem 1.3.18 (Automatic presentations as subset interpretations)   *There are effective transformations establishing the following equivalences.*

 (i) $\mathfrak{A} \in$ S-AutStr *if, and only if, $\mathfrak{A} \leq_{\mathrm{fset}} \Delta_1$*
 (ii) $\mathfrak{A} \in \omega$S-AutStr *if, and only if, $\mathfrak{A} \leq_{\mathrm{set}} \Delta_1$*
(iii) $\mathfrak{A} \in$ T-AutStr *if, and only if, $\mathfrak{A} \leq_{\mathrm{fset}} \Delta_2$*
(iv) $\mathfrak{A} \in \omega$T-AutStr *if, and only if, $\mathfrak{A} \leq_{\mathrm{set}} \Delta_2$*

Equivalently, one may formulate universality with respect to FO interpretations. Following [47] we define the *(finite) subset envelope* $\mathcal{P}_{(f)}(\mathfrak{A})$ of a structure $\mathfrak{A}$ by adjoining to $\mathfrak{A}$ its (finite) subsets as new elements ordered by set inclusion.

Definition 1.3.19   Given $\mathfrak{A} = (A, \{R_i\})$ write $P(A)$ for the set of all subsets of A. The *subset envelope* $\mathcal{P}(\mathfrak{A})$ is the structure with domain $P(A)$ and relations $R_i' := \{(\{a_1\}, \ldots, \{a_n\}) \mid (a_1, \ldots, a_n) \in R_i\}$ and the subset relation $\subseteq$ defined on $P(A)$. The *finite-subset envelope* $\mathcal{P}_f(\mathfrak{A})$ is the substructure of $\mathcal{P}(\mathfrak{A})$ whose domain is the set of finite subsets of A.

It is immediately clear that

$$\mathfrak{B} \leq_{(f)\text{set}} \mathfrak{A} \iff \mathfrak{B} \leq_{\text{FO}} \mathcal{P}_{(f)}(\mathfrak{A})$$

In particular, this yields natural universal structures, with respect to FO-interpretations, for each of the four classes of automatic structures.

Corollary 1.3.20 (i) $\mathcal{P}_f(\Delta_1)$ *is universal for* S-AutStr.
(ii) $\mathcal{P}(\Delta_1)$ *is universal for* ωS-AutStr.
(iii) $\mathcal{P}_f(\Delta_2)$ *is universal for* T-AutStr.
(iv) $\mathcal{P}(\Delta_2)$ *is universal for* ωT-AutStr.

VRS-Equational structures

Recall that the VRS-algebra of graphs extends the VR-algebra with the synchronous product operation and that VRS-equational systems define exactly the finite-tree automatic graphs (see Section 1.2.4 and Theorem 1.2.26).

A finite VRS-equational system whose unfolding is a linear VRS-term specifies a structure in S-AutStr. This happens if in the defining equations one of the arguments of each occurrence of $\oplus$ and of $\otimes_S$ is a finite graph (and so these act like unary operations). Conversely, for word-automatic presentations Equation (1.7) reduces to the following form:

$$X = \vartheta \left(\bullet^\perp \oplus (\vartheta_0 \otimes \vartheta_1(X)) \right) \tag{1.8}$$

This scheme matches the following type definition obtained by restricting (1.6) to words:

$$\mathcal{T} = \perp \oplus (\{0, 1\} \otimes \mathcal{T}) \tag{1.9}$$

This recursive definition of the set of words has the same advantage over (1.2) as (1.6) has over (1.4) when it comes to defining binary relations over words via structural induction, e.g. via finite automata. Over words we have the following special case of Theorem 1.2.26.

Theorem 1.3.21 (Colcombet [50])
For every countable structure $\mathfrak{A}$ the following are equivalent

(1) $\mathfrak{A}$ *is isomorphic to a word-automatic graph.*
(2) $\mathfrak{A}$ *is the restriction of some $\mathfrak{B}$ to its elements of a certain colour, where $\mathfrak{B}$ can be specified by a VR-equation $Z = \pi(X)$, where π simply forgets some of the structure of X, together with a VRS-equation for X of the form (1.8);*
(3) $\mathfrak{A}$ *is finite-subset interpretable in* $(\mathbb{N}, \text{suc})$.

The equivalence of the first and the third item is a direct consequence of the classical correspondence of automata on words and monadic second-order logic of one successor and was already stated in Theorem 1.3.18. Nonetheless, this can also be inferred from the fact that the solution term obtained by unfolding (1.8) is (essentially) a periodic linear VRS-term that evaluates, via a finite-subset interpretation, to the word-automatic structure specified by equation (1.8).

More generally, let VRS$^-$ denote the extension of VR with unary operations $X \mapsto G_0 \otimes_s X$ where G_0 is any finite graph. Moreover let us call a *chain interpretation* a subset interpretation in a tree where each of the subsets representing an element is linearly ordered by the ancestor relation of the tree. It is not hard to see that solutions of finite systems of VRS$^-$-equations are finite-chain interpretable in a regular tree and that these in turn are word automatic [50].

1.3.5 Rational graphs

If we allow the more general *asynchronous automata* in the definition of an automatic presentation of a graph we get the notion of a rational graph. Thus vertices are labelled with finite words of a rational language over some finite alphabet Σ, and the edge relations are required to be rational subsets of $\Sigma^* \times \Sigma^*$.

With no aim for completeness we list below some results on rational graphs (asynchronous) in comparison with automatic graphs (synchronous). For a comprehensive treatment the reader is referred to [105].

The class of rational graphs strictly includes that of finite-word automatic graphs. In their seminal paper [87] Khoussainov and Nerode also introduced asynchronous automatic structures. As an example they gave an asynchronous automatic presentation of ω^ω, which is not in S-AutStr (see Theorem 1.4.12). Asynchronous automatic presentations of Cayley-graphs of finitely generated groups have also been considered as generalisations of 'automatic groups' [31].

The price of increasing expressiveness is a loss of tractability: in general, rational graphs do not have a decidable first-order theory. This renders rational graphs useless for representing data, let alone programs. However, in the context of formal language theory rational graphs seem to fill a gap. Considering rational graphs as infinite automata, i.e. as acceptors of languages, Morvan and Stirling have shown that they trace exactly the context-sensitive languages [108, 107] (see also [34] for a simplified approach). Rispal and others [123, 107, 34] have subsequently observed that this holds true for automatic graphs as well.

Although first-order queries on rational graphs are in general intractable there are some interesting decidable subclasses.

Morvan observed that by a result of Eilenberg and Schützenberger, graphs defined by rational relations over a *commutative* monoid have a decidable first-order theory. In particular, over the unary alphabet the monoid structure is isomorphic to $(\mathbb{N}, +)$ whence the unary rational graphs are those first-order definable in $(\mathbb{N}, +)$ [105]. Similarly, rational graphs over $(\mathbb{N}, +)^d$ are those having a d-dimensional first-order interpretation in $(\mathbb{N}, +)$.

Carayol and Morvan showed that on rational graphs that also happen to be trees (this is an undecidable property) first-order logic is decidable [36, 106]. The decision method is based on locality of FO as formulated by Gaifman and uses a compositional technique. The authors also exhibit a rational graph that is a finitely branching tree but is not finite-word automatic.

1.3.6 Generalisations

Automata with oracles

Consider an expansion Δ_i^O of $\Delta_i := ([i]^*, \mathrm{suc}_0, \ldots, \mathrm{suc}_{i-1})$ by a unary predicate $O \subset [i]^*$. Every MSO formula (with free MSO variables) of the expanded structure corresponds to a tree automaton with *oracle* O. An automaton with oracle is one that, while in position $u \in [i]^*$, can decide on its next state using the additional information of whether or not $u \in O$. Thus for automata working on infinite words/trees the oracle O is simply read as part of the input. In the case of automata working on finite words/trees, the entire oracle is scanned, and so the acceptance condition should be taken appropriately (eg. Muller/Rabin).

Call a set O *decidable* if $\mathrm{MSO}(\Delta_i^O)$ is decidable, and *weakly decidable* if $\mathrm{wMSO}(\Delta_i^O)$ is decidable. Early work on decidable oracles used the contraction method to show that certain oracles on the line, such as $\{n! \mid n \in \mathbb{N}\}$, are decidable [67]. This was extended to the effectively profinitely ultimately periodic words [38], which it turns out capture all the decidable unary predicates on the line [119, 120]. Nonetheless, it is still of interest to produce explicit examples of decidable oracles, see for instance [38, 74, 75, 7].

Definition 1.3.22 If in the definition of automatic presentation (1.3.2) we replace □-automata with □-automata with oracle O, we get a notion of □-*automatic presentation with oracle* O. A structure is called *automatic with oracle* if it has a □-automatic presentation with some oracle.

Example 1.3.23 The group of rationals $(\mathbb{Q}, +)$ has recently been shown to have no word-automatic presentation [136]. However it is finite-word automatic with oracle #2#3#4 $\cdots$. This is based on the idea, independently found by Frank

Stephan and Joe Miller and reported in [114], that there is a presentation of $([0, 1) \cap \mathbb{Q}, +)$ by finite words in which $+$ is regular, but the domain is not: every rational in $[0, 1)$ can be expressed as $\sum_{i=2}^{n} \frac{a_i}{i!}$ for a unique sequence of natural numbers a_i satisfying $0 \le a_i < i$. The presentation codes this rational as $\#a_2\#a_3\#a_4 \cdots$ where a_i is written in decimal notation (and hence has length less than the length of i written in decimal notation). Addition is performed with the least significant digit first, based on the fact that

$$\frac{a_i + b_i + c}{i!} = \frac{1}{(i-1)!} + \frac{a_i + b_i + c - i}{i!}$$

where $c \in \{0, 1\}$ is the carry in.

We immediately have that a structure is (finite-)word/tree automatic with oracle O if and only if it is (finite) set interpretable in Δ_1^O / Δ_2^O. Hence we have the following generalisation of the Fundamental Theorem and its corollaries (1.3.4).

Theorem 1.3.24   (i) Definability: *Say $(\mathfrak{d}, f)$ is a $\square$-automatic presentation with oracle O of a structure $\mathfrak{A}$ and $\varphi(\overline{x})$ is a FO-formula in the signature of $\mathfrak{A}$ defining a k-ary relation R over $\mathfrak{A}$. Then the relation $f^{-1}(R)$ is recognised by an $\square$-automaton with oracle O.*

 (ii) Interpretations: *The class of $\square$-automatic structures with oracle O is closed under FO-interpretations.*

(iii) Decidability: *The previous statements can be made effective under the following conditions.*

 1 For $\square \in \{\text{word, tree}\}$ we require that $\text{wMSO}(\Delta_i^O)$ be decidable.

 2 For $\square \in \{\omega\text{-word}, \omega\text{-tree}\}$ we require that $\text{MSO}(\Delta_i^O)$ be decidable.

 In particular, under these conditions, every $\mathfrak{A}$ that is $\square$-automatic with oracle O has decidable FO-theory.

Of course Δ_i^O can be viewed as a coloured tree. As in Corollary 1.3.20 we have universal structures with respect to FO-definability. For instance $\mathcal{P}(\Delta_2^O)$ is universal for $\omega\text{T-AutStr}$ with oracle O. The following result concerns finite-set interpretations in arbitrary trees.

Theorem 1.3.25 ([47])   *To every finite set interpretation $\mathcal{I}$ one can effectively associate a wMSO interpretation $\mathcal{J}$ such that for every tree t and structure $\mathfrak{A}$ if $\mathcal{P}_f(\mathfrak{A}) \cong \mathcal{I}(t)$ then $\mathfrak{A} \cong \mathcal{J}(t)$.*

This can be used to show that certain structures, such as the random graph, are not finite-tree automatic in the presence of any oracle [47].

1.3.7 Subclasses

In this section we restrict the complexity of the regular domains in automatic presentations to yield some of the more robust subclasses of S-AutStr and T-AutStr.

Polynomial domain

The most natural restriction is to consider presentations where the words and trees take labels from a unary alphabet $|\Sigma| = 1$. Word-automatic presentations over a unary alphabet were introduced and studied by Blumensath [20] and Rubin [89, 124].

The *density* of a language $L \subset \Sigma^*$ is the function $n \mapsto |L \cap \Sigma^n|$.

Definition 1.3.26 A structure is *unary automatic* if it has an injective word-automatic presentation in which the domain consists of words from a unary alphabet. A structure is *p-automatic* if it has an injective word-automatic presentation in which the domain has polynomial density. Let 1-AutStr and P-AutStr denote these respective classes of structures.

Regular sets of polynomial density were characterised by Szilard et al. [131] as being a finite union of the form

$$D = \bigcup_{i < N} u_{i,1} v_{i,1}^* u_{i,2} \ldots u_{i,n_i} v_{i,n_i}^* u_{i,n_i+1} \tag{1.10}$$

where the degree of the polynomial of the density function is equal to the maximum of the n_i's. In [6] it was demonstrated that every finite-word-automatic presentation over a domain as in (1.10) can be transformed into an equivalent one (cf. Section 1.4.4) over a domain that is a regular subset of

$$a_1^* a_2^* \ldots a_n^*$$

where n is equal to the maximum of the n_i's. In particular, word-automatic presentations over a domain of linear density are unary automatic. This transformation yields a kind of normal-form of word-automatic presentations over a polynomially growing domain.

Theorem 1.3.27 ([6]) *A structure* $\mathfrak{A}$ *has an automatic presentation over a domain of density* $\mathcal{O}(n^d)$ *if, and only if, it has a d-dimensional interpretation in* $\mathfrak{M} := (\mathbb{N}, <, \{\equiv_{(\bmod m)}\}_{m>1})$ *if, and only if, it is finite-subset interpretable in* $\Delta_1 := (\mathbb{N}, \mathrm{suc})$ *with subsets of size at most d.*

Corollary 1.3.28 ([113],[20]) *A structure* $\mathfrak{A}$ *is unary automatic if, and only if, it is first-order definable in* $\mathfrak{M}$ *if, and only if, it is MSO-interpretable in* Δ_1.

Unary automatic structures form a very restricted subclass of VR-equational structures and have a decidable MSO-theory. Using pumping arguments one can show that Presburger arithmetic $(\mathbb{N}, +)$ has no p-automatic presentation [20, 121]. On the other hand, the infinite grid is p-automatic but not unary automatic. Thus we have

$$1\text{-AutStr} \subsetneq \text{P-AutStr} \subsetneq \text{S-AutStr}.$$

The expansion of $\mathfrak{M}$ with the successor function suc and a constant for 0 admits quantifier elimination. Hence, every p-automatic structure can be interpreted in $(\mathbb{N}, 0, \text{suc}, <, \{\equiv_{(\bmod m)}\}_{m>1})$ using quantifier-free formulas.

Every p-automatic structure inherits the PSPACE upper-bound on the complexity of its first-order theory from $\mathfrak{M}$. This is as low as possible since FO model-checking is PSPACE-hard for any structure with at least two elements. Adding even the simplest form of iteration to FO leads to undecidability. For every k-counter machine it is straightforward to construct a p-automatic presentation of its configuration graph where each configuration $(q, n_1, \ldots, n_k)$ is represented by the word $q c_1^{n_1} \cdots c_k^{n_k}$. It follows that the first-order theory with reachability FO[**R**] of a p-automatic structure is undecidable in general. In comparison, while unary automatic structures have a decidable MSO-theory, the FO(DTC) theory of $(\mathbb{N}, \text{succ})$ interprets full first-order arithmetic and is therefore highly undecidable [20].

Observe, that graphs having rational presentation over a finitely generated commutative monoid (cf. Section 1.3.5) can be seen as analogues of p-automatic graphs. Indeed, every monoid element is represented by some word $g_1^{r_1} g_2^{r_2} \ldots g_n^{r_n}$ over the generators.

Finite-rank tree-automatic presentations

The analogue of p-automatic to tree-automatic structures is restricting to presentations involving trees of bounded rank. Intuitively the rank of a tree corresponds to its branching degree (which can be measured in terms of the Cantor-Bendixson rank).

Recall a Σ-labelled n-ary tree T is a function from a prefix-closed subset of $[n]^*$ to Σ. We say that T has *rank* k if its domain has polynomial density of degree at most k.

A finite-tree automatic presentation is called of *rank* k if for some regular language D of polynomial density of degree at most k the domain of every tree in the presentation is a subset of D. Collectively we speak of *bounded-rank tree-automatic presentations*. The class of structures with rank k presentations is denoted k-T-AutStr.

Example 1.3.29 The ordinal ω^{ω^k} has a rank $k + 1$ tree-automatic presentation.

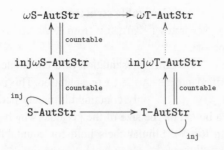

Figure 1.2 Relationship of classes of automatic structures

Let $\mathcal{T}_k$ denote the structure corresponding to the unlabelled k-ary tree with domain $0^*1^* \cdots (k-1)^*$. Note that $\mathcal{T}_k$ is wMSO-interpretable in the ordinal ω^k (in the signature of order), and vice-versa.

Proposition 1.3.30 *The following are equivalent.*

- $\mathfrak{A}$ *is in* k-T-AutStr,
- $\mathfrak{A}$ *is finite-set interpretable in* $\mathcal{T}_k$ *(or equivalently in the ordinal* ω^k*),*
- $\mathfrak{A}$ *is the solution of a finite system of VRS-equations whose unfolding is a term-tree of rank k.*

The hierarchy is strict:

$$\text{S-AutStr} = \text{1-T-AutStr} \subsetneq \text{2-T-AutStr} \subsetneq \cdots \subsetneq \text{T-AutStr}.$$

Indeed, if k+1-T-AutStr = k-T-AutStr for some k then the finite-subset envelope $\mathcal{P}_f(\omega^{k+1})$ would be finite-set interpretable in ω^k. But by Theorem 1.3.25 then ω^{k+1} is wMSO interpretable in ω^k, which is known not to be possible [98, Lemma 4.5].[15]

1.3.8 Comparison of classes

Since words are special cases of trees, and finite ones special cases of infinite ones, one immediately sees the inclusions indicated by the arrows in the figure. All the arrows except for the dotted one are known to be strict inclusions. We now discuss the separating examples as well as the double lines indicating equality of the classes when restricted to countable structures. Since ωS-AutStr and ωT-AutStr contain uncountable structures while S-AutStr and T-AutStr do not, we split our discussion along these lines.

---

[15] We thank Łukasz Kaiser for discussions on the notions of this section and Alex Rabinovich for providing the latter reference.

Countable structures

The structure $(\mathbb{N}, \times)$ separates T-AutStr from S-AutStr (see [20], or [88] for an alternative proof).

Every injective ωS-AutStr presentation of a countable structure can be effectively transformed into a S-AutStr presentation. This is because a countable ω-regular set $X \subseteq \{0, 1\}^{\omega}$ only contains ultimately periodic words, and moreover there is a bound on the size of the periods (which can be computed from an automaton for X). Similar facts hold for countable regular sets of infinite trees [115].

The next theorem generalises this in the word case:

Theorem 1.3.31 ([85])   *(i) The countable structures in ωS-AutStr are precisely those in S-AutStr.*

(ii) Given a (not necessarily injective) automatic presentation of some $\mathfrak{A} \in \omega$S-AutStr it is decidable whether $\mathfrak{A}$ is countable or not, and if it is, an automatic presentation of $\mathfrak{A}$ over finite words can be constructed.

On the other hand, we do not know whether every countable structure in ωT-AutStr is in T-AutStr.

Uncountable structures

The only known non-trivial methods dealing with uncountable structures appear in [82]:

(i) The algebra $(\mathcal{P}(\{0, 1\}^*), \cap, \cup, \cdot^c, \mathcal{F})$ from example 1.3.12(vi) is an uncountable structure separating ωT-AutStr from ωS-AutStr.

(ii) Recall Example 1.3.12(iv) consisting of the algebra of subsets of $\mathbb{N}$ (call it $\mathcal{A}$) quotiented by having finite symmetric difference (call it $\approx$). Construct a variant structure as the disjoint union of $\mathcal{A}$ and $\mathcal{A}/_{\approx}$, with a unary predicate U identifying the elements of $\mathcal{A}$ and a binary relation R relating $a \in A$ to its representative in $\mathcal{A}/_{\approx}$. This uncountable structure separates ωS-AutStr from injωS-AutStr.

1.4 More on word-automatic presentations

1.4.1 Beyond first-order logic

The Fundamental Theorem can be strengthened to include order-invariant definable formulas as well as certain additional quantifiers.

Generalised quantifiers

We briefly recall the definition of generalised quantifiers as introduced by Lindström.

Definition 1.4.1 Fix a finite signature $\tau = (R_i)_{i \leq k}$, where R_i has associated arity r_i. A *quantifier* Q is a class of τ-structures closed under isomorphism. Let σ be another signature. Given σ-formulas $\Psi_i(\bar{x}_i, \bar{z})$ with $|\bar{x}_i| = r_i$ ($i \leq k$), the syntax $Q\bar{x}_1, \ldots, \bar{x}_n(\Psi_1, \ldots, \Psi_k)$ has the following meaning on a σ-structure $\mathcal{A}$:

$$(\mathcal{A}, \bar{a}) \models Q\bar{x}_1, \ldots, \bar{x}_k(\Psi_1, \ldots, \Psi_k) \text{ iff } (A; \Psi_1^{\mathcal{A}}(\cdot, \bar{a}), \ldots, \Psi_k^{\mathcal{A}}(\cdot, \bar{a})) \in Q,$$

where $\Psi^{\mathcal{A}}(\cdot, \bar{a})$ is the relation defined in $\mathcal{A}$ by Ψ with parameters $\bar{a}$. The *arity* of a quantifier is the maximum of the r_is. A quantifier is *n-ary* if its arity is at most n.

The extension of first-order logic by a collection **Q** of generalised quantifiers will be denoted FO[**Q**].

Examples 1.4.2   (i) The unary quantifier $\{(A; X) \mid \emptyset \neq X \subset A\}$ is 'there exists'.

 (ii) The unary quantifier 'there exist infinitely many', written $\exists^{\infty}$, is the class of structures $(A; X)$ where X is an infinite subset of A.

(iii) The unary *modulo quantifier* 'there are k modulo m many' (here $0 \leq k < m$), written $\exists^{(k,m)}$, is the class of structures $(A; X)$ where X contains k modulo m many elements. Write $\exists^{\mathsf{mod}}$ for the collection of modulo quantifiers.

(iv) The unary *Härtig quantifier* is the class of structures $(A; P, Q)$ where $P, Q \subset A$ and $|P| = |Q|$.

 (v) Every set $C \subset (\mathbb{N} \cup \{\infty\})^n$ induces the unary *cardinality quantifier* $Q_C = \{(A; P_1, \ldots, P_n) \mid (|P_1|, \ldots, |P_n|) \in C\}$. In fact, a given unary quantifier over signature $(R_i)_{i \leq k}$ is identical to some cardinality quantifiers with $n = 2^k$.

(vi) The binary *reachability quantifier* is the class of structures of the form $(A; E, \{c_s\}, \{c_f\})$ where $E \subset A^2$, $c_s, c_f \in A$, and there is a path in the directed graph $(A; E)$ from c_s to c_f.

(vii) The k-ary *Ramsey quantifier* $\exists^{k\text{-ram}}$ is the class of structures $(A; E)$, $E \subset A^k$, for which there is an infinite $X \subset A$ such that for all pairwise distinct $x_1, \ldots, x_k \in X$, $E(x_1, \ldots, x_k)$.

The following general definition will allow us to compare the expressive strength of quantifiers.

Definition 1.4.3   Let Q be a quantifier, **Q** a collection of quantifiers, and τ the signature of Q. Say that Q *is definable in* **Q** if there is a sentence θ over the signature τ in the logic FO[**Q**] with $Q = \{\mathcal{A} \mid \mathcal{A} \models \theta\}$.

For instance, a structure $(A; X)$ satisfies $\exists^{(0,2)} z\, X(z) \;\vee\; \exists^{(1,2)} z\, X(z)$ if and only if X is finite. Hence $\exists^{\infty}$ is definable in $\{\exists^{(0,2)}, \exists^{(1,2)}\}$.

Of course the generalised quantifiers that interest us most are the ones, like $\forall$ and $\exists$, that preserve regularity.

Definition 1.4.4   Fix class $\mathcal{C}$ as one of S-AutStr, T-AutStr, ωS-AutStr, or T-AutStr. Let Q be a quantifier with signature $\tau = (R_i)_{i \leq k}$, where R_i has associated arity r_i. Say that quantifier Q *preserves regularity for the class* $\mathcal{C}$ if for every $n \in \mathbb{N}$, and every automatic presentation μ of a structure $\mathcal{A} \in \mathcal{C}$, every formula

$$Q\overline{x}_1, \ldots, \overline{x}_k(\Psi_1^{\mathcal{A}}(\overline{x}_1, \overline{z}), \ldots, \Psi_k^{\mathcal{A}}(\overline{x}_k, \overline{z}))$$

defines a relation R in $\mathcal{A}$ with $\mu^{-1}(R)$ regular (here $\overline{z} = (z_1, \ldots, z_n)$ and the Ψ_i are first-order $\mathcal{A}$-formulas).

Say that Q *preserves regularity effectively* if an automaton for $\mu^{-1}(R)$ can effectively be constructed from the automata of the presentation and the formulas Ψ_i.

Since not every structure is injectively presentable, we may restrict this definition to the class $\mathcal{C}$ of injectively presentable structures from ωS-AutStr (or ωT-AutStr). For this, replace 'automatic presentation' with 'injective automatic presentation' in the above definition.

Example 1.4.5   The reachability quantifier is not regularity preserving (for any of the classes). For otherwise, by Example 1.3.13, the set of starting configurations that drive a given Turing Machine to a halting state would be regular, and hence computable.

The first steps have been taken in exploring those quantifiers that preserve regularity.

Theorem 1.4.6   *Let $\mathcal{C}$ be any of the following classes of structures* inj-ωT-AutStr, ωS-AutStr, T-AutStr, S-AutStr.

(i) *The following unary quantifiers preserve regularity effectively for $\mathcal{C}$:* $\exists^{\infty}$, $\exists^{mod}$, $\exists^{\leq \aleph_0}$, $\exists^{> \aleph_0}$ [20, 90, 94, 85, 9].

(ii) *Every unary quantifier that preserves regularity for the class* S-AutStr *is already definable from* $\exists^{mod}$, $\exists^{\infty}$ [125].

The second item also implies that every unary quantifier that preserves regularity for the class $\text{inj-}\omega\text{S-AutStr}$ is already definable from $\exists^{\text{mod}}$, $\exists^{\infty}$, $\exists^{\leq\aleph_0}$, $\exists^{>\aleph_0}$. This is because for an ω-regular relation $R(\overline{x}, \overline{z})$ the cardinality of the set $R(-, \overline{c})$ (for any fixed parameter $\overline{c}$) is finite, countable or has size continuum [94].

Theorem 1.4.7 (see [125]) *Each k-ary Ramsey quantifier preserves regularity effectively for the class* S-AutStr.

Kuske and Lohrey observed that the proof of this theorem can be generalised to quantifiers of the form 'there exists an infinite set X satisfying θ', where θ is a property of sets closed under taking subsets. They use this to show that certain problems, while Σ_1^1-complete for recursive graphs, are decidable on automatic graphs [96].

Order-invariance

Definition 1.4.8 Fix a signature τ and a new symbol $\leq$. A formula $\phi(\overline{x})$ in the signature $\tau \cup \{\leq\}$ is called *order invariant* on a τ-structure $\mathcal{A}$ if for all tuples $\overline{a}$ from A and all linear orders $\leq_1$ and $\leq_2$ on A, we have that $(\mathcal{A}, \leq_1) \models \phi(\overline{a})$ if and only if $(\mathcal{A}, \leq_2) \models \phi(\overline{a})$. The *relation defined by the order invariant ϕ in $\mathcal{A}$* is the set of tuples $\overline{a}$ from A such that $(\mathcal{A}, \leq) \models \phi(\overline{a})$ for some (and hence all) linear orders $\leq$ on A.

The Fundamental Theorem can be extended on injective presentations to include order-invariant formulas in those cases where there is a regular linear ordering of the set $f^{-1}(A)$. On finite-words, finite-trees and ω-words there are regular linear orderings. However, we do not know if there is a regular linear ordering on the set of all ω-trees. On the other hand, certain separating examples from finite model theory are adaptable to the automatic world.

Proposition 1.4.9 ([5]) *There exists a structure $\mathfrak{B} \in \text{S-AutStr}$ and an order-invariant definable relation S^* in $\mathfrak{B}$ that is not definable in $\mathfrak{B}$ using any extension of* FO *with only unary quantifiers.*

1.4.2 Complexity of some problems

First-order theories

By Theorem 1.3.4 query-evaluation and model-checking for first-order formulas are effective on automatic structures. However, the complexity of these problems is in general non-elementary, i.e. it exceeds any fixed number of iterations of the exponential function. For instance the first-order theories of the

	Structure-Complexity[a]	Expression-Complexity
Model-Checking		
Σ_0	LOGSPACE-complete	ALOGTIME-complete
Σ_0 + func	NLOGSPACE	in quadratic time[16]
		and PTIME-complete
Σ_1	PTIME[17]	PSPACE-complete
		(EXPTIME-c. for T-AutStr)
Σ_2	PSPACE-complete[17]	EXPSPACE-complete
		(2EXPTIME-c. for T-AutStr)
Query-Evaluation		
Σ_0	LOGSPACE	PSPACE
Σ_1	PSPACE	EXPSPACE

Figure 1.3 Complexity of fragments of FO on automatic structures
[a] Structure complexity is measured in terms of the size of the largest deterministic automaton in the input presentation.

universal structures $\mathcal{N}_k$ and $\mathcal{S}_{[k]}$ ($k \geq 2$) have non-elementary complexity [77] (cf. also the remark after Example 1.4.39).

There are various sensible ways of measuring model-checking complexity. First, one may fix a formula and ask how the complexity depends on the input structure. This measure is called *structure complexity*. On the other hand, *expression complexity* is defined relative to a fixed structure in terms of the length of the formula. Finally, one can look at the combined complexity where both parts may vary.

In [25] Blumensath and Grädel studied the expression and structure complexity of model-checking and query evaluation for quantifier-free and existential first-order formulas both in a relational signature and allowing terms in quantifier-free formulas. Their results are complemented by those of Kuske and Lohrey [95] on the expression complexity of Σ_1 (existential) and Σ_2 formulas of a relational signature over arbitrary word- and tree-automatic structures. Figure 1.3 provides a summary.

On certain subclasses of automatic structures there is better complexity. In section 1.3.7 above we have mentioned that the first-order theory of each structure allowing a word-automatic presentation of polynomial density is

---

[16] This is a generalisation of the quadratic solution of the word problem in automatic groups [31] (see Section 1.4.5).
[17] Model checking with a fixed Σ_1 formula reduces to a membership or non-emptiness test for an NFA. For fixed Π_2 formulas the problem is polynomially equivalent to the universality problem of NFAs, and thus PSPACE-complete. (We thank Anthony To for pointing out the error in [25].)

decidable in PSPACE. Kuske and Lohrey [101, 95] studied automatic structures whose Gaifman graphs are of *bounded degree*. Relying on locality of first-order logic they have identified the expression complexity of FO model checking on word-automatic and tree-automatic structures of bounded degree to be 2ExpSpace-complete and 3ExpTime-complete, respectively. The combined complexity remains 2ExpSpace for word-automatic presentations and is in 4ExpTime for tree-automatic presentations. For finer results we refer to [95].

Beyond first-order

A fundamental problem in verification is deciding *reachability*: whether there is a path between specified source and target nodes. Since the configuration space of an arbitrary Turing machine is finite-word automatic, the halting problem can be reduced to the reachability problem on the configuration graph of a universal Turing-machine. Similar reductions show the undecidability, over (finite-word) automatic structures, of connectivity, isomorphism, bisimulation and hamiltonicity [25, 96].

On the other hand there are natural classes of automatic structures for which these problems become decidable (see Figure 1.1). For instance, VRA-equational graphs have a decidable FO-theory with reachability and are finite-tree automatic. Reachability and connectivity in locally-finite unary-automatic graphs are in fact decidable in PTIME. Bisimulation equivalence of HR-equational graphs of finite out-degree is decidable [128] (see section 1.2.2).

Finally we mention some cases where full MSO is decidable. Prefix recognisable structures (which include the unary automatic structures) are finite-word automatic. A structure of the form $(\mathbb{N}, <, C_1, \ldots, C_k)$ is called a colouring of the line. Every known finite-word automatic colouring of the line, and this includes every morphic sequence, has decidable MSO-theory (cf. Theorem 1.4.38 and see [7]). Furthermore, every word-automatic equivalence relation has a decidable MSO-theory. This follows from the above and the observation (Proposition 1.4.40) that if there are only finitely many infinite classes then the equivalence relation is FO-definable in some word-automatic colouring of the line [7].

Isomorphism problem

A measure of the complexity of a class of structures is the *isomorphism problem*, namely the problem of deciding, given two □-automatic presentations $\eth$ and $\eth'$, whether or not the structures they present are isomorphic.

The characterisations of the finite-word automatic Boolean algebras and ordinals [88, 63] imply that the isomorphism problem for each of these classes is decidable. Also, as noted, the isomorphism problem for equational graphs is decidable 1.2.10.

Configuration spaces of Turing machines are locally finite and the complexity of the isomorphism problem for locally-finite directed graphs in S-AutStr is Π_3^0-complete [124]. However, by massaging the configuration spaces we get that the isomorphism problem for automatic graphs is as hard as possible: Σ_1^1-complete. This is done by reducing the isomorphism problem for computable structures, known to be Σ_1^1-complete, to that of automatic structures.

Theorem 1.4.10 ([124])   *The complexity of the isomorphism problem for each of the following classes of* S-AutStr *structures is* Σ_1^1*-complete: (i) undirected graphs, (ii) directed graphs, (iii) successor trees, and (iv) lattices of height* 4.

Problem 1.4.11   What is the exact complexity of the isomorphism problem for the following classes:[18]

(i) Automatic equivalence structures (easily seen to be Π_1^0).
(ii) Automatic linear orders.

Traces

Infinite edge-labelled graphs, when viewed as infinite automata, can accept non-regular languages. Naturally, context-free graphs accept precisely the context-free languages. Though prefix-recognisable graphs form a structurally much richer class they have the same language accepting power as context-free graphs (cf. Theorem 1.2.11 items (1) and (6)). Graphs in the Caucal hierarchy have the same accepting power as higher-order pushdown automata (see Theorem 1.2.16) tracing languages on the corresponding levels of the OI-hierarchy of [62]. The traces of GTRS-graphs form a language class in between the context-free and context-sensitive classes of the Chomsky hierarchy [99]. Rational graphs accept precisely the context-sensitive languages [108]. All context-sensitive languages can in fact be accepted by word-automatic graphs [123], cf. also [35] for a more accessible proof and finer analysis. Meyer proved that the traces of tree-automatic graphs are those languages recognisable in Etime, i.e. in $2^{\mathcal{O}(n)}$ time [103].

---

[18] While this work has been in print, Kuske, Liu and Lohrey have greatly contributed to settling these and related questions. We refer to their forthcoming paper.

1.4.3 Non-automaticity via pumping and counting

It is usually quite simple to show that a structure has an automatic presentation (if indeed it does have one!). On the other hand, there are only a handful of elementary techniques for showing that a structure has no automatic presentation. Most rely on the pumping lemma of automata theory.

Sometimes we can provide a full characterisation of classes of automatic structures. The first non-trivial characterisation was for the word-automatic ordinals (in the signature of order).

Theorem 1.4.12 (Delhommé [63])

(i) *An ordinal α is in* S-AutStr *if, and only if, $\alpha < \omega^\omega$.*
(ii) *An ordinal α is in* T-AutStr *if, and only if, $\alpha < \omega^{\omega^\omega}$.*

A relation R is $(n + m)$ *locally finite* if for every $(x_1, \ldots, x_n)$ there are only finitely many $(y_1, \ldots, y_m)$ such that $R(\overline{x}, \overline{y})$ holds. Obviously, every functional relation $f(\overline{x}) = y$ is locally finite. Other examples of locally finite relations are equal-length el, length comparison $|y| < |x|$, and the prefix relation $y \prec_{\texttt{prefix}} x$. Note that local finiteness depends on the partitioning of the variables, e.g. $x \prec_{\texttt{prefix}} y$ is not locally finite.

A simple pumping argument gives the following important tool.

Proposition 1.4.13 (Elgot and Mezei [66])   *Let $R \subseteq (\Sigma^*)^{n+m}$ be a regular and locally finite relation. Then there is a constant k such that for all $\overline{x}, \overline{y}$ satisfying R, $max_j |y_j| \leq max_i |x_i| + k$. In particular, if f is a regular function then there is a constant k such that for every $\overline{x}$ in its domain we have $|f(\overline{x})| \leq max_i |x_i| + k$.*

Growth of generations

Consider a structure $\mathfrak{A}$ with functions $\mathcal{F} = \{f_1, \ldots, f_s\}$ and a sequence $E = \{e_0, e_1, e_2, \ldots\}$ of elements of $\mathfrak{A}$. The generations of E with respect to $\mathcal{F}$ are defined recursively as follows.

$$G_{\mathcal{F}}^0(E) = \{e_0\}$$
$$G_{\mathcal{F}}^{n+1}(E) = G_{\mathcal{F}}^n(E) \bigcup \{e_{n+1}\}$$
$$\bigcup \{f(\overline{a}) \mid f \in \mathcal{F},\ a_i \in G_{\mathcal{F}}^n(E) \text{ for each } i \leq |\overline{a}|\}$$

We are interested in how fast $|G_{\mathcal{F}}^n(E)|$ grows as a function of n.

Example 1.4.14   (i) Free semigroup on m generators: here $\mathcal{F} = \{\cdot\}$ and $E = \{e_1, \ldots, e_m\}$. For $m \geq 2$, since $G_{\mathcal{F}}^m(E) \supset E$, the set $G_{\mathcal{F}}^{m+n}(E)$ includes all strings over E of length at most 2^n; thus the cardinality of $G_{\mathcal{F}}^{m+n}(E)$ is at least a double exponential in n.

(ii) If $p : D \times D \to D$ is injective then for $\mathcal{F} = \{p\}$ and $E = \{e_1, e_2\}$ (distinct elements of D) $|G_{\mathcal{F}}^n(E)|$ is at least a double exponential.

We now iterate Proposition 1.4.13.

Proposition 1.4.15 ([87], [20, 25]) *Let $\mathfrak{A} \in$ S-AutStr and consider an injective presentation $\mathfrak{d}$ with naming function f. Let $\mathcal{F}$ be a finite set of functions FO-definable in $\mathfrak{A}$ and $E = \{e_0, e_1, \ldots\}$ a definable set of elements ordered according to length in $\mathfrak{d}$, i.e. $|f^{-1}(e_0)| \leq |f^{-1}(e_1)| \leq \cdots$. Then there is a constant k such that for every n and for every $a \in G_{\mathcal{F}}^n$ $|f^{-1}(a)| \leq kn$. In particular, $|G_{\mathcal{F}}^n| = 2^{\mathcal{O}(n)}$.*

In other words, the number of elements that can be generated using functions is at most a single exponential in the number of iterations. Continuing the previous examples, neither the free semigroup nor any bijection $f : D \times D \to D$ (also called a pairing function) is word-automatic. It is trickier to apply the proposition to show that Skolem arithmetic $(\mathbb{N}, \times)$ is not word-automatic (see [20, 25]). It is nevertheless tree-automatic, cf. Example 1.3.9.

The application of propositions 1.4.13 and 1.4.15 has been pushed to their limits:

Proposition 1.4.16 (i) *If a group $(G, \cdot)$ is word-automatic then every finitely generated subgroup is virtually Abelian (has an Abelian subgroup of finite index). In particular, a finitely generated group is in* S-AutStr *if, and only if, it is virtually Abelian [116, 114].*

(ii) *A Boolean Algebra (in the signature $(\cup, \cap, \cdot^c, \bot, \top)$) is in* S-AutStr *if, and only if, it is finite or a finite power of the Boolean Algebra of finite or co-finite subsets of $\mathbb{N}$ [88]. In particular, the countable atomless Boolean Algebra is not in* S-AutStr.

(iii) *There is no infinite integral domain in* S-AutStr *[88].*

(iv) *No word-automatic structure (D, R) has a subset $N \subset D$ such that (N, R) is isomorphic to $(\mathbb{N}, \cdot)$, cf. [114].*

The proof of the first item starts with the observation that every finitely-generated group $G \in$ S-AutStr has polynomial density - that is, for every finite set $A = \{a_1, \ldots, a_k\}$ the function

$$\gamma(n) = |\{\prod_{i < n} c_i^{\sigma_i} \mid \forall i < n : c_i \in A, \ \sigma_i \in \{1, -1\}\}|$$

is bounded by a polynomial (this exploits associativity of the group operation). The rest of the proof uses powerful theorems of Gromov and Ershov (see [114] for a survey of word-automatic groups).

Number of definable subsets

Various countable random structures, such as the random graph, do not have word- or tree-automatic presentations [88, 63]. The approach to proving these facts has a model-theoretic flavour: for a purported automatic presentation, it involves counting the number of definable subsets of elements represented by words of bounded length.

Consider the usual definition of a set defined by φ with parameter b that remains fixed:

$$\varphi(-, b)^{\mathfrak{A}} = \{a \in \mathfrak{A} \mid \mathfrak{A} \models \varphi(a, b)\} \, .$$

A finite set $X \subset A$ is *fully shattered by* φ if the cardinality of the family

$$\{\varphi(-, b)^{\mathfrak{A}} \cap X \mid b \in A\}$$

is as large as possible, namely $2^{|X|}$. For instance, Benedikt et al. [16] observe that in $\mathcal{S}_{[2]}$ each of the sets $\{0, 00, \ldots, 0^n\}$ can be fully shattered by the formula $\varphi(x, b) = \exists z(\texttt{suc}_1 z \prec_{\texttt{prefix}} b \wedge \texttt{el}(z, x))$.

By contrast, in every automatic presentation with naming function f and domain $D \subseteq \Sigma^*$, the image under f of each $D_{\leq n} := D \cap \Sigma^{\leq n}$ can only be linearly shattered by definable families.

Proposition 1.4.17 ([88, 63]) *In every automatic presentation of a structure $\mathfrak{A}$ with naming function f and for every formula φ:*

$$|\{\varphi(-, b)^{\mathfrak{A}} \cap f(D_{\leq n}) \mid b \in A\}| = \mathcal{O}(|f(D_{\leq n})|) \, .$$

As an application recall that the random graph is characterised by the property that for every partition of a finite set X of vertices into sets U and V, there is a vertex b connected to all elements of U and to no element of V. In other words, every finite set X of vertices is fully shattered by the edge relation as the parameter b is varied. So by Proposition 1.4.17 the random graph has no word-automatic presentation. Similar reasoning yields the following.

Proposition 1.4.18 ([88, 63]) *The following are not in* S-AutStr: *the random graph, the random partial order, the random K_n-free graph.*

Using Theorem 1.3.25 one can established non-automaticity of the random graph in a far more general sense.

Theorem 1.4.19 ([47]) *Neither the random graph nor the the free monoid on two generators is finite-tree automatic with any oracle.*

In fact neither is ω-word automatic with any oracle, as witnessed by the following theorem which follows from the proof of Theorem 1.3.31.

Theorem 1.4.20   *If a* countable *structure is ω-word automatic with oracle, then it is also finite-word automatic with (the same) oracle.*

1.4.4 Comparing presentations

When we think of an automatic structure we frequently have a particular automatic presentation in mind. Some structures have *canonical* presentations. For instance, $(a^*, <_{\text{len}})$ is arguably the canonical presentation of $(\mathbb{N}, <)$ and $(\{0, 1\}^*, \text{suc}_0, \text{suc}_1, \prec_{\text{prefix}}, \text{el})$ is the canonical presentation of itself. Some well-known structures have *natural* presentations, none of which can be indisputably called canonical. The base $k \in \mathbb{N}$ $(k > 1)$ presentations of $(\mathbb{N}, +)$ can be considered equally natural; but then what about the Fibonacci numeration system? The field of *regular numeration systems*, though using a somewhat different terminology, investigates automatic presentations of $(\mathbb{N}, +)$ and ω-word automatic presentations of $(\mathbb{R}, +)$. Finally, there are *pathological* presentations that are used to pin down the relationship between definability in a structure and regularity in its presentations [90].

How are we to compare different automatic presentations of the same structure? What are the crucial aspects of a presentation that distinguish it from others?

Canonical representations of context-free graphs were investigated by Sénizergues. In [127] a p-structure for a graph G is a PDA $\mathcal{A}$ (having no ϵ-transitions) together with an isomorphism between the configuration graph of $\mathcal{A}$ and G. Furthermore, a p-structure for G is *P-canonical* if the distance in G between a vertex v and the root is equal to the stack height of the configuration representing v (cf. [112]'s notion of a canonical automaton for a context-free graph; and [41, 44]). For a fixed graph G Sénizergues considers two p-structures equivalent if there is a rational isomorphism between them, and shows that every equivalence class of p-structures contains a P-canonical one [127].

An example from the theory of numeration systems is provided by the celebrated result of Cobham and Semenov. Recall that naturals p and q are called *multiplicatively independent* if they have no common power (ie. $p^k \neq q^l$ for all $k, l \geq 1$) and *multiplicatively dependent* otherwise.

Theorem 1.4.21 (Cobham-Semenov[19], cf. [26, 19, 109])
The following dichotomy holds for $p, q \geq 2$.

---

[19] Cobham proved it for sets; Semenov later extended it to arbitrary relations.

(i) If p and q are multiplicatively dependent then a relation $R \subseteq \mathbb{N}^r$ is regular when coded in base p iff it is regular when coded in base q.

(ii) If p and q are multiplicatively independent then a relation $R \subseteq \mathbb{N}^r$ is regular in both base p and base q iff R is FO-definable in $(\mathbb{N}, +)$.

The meaning of (i) is that, for instance, bases 2^l and 2^k are expressively equivalent. There is a very simple coding translating numerals between these bases, which bijectively maps blocks of k digits in the first system to blocks of l digits in the second system. Every pair of multiplicatively dependent numeration systems are linked by similar translations.

According to (ii) the base 2^k presentation is as different as it can be from, say, the base 3 presentation. This point is further stressed by the following result of Bés based on the work of Michaux and Villemaire.

Theorem 1.4.22 ([18]) *Let p and q be multiplicatively independent, and $R \subseteq \mathbb{N}^r$ regular when coded in base q, but not first-order definable in $(\mathbb{N}, +)$. Then the first-order theory of $(\mathbb{N}, +, |_p, R)$ is undecidable.*

On a similar note we introduce the following general notions.

Definition 1.4.23 (Subsumption and equivalence)
Consider two □-automatic presentations of some structure $\mathfrak{A}$ with naming functions f and g, respectively. We say that f *subsumes* g ($g \preccurlyeq f$) if for every relation R over the domain of $\mathfrak{A}$, if $g^{-1}(R)$ is □-regular then $f^{-1}(R)$ is □-regular. If both $f \preccurlyeq g$ and $g \preccurlyeq f$ then we say that the two presentations are *equivalent* and write $f \sim g$. Moreover, we say that a □-automatic presentation of $\mathfrak{A}$ is *prime* if it is subsumed by all other □-automatic presentations of $\mathfrak{A}$.

Word-automatic presentations

The definition of equivalence of automatic presentations is modelled on case (i) of Theorem 1.4.21. In [5] it has been shown that two finite-word automatic presentations are equivalent if and only if the transduction translating names of elements from one presentation to the other is computable by a *semi-synchronous transducer*: a two-tape finite automaton processing its first tape in blocks of k letters and its second tape in blocks of l letters for some fixed positive k and l. (Note that, except in trivial cases, k/l is uniquely determined [5].)

Theorem 1.4.24 ([5]) *Two finite-word automatic presentations of some $\mathfrak{A} \in$ S-AutStr with naming functions $f_i : D_i \to A$, $i \in \{1, 2\}$, are equivalent if, and only if, the transduction $T = \{(x, y) \in D_1 \times D_2 \mid f_1(x) = f_2(y)\}$*

translating names of elements from one presentation to the other is semi-synchronous rational.

Corollary 1.4.25   *Let f_1 and f_2 be naming functions of equivalent automatic presentations of $\mathfrak{A}$. Then there is a constant C such that for every n-ary relation R over $dom(\mathfrak{A})$ and for every automaton $\mathcal{A}_1$ recognising $f_1^{-1}(R)$ there is an automaton $\mathcal{A}_2$ of size $|\mathcal{A}_2| \leq C^n \cdot |\mathcal{A}_1|$ recognising $f_2^{-1}(R)$, and vice versa.*

Let $\mathfrak{U}$ be one of the universal finite-word automatic structures $\mathcal{S}_\Sigma$ (for $|\Sigma| >$ 1), $\mathcal{P}_f(\Delta_1)$, or $(\mathbb{N}, +, |_k)$ (for $k > 1$). Using semi-synchronous translations one can establish the following.

Theorem 1.4.26 ([5, 6])   *The universal structure $\mathfrak{U}$ has only a single word-automatic presentation up to equivalence.*

The assertion of the theorem can be reformulated as follows.

Corollary 1.4.27   *For a relation R, the expansion $(\mathfrak{U}, R)$ is in* S-AutStr *if, and only if, R is* FO-*definable in $\mathfrak{U}$.*

The prime presentation of a structure, if one exists, is unique up to equivalence, hence may as well be called canonical. The unary presentation of $(\mathbb{N}, <)$ is a prime word-automatic presentation. It is, however, not a prime presentation of $(\mathbb{N}, \text{suc})$, which allows, for every $m > 1$ a word-automatic presentation in which divisibility by m is not regular [90]. It can be inferred that $(\mathbb{N}, \text{suc})$ has no prime presentation.

Recall Theorem 1.3.27 stating that each word-automatic presentation, of structure $\mathfrak{A}$, over a domain of polynomial density of degree d directly corresponds to a d-dimensional interpretation of $\mathfrak{A}$ in the structure $\mathfrak{M} = (\mathbb{N}, <, \{\equiv_{(\bmod m)}\}_{m>1})$, and hence also in $(\mathbb{N}, +)$. So every p-automatic structure has infinitely many pairwise incomparable word-automatic presentations 'inherited' from $(\mathbb{N}, +)$, namely, based on different numeration systems.

In fact, $\mathfrak{M}$ allows a non-trivial 2-dimensional interpretation in itself. Simply consider the lexicographic ordering of all pairs (n_1, n_2) such that $n_1 \geq n_2$ as an interpretation of $(\mathbb{N}, <)$ and observe that moduli of positions within the lexicographic ordering of tuples can be expressed in terms of moduli of their components. Thus, by composing interpretations, every p-automatic presentation of $\mathfrak{M}$ is properly subsumed by other p-automatic presentations with domains of asymptotically greater polynomial densities. This carries over to all p-automatic structures.

In contrast, from results of [5, 8] it follows that $g \preccurlyeq f$ implies $g \sim f$ for any two word-automatic presentations of a given structure, provided that either both f and g have domains of exponential density, or both have a domain of polynomial density of the same degree.

Therefore, the height of the partial order of word-automatic presentations of $\mathfrak{A}$ under subsumption and modulo equivalence is ω if $\mathfrak{A}$ is p-automatic and 1 if $\mathfrak{A}$ is not p-automatic. It is not known whether the width of the subsumption order modulo equivalence is always one or infinite for word-automatic structures that are not p-automatic.

Tree-automatic presentations

Colcombet and Löding [47] investigated the power of finite-subset interpretations applied to arbitrary trees. In our terminology these are tree-automatic presentations with arbitrary oracles.

In the tree-automatic model the analogue of Theorem 1.4.26 does not hold. A tree-automatic presentation of $\mathcal{P}_f(\Delta_2)$ incomparable with the natural one can be forged simply by 'folding each tree in half about the vertical axis', i.e. taking the mirror image of the subtree below the right child of the root and smoothly combing it together with the untouched left half, e.g. as in Example 1.3.11(iii). Despite this, the fact concerning primality of the natural presentation of the universal structure holds in an even stronger sense.

Proposition 1.4.28 ([47, Lemma 5.6]) *The natural tree-automatic presentation with oracle O and with the identity naming function of the finite-subset envelope $\mathcal{P}_f(\mathfrak{T}_O)$ of the oracle tree $\mathfrak{T}_O$ is a prime presentation with respect to tree-automatic presentations with arbitrary oracle.*

In particular, 'the' word-automatic presentation of $\mathcal{P}_f(\Delta_1)$ and the natural tree-automatic presentation of $\mathcal{P}_f(\Delta_2)$ are both prime even among tree-automatic presentations with arbitrary oracles. This is complemented by the following result of [47].

Theorem 1.4.29 *All tree-automatic presentations of $\mathcal{P}_f(\Delta_1)$ are equivalent.*

Therefore, the same holds true for all of the universal structures from Theorem 1.4.26.

1.4.5 Other notions of automaticity

Specific automatic presentations have been employed in other mathematical fields: computational group theory [31], symbolic dynamics [13], numeration systems (of integers or reals) [76], and infinite sequences represented in natural numeration systems [2, 26, 4]. In this section we survey natural presentations

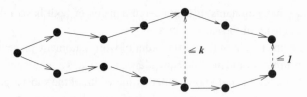

Figure 1.4 k-fellow traveler property.

of certain structures that have mostly been considered independently of the general theory of automatic structures.

Automatic groups

Thurston (1986) motivated by work of Cannon on hyperbolic groups introduced the notion of automatic groups. A finitely generated group G is *automatic* in this sense if for some set of semigroup generators S and associated canonical homomorphism $f : S^* \to G$

(i)   there is a regular language $W \subset S^*$ so that f restricted to W is surjective,
(ii) for every s a generator from S or the group identity, the following binary relation over W is regular:

$$\{(u, v) \mid f(u) = f(v)s\}.$$

This is in fact an algebraic notion: it does not depend on the particular choice of generators. From the automata presenting the group one can extract a finite presentation of the group, and a quadratic-time algorithm deciding the word problem.

Proposition 1.4.30 (*k-fellow traveler property*)   *A group G with semigroup generators $S = \{s_1, \ldots, s_r\}$ is automatic if, and only if, there exists a regular set $W \subseteq S^*$ and $k \in \mathbb{N}$ such that $f \mid_W$ is surjective and W satisfies the k-*fellow traveler property:*

$$\forall u, v \in W \text{ with } d(u, v) \le 1 \; \forall i \le \max\{|u|, |v|\} : d(u_1 \ldots u_i, v_1 \ldots v_i) \le k$$

where $d(u, v)$ denotes the length of the shortest path between u and v in the Cayley graph of G with generators S.

Virtually Abelian groups and Gromov's word hyperbolic groups constitute important examples of automatic groups in this sense. Major results of this programme are presented in [31] (see also the introductions by Farb [71] and by Choffrut [46]).

More recently, this notion has been extended to semigroups [29, 30, 84, 28] and monoids [83, 129, 102].

Let us compare the following three notions: (i) groups whose multiplication function admits a word-automatic presentation, (ii) finitely generated automatic groups, and (iii) finitely generated groups with a Cayley graph admitting a word-automatic presentation. It is known [116] that a finitely generated group allows a word-automatic presentation of type (i) iff it is virtually Abelian. All virtually Abelian finitely generated groups are automatic in the sense of this subsection. Hence (i) implies (ii) for finitely generated groups. Furthermore, by definition, the Cayley graph of every automatic group has a word-automatic presentation. Hence (ii) implies (iii), but the converse fails. As Sénizergues has pointed out the Heisenberg group is not automatic even though its Cayley graph has an automatic presentation. For further reading we recommend the survey by Nies [114].

Generalised numeration systems

The theory of *generalised numeration systems* [76] is concerned with representations of $\mathbb{N}$ and $\mathbb{R}$ in various bases and using different (possibly negative) digits. In general, the basis $U_0 < U_1 < U_2 < \ldots$ of the system does not have to be the sequence of powers of a natural. One considers bases satisfying appropriate linear recursions, or alternatively powers of a base β which is the greatest root of a polynomial of a certain type. The study of generalised numeration systems goes back to Rényi who in 1957 introduced β-expansions.

Without going into the particulars of this very rich field we point out that a number may have more than one representation in a given numeration system. Thus from a practical perspective one is interested in *normalised* numerals obtained via the *greedy* algorithm. Normalised numerals are ordered according to $<_{llex}$ (length and then lexicographically, most significant digit first). A regular set of (normalised) numerals $N \subseteq [d]^*$ over the set of digits $0, \ldots, d-1$ is simply an automatic copy of $(\mathbb{N}, <)$ of the form $(N, <_{llex})$.

A fundamental question in this context asks under which circumstances addition can be computed by a synchronous finite automaton. When this is the case one speaks of a *regular numeration system*. On this matter we refer to [76] and the references therein.

Example 1.4.31 The *Fibonacci numeration system* is a prominent example of a regular numeration system. It has the Fibonacci numbers $1, 2, 3, 5, 8, \ldots$ as its basis, and the binary digit set. The normalised numerals delivered by the greedy algorithm are $\varepsilon, 1, 10, 100, 101, 1000, 1001, 1010, 10000, 10001, \ldots$ in the length-lexicographic ordering. They are the binary strings avoiding 11 as a factor since greedy normalisation prefers 100 to 11. Naturally, 10^n represents the nth Fibonacci number.

More generally we ask how can one classify the word-automatic presentations of $(\mathbb{N}, +)$? Or those of $(\mathbb{N}, <)$? Below we survey known classes of automatic presentations of expansions of $(\mathbb{N}, <)$ by unary predicates, i.e. infinite sequences.

Automatic sequences

The theory of *automatic sequences* [2] studies ω-words representable in more-or-less standard numeration systems. Presentations of primary concern are those of base $k \in \mathbb{N}$, or of base $-k$, and possibly involving negative digits.

Definition 1.4.32   A sequence $s : \mathbb{N} \to \Sigma$ is *k-automatic* if for every $a \in \Sigma$ the set N_a of numerals in the standard base k numeration system representing all positions n such that $s(n) = a$ constitutes a regular language.

These *k-automatic sequences* have been characterised in both algebraic and logical terms. In order to formulate another characterisation some notions are required. A morphism $\varphi : \Gamma^* \to \Sigma^*$ is said to be *k-uniform* if $|\varphi(a)| = k$ for each $a \in \Gamma$. *Codings* are 1-uniform morphisms. A morphism $\varphi : \Gamma^* \to \Gamma^*$ is *prolongable* on some $a \in \Gamma$ if a is the first symbol of $\varphi(a)$. In this case the sequence $(\varphi^n(a))_{n \in \mathbb{N}}$ converges to either a finite or infinite word, which is a fixed point of φ, denoted $\varphi^\omega(a)$.

Theorem 1.4.33 ([26, 2])   *For any sequence $s : \mathbb{N} \to \Sigma$ the following are equivalent:*

(1) s is k-automatic;
(2)   the k-kernel of s: $\{(s_{nk^m+r})_n \mid r, m \in \mathbb{N}, \ r < k^m\}$ is finite;
(3)   the sets $s^{-1}(a)$ are FO-definable in $(\mathbb{N}, +, |_k)$ for each $a \in \Sigma$;
(4) $s = \sigma(\tau^\omega(a))$ for some k-uniform morphism τ on some Γ^* and a coding $\sigma : \Gamma \to \Sigma$;
(5)   (assuming k is a prime and $\Sigma \subseteq \{0, \ldots, k-1\}$): the formal power series $S(x) = \sum_n s_n x^n \in \mathbb{F}_k[[x]]$ is algebraic over $\mathbb{F}_k[x]$.

For example, consider the morphism $\tau : 0 \mapsto 01, 1 \mapsto 10$. Its fixed point $\tau^\omega(0)$ is the *Thue-Morse sequence* $t = 0110100110010110 1001 \ldots$. This is a truly remarkable sequence bearing a number of characterisations and combinatorial properties [3]. For instance, its nth digit is 1 if, and only if, the binary numeral of n contains an odd number of 1's. The 2-kernel of t is $\{t, \bar{t}\}$, where $\bar{t}$ is obtained from t by flipping every bit.

Morphic words

One obtains a definition of *morphic words* by relaxing characterisation (4) of the above theorem. Morphic words thus constitute a generalisation of automatic sequences. They and their relatives have been extensively studied in the context of formal language theory, Lindenmayer systems and combinatorics on words.

Definition 1.4.34 *Morphic words* are those of the form $\sigma(\tau^{\omega}(a))$ for arbitrary homomorphism τ prolongable on a and arbitrary homomorphism $\sigma : \Gamma^* \to \Sigma^*$ extended to ω-words in the obvious way.

Example 1.4.35 Consider $\tau : a \mapsto ab, b \mapsto ccb, c \mapsto c$ and $\sigma : a, b \mapsto 1, c \mapsto 0$ both homomorphically extended to $\{a, b, c\}^*$. The fixed point of τ starting with a is the word $abccbccccbc^6b\ldots$, and its image under σ, $110010^410^610^81\ldots$, is the characteristic sequence of the set of squares. In general, for every strictly positive $\mathbb{N}$-rational sequence (s_k) the characteristic sequence of the set $\{\sum_{k=0}^{n} s_k \mid n \in \mathbb{N}\}$ is morphic [38]. This result also follows from Proposition 1.4.37.

While k-automatic sequences allow automatic presentations over the set of standard base k numerals, the above example suggests that morphic words may need generalised numeration systems. Indeed, every morphic word is automatically presentable in the following sense.

Consider a finite ordered alphabet $\Gamma = \{a_1 < a_2 < \ldots < a_r\}$. In the induced *length-lexicographic order*, denoted $<_{\text{llex}}$, words over Γ are ordered according to their length first, while words of the same length are ordered lexicographically. Thus $(D, <_{\text{llex}})$ provides an automatic presentation of $(\mathbb{N}, <)$ for every infinite regular language D over Γ. Base k as well as so called generalised numeration systems are special cases of this scheme. The following notion thus generalises Definition 1.4.32.

Definition 1.4.36 We say that an ω-word $w : \mathbb{N} \to \Sigma$ is *length-lexicographically presentable* if there is an automatic presentation $(D, <_{\text{llex}})$ of $(\mathbb{N}, <)$ with naming function $f : D \to \mathbb{N}$ such that the sets $f^{-1}(w^{-1}(a))$ are regular for each $a \in \Sigma$.

It is not hard to see that an ω-word is length-lexicographically presentable if and only if it is morphic. There is a perfectly natural correspondence between the morphisms generating a word and the automaton recognising the set of 'numerals', which, when length-lexicographically ordered, give an automatic presentation of the morphic word.

Proposition 1.4.37 ([122])   *An ω-word w is length-lexicographically presentable if, and only if, w is morphic.*

We illustrate the transformation from one formalism to the other on the characteristic sequence of squares from Example 1.4.35. Recall that it is generated by the following morphism τ and final substitution σ

$$\tau : \quad a \mapsto ab \quad b \mapsto ccb \quad c \mapsto c$$
$$\sigma : \quad a \mapsto 1 \quad\quad b \mapsto 1 \quad\quad c \mapsto 0$$

The idea is to interpret symbols $\{a, b, c, 0, 1\}$ as states. Without loss of generality, the alphabets of the ranges of σ and τ are disjoint. The alphabet Γ of the automatic presentation consists of digits ranging from 0 to $|\tau| + |\sigma| - 1$, where $|\tau|$ is the maximum of $|\tau(x)|$ with $x \in \{a, b, c\}$ and $|\sigma|$ is defined similarly. Letters of the alphabet, ordered as usual, are used to index positions within the right-hand side of a τ-rule, or, when larger, positions inside the right-hand side of a substitution via σ.

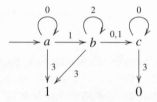

The domain D of the presentation is recognised by the above automaton with both 1 and 0 as final states. With only 1 as a terminal state, the automaton recognises the numerals representing a square relative to the length-lexicographic enumeration of D. Starting with a deterministic automaton this transformation can be reversed producing a morphism τ representing the transition function linearised according to the ordering on the alphabet and with σ identified by the terminal states.

The MSO-theory of the structure $(\mathbb{N}, <, (w^{-1}(a))_a)$ for morphic w is decidable [38]. Moreover, the class of morphic words is closed under MSO-definable recolourings, i.e. under *deterministic generalised sequential mappings* [118]. These results are generalised by the following one, which can be seen as an extension of the Fundamental Theorem 1.3.4.

Theorem 1.4.38 ([7])   *Let $\mathfrak{d} = (D, <_{\text{llex}}, \overline{P})$ be a length-lexicographic presentation of a morphic word w and let $\varphi(\overline{x})$ be an MSO$[<, \overline{P}]$-formula having only first-order variables free. Then there is an automaton $\mathcal{A}$, computable from $\mathfrak{d}$ and φ and such that $(\mathfrak{d}, \mathcal{A})$ is a word-automatic presentation of w expanded by the relation defined by φ.*

Caucal has shown that morphic sequences can be constructed as graphs on the second level of the pushdown hierarchy (cf. Definition 1.2.14) [43]. However, there are automatically presentable ω-words on higher levels as well.

Higher-order morphic words

Higher-order morphic words were introduced in [4, 7]. Morphic words of order k can be defined either in the style of Definition 1.4.34 based on a notion of 'morphisms of order-k stacks' or similar rules, or as in Definition 1.4.36 as those having an automatic presentation using the 'k-fold nested length-lexicographic order' induced by an ordered alphabet. Theorem 1.4.38 extends to these automatic presentations of higher-order morphic words. The classes of order k morphic words form an infinite hierarchy, and are constructible on the $2k$-th level of the pushdown hierarchy [7].

Example 1.4.39 As an example we mention the Champernowne word (cf. Example 1.3.23) obtained by concatenating decimal numerals in their usual order:

$$C = 1234567891011121314\ldots$$

It is on the second level of this hierarchy (and on the fourth level of the pushdown hierarchy). Consider the level 2 morphism Δ given by the following intuitive production rules

$$S_x \rightarrow S_x A_{\tau_1(x)} \ldots A_{\tau_9(x)}$$
$$A_x \rightarrow A_{\tau_0(x)} A_{\tau_1(x)} \ldots A_{\tau_9(x)}$$

where each τ_i is a (level 1) morphism of words in the usual sense mapping each digit $d \in \{0, \ldots, 9\}$ to d and # to i#. Applying Δ repeatedly to the initial level 2 stack $S_\#$ yields the following converging sequence

$$S_\# \rightarrow S_\# A_{1\#} A_{2\#} \ldots A_{9\#}$$
$$\rightarrow S_\# A_{1\#} A_{2\#} \ldots A_{9\#} A_{10\#} \ldots A_{19\#} \cdots\cdots A_{90\#} \ldots A_{99\#}$$
$$\rightarrow \cdots$$

Hence C can be specified as $C = \sigma(\Delta^\omega(S_\#))$ with the morphism σ erasing all #'s while preserving the other (level 1) symbols.

To give a word-automatic presentation we take the domain D to be comprised of all words of the form $d_1 m_1 d_2 m_2 \ldots d_s m_s$ with $d_1 d_2 \ldots d_s$ a conventional decimal numeral and $m_1 m_2 \ldots m_s = o^i x o^{s-i-1}$ a marker indexing the ith digit of this numeral. Elements of the domain are ordered using the length-lexicographic ordering in a nested fashion: comparing numerals (i.e. odd positions) first, and then according to the position of the marker x.

The Champernowne word contains every finite word over $\{0, 1, \ldots, 9\}$ as a factor. The satisfiability problem of first-order logic on finite words, known to be non-elementary [79], is thus expressible in the FO theory of the Champernowne word, which is therefore also non-elementary. For the same reason the Champernowne word is not morphic. Every morphic word is MSO-definable in the Champernowne word, and every word-automatic equivalence structure having only finitely many infinite equivalence classes is interpretable in a second-order morphic word [7].

Proposition 1.4.40 *Consider* $\mathfrak{A} = (A, E)$ *with* E *an equivalence relation having, for each* $n > 0$, $f(n) \in \mathbb{N}$ *many equivalence classes of size* n, *and no infinite classes. Then* $\mathfrak{A} \in$ S-AutStr *if, and only if, there is a second-order morphic word* $w = 0^{m_0} 1 0^{m_1} 1 0^{m_2} 1 \ldots$ *such that* $f(n) = |\{i \mid m_i = n\}|$.

It remains open whether the decidability and definability results for MSO hold for all word-automatic infinite sequences. We are intrigued whether the isomorphism problem of automatic ω-words, or more broadly for automatic scattered linear orders, is decidable. Already for morphic words this is a notorious long-standing open problem.

1.5 Automatic Model Theory

We may reformulate the original problem – we seek a class of finitely-presentable structures $\mathcal{C}$ that has an interesting model theory and lies somewhere between the finite structures (finite model theory) and all structures (classical model theory).

The richest and oldest class consists of the computable structures – these are structures whose domain and atomic relations are computable by Turing machines [70]. In computable model theory, a common theme is to take classical results from mathematics and model theory and to see to what extent they can be made effective. Here are two illustrative observations:

(i) A computable (consistent) first-order theory has a computable model. Indeed, Henkin's construction can be seen as an algorithm computing the domain and atomic relations.

(ii) Every two computable presentations of the rational ordering $(\mathbb{Q}, <)$ are computably isomorphic. Again, the standard back-and-forth argument can be seen as an algorithm building the isomorphism.

The program of feasible mathematics in the 1980's included the development of polynomial-time model theory [45]. However, every relational computable structure is isomorphic (in fact computably isomorphic) to a polynomial-time structure. Automatic structures can be seen as a further restriction of this class, and in fact this is the motivation in [87]. In this section we discuss some aspects of the model theory of automatic structures, a subject still in its infancy.

We split our discussion along two lines: model theory of the class S-AutStr, and model theory of the particular universal structure $S_{[2]}$ (cf. Theorem 1.3.17).

1.5.1 Model theory restricted to the class of word-automatic structures

Blumensath shows that, as expected, certain notions of model theory fail when restricted to the class of automatic structures.

Proposition 1.5.1 (i) *It is undecidable whether an* FO-*formula has a word-automatic model.*

(ii) *The following properties fail on the class of word automatic structures: compactness, Beth, Interpolation, and Łos-Tarski.*

The proofs are based on the observation that there is a FO formula which has automatic models of every finite cardinality but no infinite automatic models.

Löwenheim-Skolem

An automatic version of the Downward Löwenheim-Skolem Theorem would say that every uncountable ω-automatic structure has a countable elementary substructure that is also ω-automatic. Unfortunately this is false since there is a first-order theory with an ω-automatic model but no countable ω-automatic model. Indeed, consider the first-order theory of atomless Boolean Algebras. Kuske and Lohrey [94] have observed that it has an uncountable ω-automatic model (namely the algebra from Example 1.3.12.iv). However, Khoussainov et al. [88] show that the countable atomless Boolean algebra is not automatic and so, by Theorem 1.4.20, not ω-automatic either.

Here is the closest we can get to an automatic Downward Löwenheim-Skolem Theorem for ω-automatic structures.

Proposition 1.5.2 ([85]) *Let* $(D, \approx, \{R_i\}_{i \leq \omega})$ *be an omega-automatic presentation of* $\mathfrak{A}$ *and let* $\mathfrak{A}_{up}$ *be its restriction to the ultimately periodic words of* D. *Then* $\mathfrak{A}_{up}$ *is a countable elementary substructure of* $\mathfrak{A}$.

Proof. Relying on the Tarski-Vaught criterion for elementary substructures we only need to show that for all first-order formulas $\varphi(\overline{x}, y)$ and elements

$\overline{b}$ of $\mathfrak{A}_{up}$

$$\mathfrak{A} \models \exists y \varphi(\overline{b}, y) \quad \Rightarrow \quad \mathfrak{A}_{up} \models \exists y \varphi(\overline{b}, y) \, .$$

By Theorem 1.3.4 $\varphi(\overline{x}, y)$ defines an omega-regular relation and, similarly, since the parameters $\overline{b}$ are all ultimately periodic the set defined by $\varphi(\overline{b}, y)$ is omega-regular. Therefore, if it is non-empty, then it also contains an ultimately periodic word, which is precisely what we needed.   ◁

An identical proposition, also independently noted by Khoussainov and Nies, holds for $\mathfrak{A} \in \omega$T-AutStr with regular trees in place of ultimately periodic words.

Consider the natural, say, binary ω-automatic presentation of $(\mathbb{R}, +)$. Its restriction to the set of elements represented by ultimately periodic ω-words is isomorphic to the additive group of the rationals $(\mathbb{Q}, +)$. Tsankov [136] has shown that there is no automatic divisible torsion-free Abelian group (DTAG). Hence the theory of DTAGs is another example of a first-order theory having an uncountable ω-automatic model but no countable (ω-)automatic models.

Automatic theorems
Kőnig's Lemma

Kőnig's Lemma says that an infinite finitely-branching tree has an infinite path. We split our discussion of automatic analogues along two lines, depending on whether the signature is that of partial order $(T, \preceq)$ or successor (T, S).

Theorem 1.5.3 ([91]) *If $\mathcal{T} = (T, \preceq)$ is an automatic copy of an infinite finitely-branching tree, then $\mathcal{T}$ has a regular infinite path. That is, there exists a regular set $P \subseteq T$ where P is an infinite path of $\mathcal{T}$.*

Proof. Define a set P as those elements x such that $\exists^{\infty} w[x \prec w]$ and for which every $y \prec x$ satisfies that

$$\forall z, z' \in S(y)[z \preceq x \Rightarrow z \leq_{llex} z'].$$

Then P is the length-lexicographically least infinite path of $\mathcal{T}$ (in the ordering induced by the finite strings presenting the tree).   ◁

However, using the 2-Ramsey quantifier we can do more.

Theorem 1.5.4 ([91]) *If $\mathcal{T} = (T, \preceq)$ is an automatic copy of a tree with countably many infinite paths, then every infinite path is regular.*

Proof. Denote by $E(\mathcal{T}) \subseteq T$ the set of elements of a tree $\mathcal{T}$ that are on infinite paths. It is definable in $\mathcal{T}$ using the 2-Ramsey quantifier, so Theorem 1.4.7 gives that $E(\mathcal{T})$ is regular. Then every isolated path of $\mathcal{T}$ is regular, since it is definable as $\{x \in E(\mathcal{T}) \mid p \preceq x\} \cup \{x \in E(\mathcal{T}) \mid x \prec p\}$, for suitable $p \in E(\mathcal{T})$. Replace $\mathcal{T}$ by its derivative $d(\mathcal{T})$, which is also automatically presentable. Since the CB-rank of $\mathcal{T}$ is finite [91] and $d^{\mathrm{CB}(\mathcal{T})}(\mathcal{T})$ is the empty tree, every infinite path is defined in this way. ◁

However, automatic successor trees behave more like computable trees:

Theorem 1.5.5 ([96]) *The problem of deciding, given automata presenting a successor tree (T, S), whether or not it has an infinite path, is Σ_1^1-complete.*

The proof consists of a reduction from the problem of whether a non-deterministic Turing machine visits a designated state infinitely often.

We compare with the computable case.[20] Fix the computable presentation of the full binary tree as consisting of the finite binary sequences with the immediate successor relation (so in fact the prefix relation is also computable). To stress this presentation, we refer to the tree as 2^ω. Similarly fix a natural computable presentation ω^ω of the ω-branching tree. A *computable subtree* of either of these trees is a computable prefix-closed subset.

(i) There is an infinite computable subtree of 2^ω with no computable infinite path.

(ii) There is a computable subtree of ω^ω with exactly one infinite path, and this path is not computable.

(iii) The set of indices of computable subtrees of the binary tree 2^ω with at least one infinite path is Π_2^0-complete.

(iv) The set of indices of computable subtrees of ω^ω with at least one infinite path Σ_1^1-complete.

Cantor's Theorems

One of Cantor's theorems says that every countable linear ordering embeds in the rational ordering $\mathbb{Q}$. The standard proof is easily seen to be effective given a computable presentation of $(\mathbb{Q}, <)$.

There are potentially a variety of automatic versions. The following proposition is the best known.

Proposition 1.5.6 [93] *Every automatic copy $\mathcal{M}$ of a linear order can be embedded into some automatic copy of $\mathbb{Q}$ by a function $f : \mathcal{M} \to \mathbb{Q}$ with the following properties:*

(i) *The function f is continuous with respect to the order topology.*

(ii) *The graph of f is regular.*

---

[20] Thanks to Frank Stephan for discussions concerning this case.

It is not known whether there is a single automatic copy of $\mathbb{Q}$ that embeds, in the sense above, all automatic copies of all automatically presentable linear orders $\mathcal{M}$.

Cantor also proved that $\mathbb{Q}$ is homogeneous: For every two tuples $x_1 < \cdots < x_m$ and $y_1 < \cdots < y_m$ there is an automorphism $f : \mathbb{Q} \to \mathbb{Q}$ with $f(x_i) = y_i$ for $i \leq m$. Again there might be a number of automatic variations. Call an automatic copy of $\mathbb{Q}$ *automatically homogeneous* if for every two tuples there is an automorphism as above that is also regular.

Proposition 1.5.7   [93] *There is an automatic copy of $\mathbb{Q}$ that is automatically homogeneous. There is an automatic copy of $\mathbb{Q}$ that is not automatically homogeneous.*

Scott ranks

Every countable structure $\mathcal{A}$ has a sentence of the infinitary logic $L_{\omega_1,\omega}$ (it allows, in addition to FO, countable disjuncts but still only finitely many free variables) that characterises $\mathcal{A}$ up to isomorphism. The *Scott rank* of $\mathcal{A}$ is the minimal quantifier rank amongst all such sentences.

Theorem 1.5.8 ([86])   *For every computable ordinal there is an automatic structure of Scott Rank at least α.*

The idea is to massage the configuration space of Turing machines presenting a computable structure (having Scott Rank α) to get an automatic structure of similar rank.

1.5.2 On the universal word-automatic structure

We conclude by highlighting some model-theoretic properties of the universal structure $\mathcal{S}_{[2]}$.

(i) $\mathcal{S}_{[2]}$ has infinite VC-dimension [15]. That is, there is a formula $\phi(x, z)$ that defines a family of sets of the form $\phi(-, z)^{\mathcal{S}_{[2]}}$ as one varies the parameter z, and this family fully shatters arbitrarily large finite sets.

(ii) $\mathcal{S}_{[2]}$ admits quantifier elimination (QE) in the expansion of all definable unary predicates and binary functions. In fact, no expansion with definable unary functions (and arbitrary predicates) admits QE [15].

Blumensath [20, p. 67] raised the question of whether there are non-standard models of the theory of the universal structure $\mathcal{S}_{[2]}$ in S-AutStr. Here we sketch an argument resting on Theorem 1.4.26 that shows that there are no word-automatic non-standard models. This result was obtained in discussions with Bakhadyr Khoussainov.

Theorem 1.5.9 $\mathcal{S}_{[2]}$ *is the only word-automatic model of its theory.*

Proof. Assume, for a contradiction, an automatic presentation of a non-standard elementary extension of $\mathcal{S}_{[2]}$. By 'component' we mean a maximal set of elements connected by successor relations. Every elementary extension of $\mathcal{S}_{[2]}$ consists of the standard component isomorphic to $\mathcal{S}_{[2]}$ (containing the root), and any number of non-standard components, that are, as unlabelled graphs, all isomorphic to one-another. The non-standard components are distinguished by the infinite sequences of 0-1 successors ascending towards the root.

(0) The set of representatives of elements of each component is regular.

Indeed, the equivalence relation of belonging to the same component is $FO + \exists^{\infty}$-definable in the model (by saying that there is a common ancestor having finite distance from both elements), hence regular in the representation.

(1) There is a non-standard element below every standard node.

This follows from the fact that the formula

$$\forall x, x', y : \mathrm{el}(x, x') \wedge x \prec y \rightarrow \exists y' : \mathrm{el}(y, y') \wedge x' \prec y'$$

being true in $\mathcal{S}_{[2]}$ must also hold in every non-standard model.

Combining observation (0) and Theorem 1.4.26 we may assume that the presentation restricted to the standard component is the natural one having the identity as naming function. The binary ω-sequence naturally associated with an infinite branch of the standard component provides a representation of the set of nodes along that branch consistent with the assumed presentation of the model. Denote by Π the set of paths with a non-standard element below them.

(2) The set Π is ω-regular.

Indeed, a Büchi-automaton is built to guess a finite word representing a non-standard element and to check, using the automata of the assumed presentation, that it is a descendant of all finite prefixes of the input path. Given that our model is countable, hence so is Π, we have the following consequence of claim (2).

(3) Every path in Π is ultimately periodic with a period of bounded length.

To close the circle, consider for each $n \in \mathbb{N}$ the sentence

$$\forall x \exists y \, |y| > |x| \wedge 0^n 1 \preceq_{\text{prefix}} y \wedge (\forall z \prec_{\text{prefix}} y)[\text{end}_1(z) \to z 0^n 1 \preceq_{\text{prefix}} y]$$

where $\text{end}_1(z)$ is shorthand for saying that the last letter of z is 1. This sentence expresses that for every length $|x|$ there is a longer word y with as many initial prefixes in $(0^n 1)^*$ as possible. In particular this sentence holds for non-standard elements x. Consequently,

(4) for every $n \in \mathbb{N}$ there is an infinite branch of the standard component with label $(0^n 1)^\omega$ and having non-standard elements below it.

This contradicts observation (3).   ◁

Therefore, by Theorem 1.4.20, there are no countable ω-word automatic non-standard models either. Furthermore, using Theorem 1.4.29 in place of Theorem 1.4.26 in the argument shows there are no non-standard finite-tree automatic models of $\mathcal{S}_{[2]}$. To prove that there are no uncountable ω-word automatic non-standard models of $\mathcal{S}_{[2]}$ one tightens (4) and exploits that all automatic presentations of non-standard components are equivalent.

References

[1] K. Aehlig, J. G. de Miranda, and C.-H. L. Ong. Safety is not a restriction at level 2 for string languages. In *FoSSaCS*, pages 490–504, 2005.

[2] J.-P. Allouche and J. Shallit. *Automatic Sequences, Theory, Applications, Generalizations*. Cambridge University Press, 2003.

[3] J.-P. Allouche and J. O. Shallit. The Ubiquitous Prouhet-Thue-Morse Sequence. In C. Ding, T. Helleseth, and H. Niederreiter, editors, *Sequences and Their Applications: Proceedings of SETA '98*, pages 1–16. Springer-Verlag, 1999.

[4] V. Bárány. A hierarchy of automatic ω-words having a decidable MSO theory. Journées Montoises '06, Rennes, 2006.

[5] V. Bárány. Invariants of automatic presentations and semi-synchronous transductions. In *STACS '06*, volume 3884 of *LNCS*, pages 289–300, 2006.

[6] V. Bárány. *Automatic Presentations of Infinite Structures*. Phd thesis, RWTH Aachen University, 2007.

[7] V. Bárány. A hierarchy of automatic ω-words having a decidable MSO theory. *R.A.I.R.O. Theoretical Informatics and Applications*, 42:417–450, 2008.

[8] V. Bárány. Semi-synchronous transductions. *Acta Informatica*, 46(1):29–42, 2009.

[9] V. Bárány, Ł. Kaiser, and A. Rabinovich. Eliminating cardinality quantifiers from MLO. Manuscript, 2007.

[10] V. Bárány, Ch. Löding, and O. Serre. Regularity problems for visibly pushdown languages. In *STACS '06*, volume 3884 of *LNCS*, pages 420–431, 2006.

[11] K. Barthelmann. On equational simple graphs. Tech. Rep. 9, Universität Mainz, Institute für Informatik, 1997.

[12] K. Barthelmann. When can an equational simple graph be generated by hyperedge replacement? In *MFCS*, pages 543–552, 1998.

[13] M.-P. Béal and D. Perrin. Symbolic Dynamics and Finite Automata. In A. Salomaa and G. Rosenberg, editors, *Handbook of Formal Languages, Vol. 2*, pages 463–503. Springer Verlag, 1997.

[14] M. Benedikt and L. Libkin. Tree extension algebras: logics, automata, and query languages. In *Proceedings of the 17th Annual IEEE Symposium on Logic in Computer Science* (LICS), pages 203–212, 2002.

[15] M. Benedikt, L. Libkin, Th. Schwentick, and L. Segoufin. A model-theoretic approach to regular string relations. In Joseph Halpern, editor, *LICS 2001*, pages 431–440. IEEE Computer Society, June 2001.

[16] M. Benedikt, L. Libkin, Th. Schwentick, and L. Segoufin. Definable relations and first-order query languages over strings. *J. ACM*, 50(5):694–751, 2003.

[17] D. Berwanger and A. Blumensath. The monadic theory of tree-like structures. In E. Grädel, W. Thomas, and T. Wilke, editors, *Automata, Logics, and Infinite Games*, number 2500 in LNCS, chapter 16, pages 285–301. Springer Verlag, 2002.

[18] A. Bès. Undecidable extensions of Büchi arithmetic and Cobham-Semënov theorem. *Journal of Symbolic Logic*, 62(4):1280–1296, 1997.

[19] A. Bès. An Extension of the Cobham-Semënov Theorem. *J. of Symb. Logic*, 65(1):201–211, 2000.

[20] A. Blumensath. Automatic Structures. Diploma thesis, RWTH-Aachen, 1999.

[21] A. Blumensath. Prefix-Recognisable Graphs and Monadic Second-Order Logic. Technical report AIB-2001-06, RWTH Aachen, 2001.

[22] A. Blumensath. Axiomatising Tree-interpretable Structures. In *STACS*, volume 2285 of *LNCS*, pages 596–607. Springer-Verlag, 2002.

[23] A. Blumensath, Th. Colcombet, and Ch. Löding. Logical theories and compatible operations. In J. Flum, E. Grädel, and T. Wilke, editors, *Logic and Automata: History and Perspectives*, Texts in Logic and Games, pages 73–106. Amsterdam University Press, 2007.

[24] A. Blumensath and E. Grädel. Automatic structures. In *LICS 2000*, pages 51–62. IEEE Computer Society, 2000.

[25] A. Blumensath and E. Grädel. Finite presentations of infinite structures: Automata and interpretations. *Theory of Comp. Sys.*, 37:641–674, 2004.

[26] V. Bruyère, G. Hansel, Ch. Michaux, and R. Villemaire. Logic and p-recognizable sets of integers. *Bull. Belg. Math. Soc.*, 1:191–238, 1994.

[27] J. R. Büchi. Weak second-order arithmetic and finite automata. *Zeit. Math. Logih Grund. Math.*, 6:66–92, 1960.

[28] A. J. Cain, E. F. Robertson, and N. Ruskuc. Subsemigroups of groups: presentations, malcev presentations, and automatic structures. *Journal of Group Theory*, 9(3):397–426, 2006.

[29] C. M. Campbell, E. F. Robertson, N. Ruskuc, and R. M. Thomas. Automatic semigroups. *Theor. Comput. Sci.*, 250(1–2):365–391, (2001).

[30] C. M. Campbell, E. F. Robertson, N. Ruskuc, and R. M. Thomas. Automatic completely-simple semigroups. *Acta Math. Hungar.*, 96:201–215, 2002.

[31] J.W. Cannon, D.B.A. Epstein, D.F. Holt, S.V.F. Levy, M.S. Paterson, and W.P. Thurston. *Word processing in groups*. Jones and Barlett Publ., Boston, MA, 1992.

[32] A. Carayol. Regular sets of higher-order pushdown stacks. In *Proceedings of Mathematical Foundations of Computer Science (MFCS 2005)*, volume 3618 of *LNCS*, pages 168–179, 2005.

[33] A. Carayol and Th. Colcombet. On equivalent representations of infinite structures. In *ICALP*, volume 2719 of *LNCS*, pages 599–610. Springer, 2003.

[34] A. Carayol and A. Meyer. Linearly bounded infinite graphs. In *MFCS*, volume 3618 of *Lecture Notes in Computer Science*, pages 180–191. Springer, 2005.

[35] A. Carayol and A. Meyer. Context-Sensitive Languages, Rational Graphs and Determinism. *Logical Methods in Computer Science*, 2(2), 2006.

[36] A. Carayol and C. Morvan. On rational trees. In Z. Ésik, editor, *CSL 06*, volume 4207 of *LNCS*, pages 225–239, 2006.

[37] A. Carayol and S. Wöhrle. The Caucal hierarchy of infinite graphs in terms of logic and higher-order pushdown automata. In *FSTTCS*, volume 2914 of *LNCS*, pages 112–123. Springer, 2003.

[38] O. Carton and W. Thomas. The monadic theory of morphic infinite words and generalizations. *Information and Computation*, 176(1):51–65, 2002.

[39] A. Caryol and Ch. Löding. MSO on the Infinite Binary Tree: Choice and Order. In *CSL*, volume 4646 of *LNCS*, pages 161–176, 2007.

[40] D. Caucal. Monadic theory of term rewritings. In *LICS*, pages 266–273. IEEE Computer Society, 1992.

[41] D. Caucal. On the regular structure of prefix rewriting. *Theor. Comput. Sci.*, 106(1):61–86, 1992.

[42] D. Caucal. On infinite transition graphs having a decidable monadic theory. In *ICALP'96*, volume 1099 of *LNCS*, pages 194–205, 1996.

[43] D. Caucal. On infinite terms having a decidable monadic theory. In *MFCS*, pages 165–176, 2002.

[44] D. Caucal. Deterministic graph grammars. In J. Flum, E. Grädel, and T. Wilke, editors, *Logic and Automata: History and Perspectives*, Texts in Logic and Games, pages 169–250. Amsterdam University Press, 2007.

[45] D. Cenzer and J. B. Remmel. Complexity-theoretic model theory and algebra. In *Handbook of Recursive Mathematics, Vol. 1*, volume 138 of *Studies in Logic and the Foundations of Mathematics*, pages 381–513. North-Holland, Amsterdam, 1998.

[46] Ch. Choffrut. A short introduction to automatic group theory, 2002.

[47] T. Colcombet and C. Löding. Transforming structures by set interpretations. *Logical Methods in Computer Science*, 3(2), 2007.

[48] Th. Colcombet. On families of graphs having a decidable first order theory with reachability. In *ICALP*, volume 2380 of *LNCS*, pages 98–109. Springer, 2002.

[49] Th. Colcombet. Equational presentations of tree-automatic structures. In Workshop on Automata, Structures and Logic, Auckland, NZ, 2004.

[50] Th. Colcombet. *Propriétés et représentation de structures infinies*. Thèse de doctorat, Université Rennes I, 2004.

[51] Th. Colcombet. A combinatorial theorem for trees. In *ICALP*, volume 4596 of *LNCS*, pages 901–912. Springer, 2007.

[52] H. Comon, M. Dauchet, R. Gilleron, F. Jacquemard, D. Lugiez, S. Tison, and M. Tommasi. Tree Automata Techniques and Applications. In preparation, draft available online at http://www.grappa.univ-lille3.fr/tata/.

[53] B. Courcelle. *Graph algebras and monadic second-order logic*. Cambridge University Press, in writing...

[54] B. Courcelle. The definability of equational graphs in monadic second-order logic. In *ICALP*, volume 372 of *LNCS*, pages 207–221. Springer, 1989.

[55] B. Courcelle. Graph rewriting: An algebraic and logic approach. In J. van Leeuwen, editor, *Handbook of Theoretical Computer Science, Volume B: Formal Models and Sematics*, pages 193–242. Elsevier and MIT Press, 1990.

[56] B. Courcelle. Recursive applicative program schemes. In J. v.d. Leeuwen, editor, *Handbook of Theoretical Computer Science, Vol. B*, pages 459–492. Elsevier and MIT Press, 1990.

[57] B. Courcelle. The monadic second-order logic of graphs ix: Machines and their behaviours. *Theoretical Computer Science*, 151(1):125–162, 1995.

[58] B. Courcelle. Finite model theory, universal algebra and graph grammars. In *LFCS '97, Proceedings of the 4th International Symposium on Logical Foundations of Computer Science*, pages 53–55, London, UK, 1997. Springer-Verlag.

[59] B. Courcelle. The Expression of Graph Properties and Graph Transformations in Monadic Second-Order Logic. In G. Rozenberg, editor, *Handbook of graph grammars and computing by graph transformations, vol. 1: Foundations*, pages 313–400. World Scientific, New-Jersey, London, 1997.

[60] B. Courcelle and J. A. Makowsky. Fusion in Relational Structures and the Verification of Monadic Second-Order Properties. *Mathematical Structures in Computer Science*, 12(2):203–235, 2002.

[61] B. Courcelle and I. Walukiewicz. Monadic second-order logic, graph coverings and unfoldings of transition systems. *Annals of Pure and Applied Logic*, 92:35–62, 1998.

[62] W. Damm. The IO- and OI hierarchies. *Theoretical Computer Science*, 20(2):95–208, 1982.

[63] C. Delhommé. Automaticité des ordinaux et des graphes homogènes. *Comptes Rendus Mathematique*, 339(1):5–10, 2004.

[64] M. J. Dunwoody. The accessibility of finitely presented groups. *Inventiones Mathematicae*, 81(3):449–457, 1985.

[65] S. Eilenberg, C. C. Elgot, and J.C. Shepherdson. Sets recognised by n–tape automata. *Journal of Algebra*, 13(4):447–464, 1969.

[66] C. C. Elgot and J. E. Mezei. On relations defined by generalized finite automata. *IBM J. Research and Development*, 9:47–68, 1965.

[67] C. C. Elgot and M. O. Rabin. Decidability and undecidability of extensions of second (first) order theory of (generalized) successor. *Journal of Symbolic Logic*, 31(2):169–181, 1966.

[68] C. C. Elgot. Decision problems of finite automata design and related arithmetics. *Trans. Amer. Math. Soc.*, 98:21–51, 1961.

[69] J. Engelfriet. Context-free graph grammars. In *Handbook of formal languages, vol. III*, pages 125–213. Springer-Verlag New York, Inc., New York, NY, USA, 1997.

[70] Y. L. Ershov, S. S. Goncharov, A. Nerode, and J. B. Remmel, editors. *Handbook of Recursive Mathematics, Vol. 1*, volume 138 of *Studies in Logic and the Foundations of Mathematics*. North-Holland, Amsterdam, 1998.

[71] B. Farb. Automatic Groups: A Guided Tour. *L'Enseignment Math.*, 38:291–313, 1992.

[72] S. Fratani. *Automates à Piles de Piles ... de Piles*. Thèse de doctorat, Université Bordeaux 1, 2005.

[73] S. Fratani. Regular sets over tree structures. Rapport Interne 1358-05, LaBRI, Université Paris 7, 2005.

[74] S. Fratani. The theory of successor extended by several predicates. Journées Montoises '06, Rennes, 2006.

[75] S. Fratani and G. Sénizergues. Iterated pushdown automata and sequences of rational numbers. *Ann. Pure Appl. Logic*, 141(3):363–411, 2006.

[76] Ch. Frougny. Numeration systems. In M. Lothaire, editor, *Algebraic Combinatorics on Words*. Cambridge University Press, 2002.

[77] E. Grädel. Simple interpretations among complicated theories. *Information Processing Letters*, 35:235–238, 1990.

[78] E. Grädel, P. G. Kolaitis, L. Libkin, M. Marx, J. Spencer, M. Vardi, Y. Venema, and S. Weinstein. *Finite Model Theory and Its Applications*. Springer-Verlag, 2007.

[79] E. Grädel, W. Thomas, and T. Wilke, editors. *Automata, Logics, and Infinite Games*, volume 2500 of *LNCS*. Springer-Verlag, 2002.

[80] A. Habel. *Hyperedge Replacement: Grammars and Languages*, volume 643 of *Lecture Notes in Computer Science*. Springer, 1992.

[81] M. Hague, A. S. Murawski, C.-H. L. Ong, and O. Serre. Collapsible pushdown automata and recursion schemes. In *LICS'08*. IEEE Computer Society, 2008.

[82] G. Hjorth, B. Khoussainov, A. Montalbán, and A. Nies. From automatic structures to Borel structures. In *23rd Symposium on Logic in Computer Science (LICS)*, 2008.

[83] M. Hoffmann, D. Kuske, F. Otto, and R. M. Thomas. Some relatives of automatic and hyperbolic groups, 2002.

[84] M. Hoffmann and R. M. Thomas. Notions of automaticity in semigroups. *Semigroup Forum*, 66:337–367 (2003).

[85] Ł. Kaiser, S. Rubin, and V. Bárány. Cardinality and counting quantifiers on ω-automatic structures. In *STACS '08*, volume 08001 of *Dagstuhl Seminar Proceedings*, pages 385–396. Internationales Begegnungs- und Forschungszentrum fuer Informatik (IBFI), Schloss Dagstuhl, Germany, 2008.

[86] B. Khoussainov and M. Minnes. Model theoretic complexity of automatic structures. *Annals of Pure and Applied Logic*, To appear, 2008.

[87] B. Khoussainov and A. Nerode. Automatic presentations of structures. In *LCC '94*, volume 960 of *LNCS*, pages 367–392. Springer-Verlag, 1995.

[88] B. Khoussainov, A. Nies, S. Rubin, and F. Stephan. Automatic structures: Richness and limitations. In *LICS'04*, pages 44–53, 2004.

[89] B. Khoussainov and S. Rubin. Graphs with automatic presentations over a unary alphabet. *Journal of Automata, Languages and Combinatorics*, 6(4):467–480, 2001.

[90] B. Khoussainov, S. Rubin, and F. Stephan. Definability and regularity in automatic structures. In *STACS '04*, volume 2996 of *LNCS*, pages 440–451, 2004.

[91] B. Khoussainov, S. Rubin, and F. Stephan. Automatic linear orders and trees. *ACM Transactions on Computational Logic*, 6(4):675–700, 2005.

[92] T. Knapik, D. Niwinski, and P. Urzyczyn. Higher-order pushdown trees are easy. In *FoSSaCS'02*, volume 2303 of *LNCS*, pages 205–222, 2002.

[93] D. Kuske. Is Cantor's theorem automatic? In *LPAR*, volume 2850 of *LNCS*, pages 332–345. Springer, 2003.

[94] D. Kuske and M. Lohrey. First-order and counting theories of ω-automatic structures. In *FoSSaCS*, pages 322–336, 2006.

[95] D. Kuske and M. Lohrey. Automatic structures of bounded degree revisited. arXiv:0810.4998, 2008.

[96] D. Kuske and M. Lohrey. Hamiltonicity of automatic graphs. In *FIP TCS 2008*, 2008.

[97] H. Lauchli and Ch. Savioz. Monadic Second Order Definable Relations on the Binary Tree. *J. of Symbolic Logic*, 52(1):219–226, 1987.

[98] S. Lifsches and S. Shelah. Uniformization and skolem functions in the class of trees. *Journal of Symbolic Logic*, 63:103–127, 1998.

[99] Ch. Löding. *Infinite Graphs Generated by Tree Rewriting*. Doctoral thesis, RWTH Aachen, 2003.

[100] Christof Löding. Reachability problems on regular ground tree rewriting graphs. *Theor. Comp. Sys.*, 39(2):347–383, 2006.

[101] M. Lohrey. Automatic structures of bounded degree. In *LPAR*, volume 2850 of *LNCS*, pages 346–360. Springer, 2003.

[102] M. Lohrey. Decidability and complexity in automatic monoids. In *Developments in Language Theory*, pages 308–320, 2004.

[103] A. Meyer. Traces of term-automatic graphs. *R.A.I.R.O. Theoretical Informatics and Applications*, 42, 2008.

[104] C. Michaux and F. Point. Les ensembles k-reconnaissables sont définissables dans $\langle \mathbf{N}, +, V_k \rangle$. *C. R. Acad. Sci. Paris Sér. I Math.*, 303(19):939–942, 1986.

[105] Ch. Morvan. *Les graphes rationnels*. Thèse de doctorat, Université de Rennes 1, Novembre 2001.

[106] Ch. Morvan. Classes of rational graphs. Journées Montoises '06, Rennes, 2006.

[107] Ch. Morvan and Ch. Rispal. Families of automata characterizing context-sensitive languages. *Acta Informatica*, 41(4-5):293–314, 2005.

[108] Ch. Morvan and C. Stirling. Rational graphs trace context-sensitive languages. In A. Pultr and J. Sgall, editors, *MFCS 01*, volume 2136 of *LNCS*, pages 548–559, 2001.

[109] A. A. Muchnik. The definable criterion for definability in Presburger arithmetic and its applications. *Theor. Comput. Sci.*, 290(3):1433–1444, 2003.

[110] D. E. Muller and P. E. Schupp. Context-free languages, groups, the theory of ends, second-order logic, tiling problems, cellular automata, and vector addition systems. *Bull. Amer. Math. Soc.*, 4(3):331–334, 1981.

[111] D. E. Muller and P. E. Schupp. Groups, the theory of ends, and context-free languages. *J. Comput. Syst. Sci.*, 26(3):295–310, 1983.

[112] D. E. Muller and P. E. Schupp. The theory of ends, pushdown automata, and second-order logic. *Theor. Comput. Sci.*, 37:51–75, 1985.

[113] A. A. Nabebin. Expressibility in a restricted second-order arithmetic. *Siberian Mathematical Journal*, 18(4):588–593, 1977.

[114] A. Nies. Describing groups. *Bulletin of Symbolic Logic*, 13(3):305–339, 2007.

[115] D. Niwiński. On the cardinality of sets of infinite trees recognizable by finite automata. In *Proceedings of the 16th International Symposium on Mathematical Foundations of Computer Science, MFCS'91*, volume 520, pages 367–376. Springer, 1991.

[116] G. P. Oliver and R. M. Thomas. Finitely generated groups with automatic presentations. In *STACS 2005*, volume 3404 of *LNCS*, pages 693–704. Springer, 2005.

[117] C.-H. L. Ong. On model-checking trees generated by higher-order recursion schemes. In *LICS*, pages 81–90. IEEE Computer Society, 2006.

[118] J.-J. Pansiot. On various classes of infinite words obtained by iterated mappings. In *Automata on Infinite Words*, pages 188–197, 1984.

[119] A. Rabinovich. On decidability of monadic logic of order over the naturals extended by monadic predicates. Unpublished note, 2005.

[120] A. Rabinovich and W. Thomas. Decidable theories of the ordering of natural numbers with unary predicates. Submitted, 2006.

[121] M. Rigo. Numeration systems on a regular language: Arithmetic operations, recognizability and formal power series. *Theoretical Computer Science*, 269:469, 2001.

[122] M. Rigo and A. Maes. More on generalized automatic sequences. *J. of Automata, Languages and Combinatorics*, 7(3):351–376, 2002.

[123] Ch. Rispal. The synchronized graphs trace the context-sensitive languages. *Electronic Notes in Theor. Comp. Sci.*, 68(6), 2002.

[124] S. Rubin. *Automatic Structures*. Phd thesis, University of Auckland, NZ, 2004.

[125] S. Rubin. Automata presenting structures: A survey of the finite-string case. *Bulletin of Symbolic Logic*, 14(2):169–209, 2008.

[126] A. L. Semenov. Decidability of monadic theories. In *Mathematical Foundations of Computer Science, Prague, 1984*, volume 176 of *LNCS*, page 162?175. Springer, Berlin, 1984.

[127] G. Sénizergues. Semi-groups acting on context-free graphs. In *ICALP '96: Proceedings of the 23rd International Colloquium on Automata, Languages and Programming*, pages 206–218, London, UK, 1996. Springer-Verlag.

[128] G. Sénizergues. The bisimulation problem for equational graphs of finite outdegree. *SIAM J. Comput.*, 34(5):1025–1106, 2005.

[129] P. V. Silva and B. Steinberg. A geometric characterization of automatic monoids. *The Quarterly Journal of Mathematics*, 55:333–356, 2004.

[130] J. Su and S. Grumbach. Finitely representable databases (extended abstract. In *In Proc. 13th ACM Symp. on Principles of Database Systems*, 1994.

[131] A. Szilard, Sh. Yu, K. Zhang, and J. Shallit. Characterizing regular languages with polynomial densities. In *MFCS*, pages 494–503, 1992.

[132] W. Thomas. Automata on infinite objects. In J. van Leeuwen, editor, *Handbook of Theoretical Computer Science, Volume B: Formal Models and Sematics*, pages 133–192. Elsevier and MIT Press, 1990.

[133] W. Thomas. Languages, automata, and logic. In G. Rozenberg and A. Salomaa, editors, *Handbook of Formal Languages, volume III*, pages 389–455. Springer, New York, 1997.

[134] W. Thomas. Constructing Infinite Graphs with a Decidable MSO-Theory. In *MFCS*, volume 2747 of *LNCS*, pages 113–124, 2003.

[135] B. A. Trahtenbrot. Finite automata and the logic of one-place predicates. Russian. *Siberian Mathematical Journal*, 3:103–131, 1962. English translation: American Mathematical Society Translations, Series 2, 59 (1966), 23–55.

[136] T. Tsankov. The additive group of the rationals is not automatic. manuscript, 2009.

[137] R. Villemaire. The theory of $\langle \mathbf{N}, +, V_k, V_l \rangle$ is undecidable. *Theoretical Computer Science*, 106:337–349, 1992.

[138] I. Walukiewicz. Monadic second-order logic on tree-like structures. *Theoretical Computer Science*, 275:311–346, 2002.

[139] S. Wöhrle and W. Thomas. Model checking synchronized products of infinite transition systems. In *LICS '04*, pages 2–11, Washington, DC, USA, 2004. IEEE Computer Society.

2
Logical aspects of spatial databases

BART KUIJPERS[a] AND JAN VAN DEN BUSSCHE[b]

2.1 Introduction

In this chapter, we consider spatial databases that are modeled as semi-algebraic sets and we present some logic-based languages to query them. We discuss various properties of these query languages, mainly concerning their expressive power.

The basic query language in this context is first-order logic over the real numbers extended with predicates to address the spatial database relations (Section 2.2). We discuss geometric properties that are expressible in this logic (Section 2.3) and then focus on first-order expressible topological properties of 2-dimensional spatial datasets. A property is called *topological* if it is invariant under homeomorphisms of the ambient space. We give a characterization of topological elementary equivalence and present a point-based language, called *cone logic* that captures exactly the topological queries expressible in first-order logic over the reals (Section 2.4 and 2.7). Next, we present another point-based language that captures the first-order queries that are invariant under affinities (Section 2.6).

The second half of this chapter is devoted to extensions of first-order logic over the reals with some form of recursion. We briefly discuss two such extensions: spatial Datalog and first-order logic extended with a while-loop (Section 2.8). We discuss in more detail extensions of first-order logic with different types of transitive-closure operators, with or without stop-conditions (Section 2.9) and investigate their expressive power (Section 2.10). The evaluation of queries expressed in transitive-closure logic with or without stop conditions may be non-terminating. In general, termination is an undecidable

[a] Hasselt University, bart.kuijpers@uhasselt.be
[b] Hasselt University, jan.vandenbussche@uhasselt.be

property, but we give examples of classes of transitive-closure queries where termination is decidable (Section 2.11).

2.2 Spatial data and first-order logic

In most general terms, a *spatial dataset* is any set $S \subset \mathbb{R}^n$ for some n. Equivalently, we can view such a set as an n-ary relation S over $\mathbb{R}$ (using Cartesian coordinates). Viewing $\mathbb{R}$ as a structure $\bar{\mathbb{R}} = (\mathbb{R}, 0, 1, +, \cdot, <)$ over the language of ordered fields, we can then use first-order logic to express properties of spatial datasets.

For example, the sentence

$$\exists a \exists b \forall x \forall y (S(x, y) \rightarrow y = a \cdot x + b)$$

expresses that $S \subset \mathbb{R}^2$ lies on a straight line.

Since the structure on $\mathbb{R}$ in this paper remains the same, we abbreviate $(\bar{\mathbb{R}}, S) \models \phi$ as $S \models \phi$.

2.3 Capturing first-order geometric properties

According to Felix Klein's Erlangen Programm, a geometric theory can be characterized by the group of transformations that preserve the fundamental geometric properties of the theory. Some examples:

geometry	group of transformations
Euclidean	similarity
affine	affinity
topology	continuous

Fix such a group G of transformations of $\mathbb{R}^n$, and consider some property ϕ of datasets in $\mathbb{R}^n$. We naturally define ϕ to be *G-geometric* if it is invariant under G, or formally:

$$\forall S \, \forall g \in G : S \models \phi \Leftrightarrow g(S) \models \phi$$

Let us see some examples:

- "S lies on a circle" is Euclidean, but not affine.
- "S lies on a straight line" is affine, but not topological.
- "S has dimension two" is topological.

A general question, for any fixed G, is: *What are the G-geometric properties expressible in first-order logic?* Can we enumerate or characterise them in some effective way?

This can easily be done when G is *first-order parameterisable*. By this we mean that there exists an injection $p : G \to \mathbb{R}^\ell$ for some ℓ such that the set

$$\{(p(g), \bar{x}, \bar{y}) \mid g \in G \text{ and } \bar{y} = g(\bar{x})\}$$

is first-order definable in $\bar{\mathbb{R}}$.

For example, the affinities in $\mathbb{R}^2$ are first-order parameterised. Indeed, each affinity g corresponds to some 6-tuple $p(g) = (a, b, c, d, e, f)$ with

$$\begin{vmatrix} a & b \\ c & d \end{vmatrix} \neq 0$$

and we have $(y_1, y_2) = g(x_1, x_2)$ iff

$$\begin{pmatrix} y_1 \\ y_2 \end{pmatrix} = \begin{pmatrix} a & b \\ c & d \end{pmatrix} \cdot \begin{pmatrix} x_1 \\ x_2 \end{pmatrix} + \begin{pmatrix} e \\ f \end{pmatrix}$$

so that is all first-order definable.

The following theorem (Gyssens et al., 1999) provides an effective characterisation of the G-geometric first-order properties, in case G is first-order parameterisable (by the injection p):

Theorem 2.1 *A property Π of sets $S \subseteq \mathbb{R}^n$, for some fixed n, is first-order expressible and G-geometric if and only if $\Pi(S)$ can be expressed by a first-order sentence of the form*

$$\phi \wedge \forall p(g) \in p(G)[\phi(S) \leftrightarrow \phi(g(S))]$$

with ϕ an arbitrary sentence over $(\bar{\mathbb{R}}, S)$.

Note that the first-order parameterisability of G guarantees that the special form of sentence in the above theorem is indeed a first-order sentence.

2.4 First-order topological properties of plain sets

When doing topology, we are only interested in properties that are invariant under "continuous transformations". More precisely, in this paper we define a property of sets in $\mathbb{R}^n$ to be *topological* if it is invariant under all *isotopies* of $\mathbb{R}^n$.

Clearly, the isotopies are not first-order parameterisable, so the easy technique from the previous section does not apply.

Figure 2.1 This set in the plane is closed, and is semi-algebraic, being defined by the real formula $x^2/25 + y^2/16 = 1 \vee x^2 + 4x + y^2 - 2y \leqslant -4 \vee x^2 - 4x + y^2 - 2y \leqslant -4 \vee (x^2 + y^2 - 2y = 8 \wedge y \leqslant -1)$.

We are still able to capture the first-order topological properties, provided we restrict our setting to the following:

1. We work in $\mathbb{R}^2$ only, i.e., sets in the real plane.
2. We consider only *semi-algebraic* sets: sets that are themselves definable in $\mathbb{R}$.
3. Moreover, we consider only *closed* sets (in the standard topological sense).

Let us call such sets "plain". Figure 2.1 shows an example of a plain set (Bochnak et al., 1998).

Let us see some examples of topological properties of plain sets, expressible in first-order logic:

- "The dimension is 0 (or 1, or 2)".
- "There is a point where three lines intersect".
- "There is a point where two 2-dimensional regions touch".

We remark that the expressibility of the third property is not obvious and uses the local conical structure of semi-algebraic sets (see later).

The following topological properties of plain sets are *not* expressible in first-order logic:

- "There is a point where an even number of lines intersect".
- "The number of points where two 2-dimensional regions touch is even".
- "The set is topologically connected".

The inexpressibility of the third property above follows from Theorem 2.3 below, but was first established in (Grumbach and Su, 1997).

So, the question is, what are the first-order expressible topological properties of plain sets? In order to formulate our answer to this question, we use the notion of a *cone* (Coste, 1982).

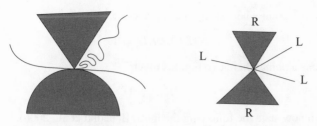

Figure 2.2 The cone of the central point of the figure is $(LLRLR)$.

Around any point on the boundary of any plain set, we always see a circular list of lines (L's) and regions (R's): this list is called the *cone* of the point. An illustration is given in Figure 2.2.

In any plain set, there may be infinitely many boundary points with cone (LL) (these are the points that lie on a line), or with cone (R) (these are the points on the boundary of a region). With the exception of these, however, there are only finitely many boundary points; this is because semi-algebraic sets have a very simple topology (Bochnak et al., 1998). In particular, there are only finitely many points with a cone different from (LL) and (R); these points are called the singular points of the set. It can be argued (Benedikt et al., 2006) that, without loss of generality, we can focus on the singular points, and we will do so from now on.

We are now ready to introduce a propositional logic, called *Cone Logic* or CL for short, designed to express properties of plain sets.

- Atomic formulas are of the form

$$|e| \geqslant n$$

  with n some natural number and e a star-free regular expression over $\Sigma = \{L, R\}$.

    The meaning of such a formula is that there are at least n singular points whose cone satisfies e. (For background on star-free regular expressions and their connection to first-order definability on strings, see (McNaughton and Papert, 1971) and (Thomas, 1997).)
- A CL-sentence is a boolean combination of atomic formulas.

Let us see some examples of properties expressed by CL formulas:

- "The dimension is 0":

$$|L\Sigma^*| = 0 \wedge |R\Sigma^*| = 0$$

- "There is a point where three lines intersect":

$$|LLLLLL| \geqslant 1.$$

- "There is a point where two regions touch":

$$|RR| \geqslant 1$$

We can now state the following theorem (Benedikt et al., 2006):

Theorem 2.2 *The first-order topological properties of plain sets are precisely those expressible in CL.*

This theorem can be proven in six steps:

1. Characterize when plain sets A and B are topologically elementary equivalent in terms of cones;
2. Define flower datasets as a normal form with respect to topological elementary equivalence;
3. Apply collapse theorems over finite structures over the reals;
4. Encode the topological content of flower datasets by abstract finite structures, called *codes*;
5. Translate topological sentences about spatial datasets into sentences about codes;
6. Establish invariance arguments over codes.

Let us go into these six steps in some more detail. For the full proof, we refer to the paper (Benedikt et al., 2006).

Topological elementary equivalence   For plain sets A and B, write $A \equiv B$ if A and B are indistinguishable by topological first-order sentences.

We have the following theorem (Kuijpers et al., 2000), which plays a crucial role in the proof of Theorem 2.2:

Theorem 2.3 $A \equiv B$ *if and only if A and B have precisely the same cones, with the same multiplicities.*

An illustration of this theorem is given in Figure 2.3.

For the full proof of this theorem we refer to the paper (Kuijpers et al., 2000), but we give an idea of the proof here. The proof is based on a transformation of plain sets into a normal form called *flower normal form*. An illustration of this transformation is given in Figure 2.4.

This transformation proceeds by the use of transformation rules, such as the "cut and paste" rule illustrated in Figure 2.5 for a 2-dimensional strip. This rule allows us to cut a 2-dimensional strip in two pieces, of which one has a

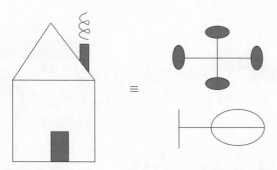

Figure 2.3 These two plain sets are topologically elementary equivalent, according to Theorem 2.3.

hole. The inverse transformation allows us to paste two strips together. There is also a stronger cut-and-paste rule that does not leave a hole as a side effect. For details we refer to (Kuijpers et al., 2000).

For both datasets at the top of Figure 2.4, first, the (strong) strip-cut transformation is used on the full strips near all the singular points of these sets. As a result these singular points have disconnected full petals. There are also some isolated 2-dimensional regions that result from the cutting.

One can show that the application of a transformation rule is indistinguishable by topological first-order sentences, using a *reduction* to first-order inexpressibility of queries on *finite structures over the reals*.

Finite structures over the reals   These are structures of the form $(\bar{\mathbb{R}}, R_1, \ldots, R_k)$ with R_i finite relations. An example of a query to such structures is *Majority:* given finite unary relations R_1 and R_2, is $\#R_1 \geqslant \#R_2$?

Let us illustrate the above-mentioned reduction for the cut-and-paste transformation. This reduction is done by writing a first-order formula $\psi(x, y)$ such that for each finite structure $D = (\bar{\mathbb{R}}, R_1, R_2)$:

- $\psi(D)$ is homeomorphic to the left-hand side of the cut-and-paste transformation if $\#R_1 \geqslant \#R_2$ in D;
- $\psi(D)$ is homeomorphic to the right-hand side of the cut-and-paste transformation if $\#R_1 < \#R_2$ in D.

We do not give this formula ψ, but illustrate it in Figure 2.6. This reduction idea is due to (Grumbach and Su, 1997).

As a consequence, if the cut-and-paste transformation were distinguishable by a topological first-order sentence, then the Majority query would be first-order expressible on finite structure over the reals. We can show the latter

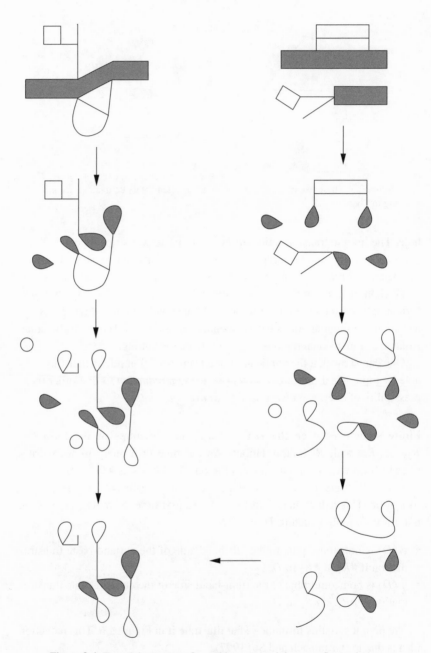

Figure 2.4 Two plain sets transformed into one and the same flower dataset.

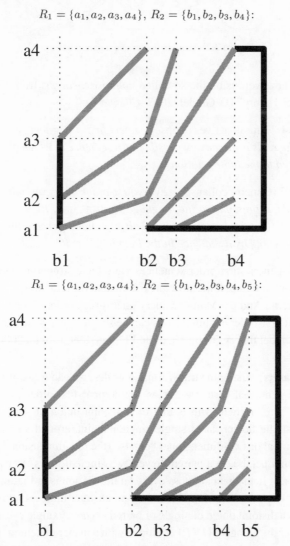

Figure 2.5 Cut and paste transformations: any local part of a spatial dataset that looks like a strip can be cut, with a hole in one of the pieces as a side effect, resulting in a topologically elementarily equivalent dataset. The converse transformation is also possible.

$R_1 = \{a_1, a_2, a_3, a_4\},\ R_2 = \{b_1, b_2, b_3, b_4\}:$

$R_1 = \{a_1, a_2, a_3, a_4\},\ R_2 = \{b_1, b_2, b_3, b_4, b_5\}:$

Figure 2.6 Reducing Majority to distinguishing the cut-and-paste transformation: the set R_1 is placed on the y-axis and the set R_2 on the x-axis. In the rectangles of the irregular raster thus formed, all diagonals from bottom left to top right are drawn with thickness. Some auxiliary lines (in darker shade) are added outside the raster. The figure thus obtained can be defined from R_1 and R_2 by a first-order formula.

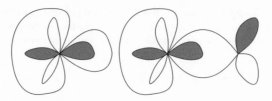

Figure 2.7 A single flower and a paired flower.

to be false, however, using the so-called *collapse theorems* from "Embedded Finite Model Theory" (E. Grädel, 2007, Chapter 5).

Theorem 2.4 (Natural–active collapse) *Every first-order query on finite structures over the reals is already expressible by a sentence in which all quantifiers are relativised to the finite relations.*

Theorem 2.5 (Generic collapse) *Every first-order query on finite structures over the reals, expressible in the language $(0, 1, +, \cdot, <, R_1, \ldots, R_k)$, that is order-generic (that is, invariant under all monotone permutations of $\mathbb{R}$) is already expressible by a sentence in the language $(<, R_1, \ldots, R_k)$.*

So, order-generic first-order sentences view finite structures over the reals just as abstract, ordered, finite structures. Note that the reduction ψ from above is order-generic. And the Majority query on abstract, ordered, finite structures is indeed not first-order expressible, as can be shown using standard arguments from finite model theory (Ebbinghaus and Flum, 1999; Libkin, 2004).

Flower datasets   Using the transformation rules, we can transform any plain dataset into a normal form (as far as topological first-order properties are concerned). This normal form is that of a disjoint union of single or paired *flowers*. A single flower has a single singular point, around which there are one-dimensional or two-dimensional "petals" (the one-dimensional petals can loop over other petals). In a paired flower, two single flowers are paired by an even number of lines that can cross over. An illustration is in Figure 2.7.

A flower dataset can be represented by an abstract finite structure called a *code:* this is a disjoint union of single or paired *cycles*. A single cycle is a word structure over the alphabet $\{L, R\}$ equipped with a planar matching. In a paired cycle, there are two words, and there is again a planar matching now on all the L's, so the matching can again cross over. An illustration is in Figure 2.8.

Translation argument   By a translation argument, we can now reduce the proof of Theorem 2.2 to an invariance question about first-order logic over

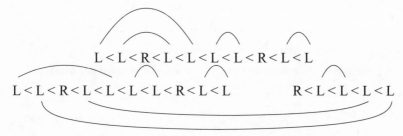

Figure 2.8 Cycle codings of the flowers of Figure 2.7.

codes. This translation argument is based on the following (Benedikt et al., 2006):

Lemma 2.6 (Drawing Lemma) *We can write an FO-formula $\delta(x, y)$ such that for any code C embedded in the reals, $\delta(C)$ is a flower dataset that is a drawing of C.*

The presence of the planar matching in codes was obviously meant to facilitate such a drawing lemma.

The lemma allows us to translate a topological sentence ϕ about flower datasets into a sentence $\psi := \phi \circ \delta$ about codes, called an *implementation* of ϕ. Using the collapse theorems, we may assume that ψ sees only an ordered version of the abstract code. This ordering $<$ is not the $\prec$ of the word structures, but without loss of generality, we can actually assume that $<$ does agree with $\prec$, so all $<$ does is shuffle the separate cycles in some order. Note, however, that ψ is *invariant* under the way this shuffling is done!

Now such $<$-invariant first-order sentences on ordered codes can be seen to be already expressible by plain first-order sentences on codes. So, we are closing in on our goal, as first-order logic on codes comes already quite close to Cone Logic. The "only" difference is that codes still contain a planar matching, which Cone Logic lacks. Crucially, however, a rather technical argument shows that ψ is *also* invariant under the particular choice of planar matching in a code. We are thus faced with one final hurdle, which is removed by the following.

Planar-matching-invariant FO on word structures Our general result concerning word structures over a finite alphabet Σ, additionally equipped with a planar matching G, is the following:

Lemma 2.7 (Main Invariance Lemma) *G-invariant first-order logic collapses to plain first-order logic on the class of word structures with a planar matching.*

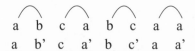

Figure 2.9 Chain matching represented as a word with alternating markers.

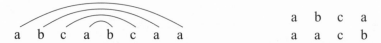

Figure 2.10 Parenthetical matching represented as a folded word.

The lemma is formulated for standard word structures, but can be adapted to cycles and cycle pairs.

The proof of the lemma focuses on two particular kinds of planar matchings: *chain matchings* and *parenthetical matchings*. A chain matching, as illustrated in Figure 2.9, can be simulated by relabeling every other position in the word with a marked letter (using a second alphabet Σ' with marked letters). A parenthetical matching, as illustrated in Figure 2.10, can be simulated by "folding" the word, moving to the alphabet Σ^2.

We can now translate first-order logic over words with chain matchings to first-order logic over words with alternating markers; and we can likewise translate first-order logic over words with parenthetical matchings to first-order logic over folded words. Both translations imply that set W of words accepted by a planar-matching-invariant sentence is surely regular, and can have only very limited kind of *counters* in the sense of (McNaughton and Papert, 1971). A final, complicated, argument then shows that $W = W' \cap (\Sigma\Sigma)^*$ with W' a counter-free regular language; since counter-free regular languages are first-order axiomatisable, the Lemma is proved.

2.5 Conclusion on first-order topological properties

A corollary of Theorem 2.2 is what can be considered a topological version, for (typically infinite) real datasets, of the earlier-mentioned collapse theorems for finite structures over the reals:

Corollary 1 (Topological Collapse) *Every topological first-order property of plain sets S is already expressible by a sentence using only $<$ and S.*

This corollary follows because Cone Logic can already be expressed in first-order logic over $(\bar{\mathbb{R}}, S)$ using only $<$ and S.

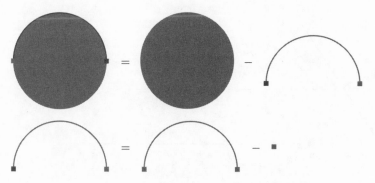

Figure 2.11 Every set is a boolean combination of closed sets.

A natural open problem is to go beyond plain sets. What about non-closed sets? We can always decompose a set in $\mathbb{R}^n$ in $n + 1$ closed sets, as illustrated in Figure 2.11. Hence, the more general question is, what about the first-order topological properties of ensembles of plain sets, as opposed to single plain sets? Grohe and Segoufin have shown that the situation there is considerably more complex (Grohe and Segoufin, 2002). In particular, just looking at cones is not enough anymore, and certain global properties such as "inside" and "outside" are expressible.

Another very natural open question is to move to higher dimensions: capturing the first-order topological properties of semi-algebraic sets in $\mathbb{R}^3$.

And, what about non-semi-algebraic sets? Consider, for example, the property "every point in the set has cone (LL)" (i.e., the dataset consists of a number of disjoint curves either closed or going to infinity). This is first-order expressible, and it is topological over semi-algebraic sets, but it is not topological over all sets (semi-algebraic sets have a very tame topology). Can one also find an example of a topological property that is first-order expressible over semi-algebraic sets, but not over all sets?

Last but not least, an obvious question is how much of our results can be generalized from semi-algebraic sets to definable sets in O-minimal structures (van den Dries, 1998).

2.6 Point-based logics for geometric queries

First-order logic over $\bar{\mathbb{R}}$ is clearly a coordinate-based logic. Cone Logic, on the other hand, is a point-based logic, as it deals directly with the (singular) points in a dataset. Can we find point-based logics for other kinds of geometric

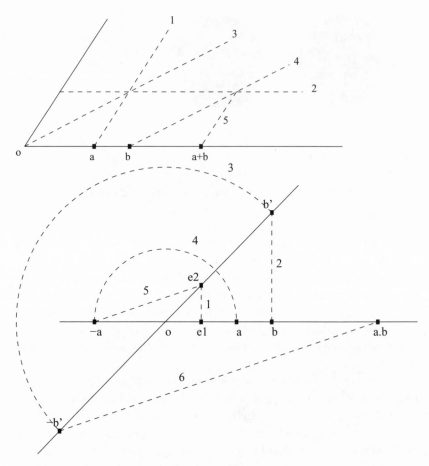

Figure 2.12 Geometric constructions of addition and multiplication: for addition, we construct straight lines in the indicated order. The construction for multiplication also involves finding the mirror image of a point with respect to o, which is indicated by circles, but can be accomplished using a parallelogram construction.

queries? The answer is affirmative, thanks to an observation made by Tarski (Schwabhäuser et al., 1983) to the effect that *the geometric constructions of addition and multiplication are first-order expressible using a single ternary predicate β ("betweenness") on points.* These constructions are illustrated in Figure 2.12.

Let us see how we can use this to obtain a point-based logic for the first-order affine queries. In this logic, we view a dataset $S \subset \mathbb{R}^2$ as a *unary* relation over the structure $(\mathbb{R}^2, \beta)$, and we use ordinary first-order logic over the universe $\mathbb{R}^2$, with ternary β (betweenness) and unary S as the only predicates (as always, β is fixed and belongs to the background, while S is variable and represents the input to the query). Note how this setup differs from the coordinate-based approach, where we view S as a *binary* relation over $(\mathbb{R}, 0, 1, +, \cdot, <)$. Let us denote the point-based logic by $FO(\beta)$, and let us denote the coordinate-based logic by $FO(\bar{\mathbb{R}})$. Both are first-order logics.

It now turns out that we can simulate $FO(\bar{\mathbb{R}})$ by $FO(\beta)$ in the following sense. Call a triple (o, e_1, e_2) of non-collinear points, a *basis*. Then for each $FO(\bar{\mathbb{R}})$-sentence ϕ there exists an $FO(\beta)$-formula $\psi(o, e_1, e_2)$ such that for every dataset S and for every basis (o, e_1, e_2):

$$S \models \psi(o, e_1, e_2) \quad \Leftrightarrow \quad \alpha(S) \models \phi$$

where α is the unique affinity that maps (o, e_1, e_2) to $((0, 0), (1, 0), (0, 1))$.

As a corollary, we obtain (Gyssens et al., 1999):

Theorem 2.8 *For each* $FO(\bar{\mathbb{R}})$-*sentence* ϕ *expressing an affine geometric query there exists an equivalent* $FO(\beta)$-*sentence* ψ *(and vice versa).*

Adding the 4-ary equidistance predicate, we can likewise capture the first-order Euclidean queries (Gyssens et al., 1999).

2.7 Plane graphs

To conclude this survey we must mention a very nice result from (Segoufin and Vianu, 2000). The topology of a semi-algebraic set in the plane can be represented by a finite data structure called a plane graph. (A plane graph is a data structure representing a planar graph embedding.) This is illustrated in Figure 2.13.

We have (Segoufin and Vianu, 2000):

Theorem 2.9 *Every topological first-order sentence about semi-algebraic sets in the plane, using only* $<$ *and* S, *can be translated to a first-order sentence about the corresponding plane graphs.*

By topological collapse (see Section 2.5), we know that (for a single plain set at least) the restriction in the above statement that 0, 1, $+$ and $\cdot$ cannot be used, is harmless.

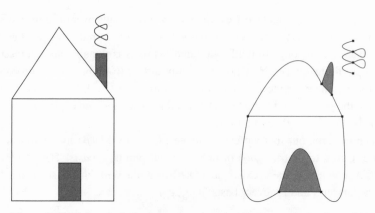

Figure 2.13 Plane graph data structure representing the topology of a semi-algebraic set.

2.8 Spatial datalog and first-order logic extended with a while-loop

Topological connectivity is a property that is important in many applications, in particular in geographical information systems (GIS) (Worboys and Duckham, 2004). As we remarked in Section 2.4, topological connectivity is not expressible in first-order logic, and several more expressive extensions of first-order logic over the reals have been proposed that do allow the expression of topological connectivity of spatial datasets. In this section, we briefly discuss two such extensions: spatial Datalog and first-order logic extended with a while-loop. In the next section, we will discuss in more detail extensions of first-order logic with different types of transitive-closure operators.

Spatial Datalog   Essentially, the query language *spatial Datalog* is Datalog extended with polynomial inequalities in the body of rules, with the understanding that: the underlying domain is $\mathbb{R}$; the only extended database predicate is S (the input spatial dataset); and relations can be infinite (to represent spatial datasets as binary relations, and auxiliary relations) (Kuijpers et al., 1996).

The following spatial Datalog program expresses linear-path connectivity of a two-dimensional spatial dataset. In the relation $Obstr(x, y, x', y')$ couples of points of S are stored that cannot be connected by a straight line segment that is entirely in S. Couples of points that are not obstructed are collected in the *Path* relation (see Figure 2.14) and next the transitive closure

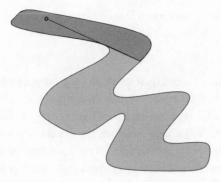

Figure 2.14 Linear path connectivity of a spatial dataset in the plane.

of *Path* is computed. Only if all pairs of points in S end up in *Path*, S is connected.

$$
\begin{aligned}
Obstr(x, y, x', y') \quad &\longleftarrow \quad \neg S(\bar{x}, \bar{y}), S(x, y), S(x', y'), \\
&\qquad \bar{x} = a_1 t + b_1, \\
&\qquad \bar{y} = a_2 t + b_2, 0 \leqslant t, t \leqslant 1, \\
&\qquad b_1 = x, b_2 = y, \\
&\qquad a_1 + b_1 = x', \\
&\qquad a_2 + b_2 = y' \\
Path(x, y, x', y') \quad &\longleftarrow \quad \neg Obstr(x, y, x', y') \\
Path(x, y, x', y') \quad &\longleftarrow \quad Path(x, y, x'', y''), \\
&\qquad Path(x'', y'', x', y') \\
Disconnected \quad &\longleftarrow \quad S(x, y), \; S(x', y'), \\
&\qquad \neg Path(x, y, x', y') \\
Connected \quad &\longleftarrow \quad \neg Disconnected.
\end{aligned}
$$

To show that this spatial Datalog program correctly tests topological *connectivity* on a class $\mathcal{C}$ of spatial datasets, we have to show that for any set S in $\mathcal{C}$, that two points in S are in the same connected component of S if and only if they can be connected by a piecewise linear curve lying entirely in S (soundness); and that the number of line segments needed to connect any such pair of points in S is bounded (termination). Termination guarantees that the transitive closure will terminate. Soundness then establishes the correctness of the test for connectivity performed by the program after the transitive closure is completed.

We have the following result (Kuijpers et al., 1996).

Theorem 2.10 *The above spatial Datalog program correctly tests connectivity of semi-linear spatial datasets (i.e., spatial datasets that can be described using addition only).*

In fact, this program correctly tests topological connectivity for a wider class of spatial datasets, called *border-visible* (Kuijpers et al., 1996). The program does not work for arbitrary semi-algebraic figures in $\mathbb{R}^2$, however. For example, when the input set is the area between the parabola given by $y = x^2$ and $y = 2x^2$, then soundness is satisfied for all points except the origin. Even when the origin is left out, termination is violated since there is no uniform bound on the number of line segments needed to connect two points. The closer one point gets to the origin, the more segments are needed.

First-order logic extended with a while-loop   Another extension is first-order logic with a while-loop. Basically, in this language first-order definable relations can be created using the input spatial dataset and previously created relations. Also, a while-loop with a first-order expressible stop-condition is allowed. These two constructs are illustrated in the following program to test linear-path connectivity.

$Seg := \{(x, y, x', y') \mid \forall \lambda (0 \leqslant \lambda \leqslant 1 \wedge \forall u \forall v ((u, v) = \lambda(x, y) + (1 - \lambda)(x', y') \rightarrow S(u, v)))\};$

$Path_1 := Seg;$

$Path_2 := \{(x, y, x', y') \mid \exists u \exists v (Path_1(x, y, u, v) \wedge Seg(u, v, x', y'))\};$

while $Path_1 \neq Path_2$

   do

   $Path_1 := Path_2;$

   $Path_2 := \{(x, y, x', y') \mid \exists u \exists v (Path_1(x, y, u, v) \wedge Seg(u, v, x', y'))\};$

   od

$R_{out} := \{() \mid \forall x \forall y \forall x' \forall y' ((S(x, y) \wedge S(x', y')) \leftrightarrow Path_2(x, y, x'x, y'))\};$

In the *Seg* relation, couples of points are collected that can be connected by a line segment that is completely in S. Next, in the while-loop, the transitive closure of this relation is computed and the output relation R_{out} reflects whether all pairs of points of S are in this transitive closure.

This example shows that topological connectivity of linear spatial datasets can be expressed in this language, but in fact we have the following more powerful result (Gyssens et al., 1999).

Theorem 2.11 *First-order logic extended with a while-loop is a computationally complete language on spatial (semi-algebraic) datasets.*

By a computationally complete query language we mean that every mapping from dataset to dataset, that is effectively computable by an algorithm (working on representations of datasets by defining formulas), is expressible by a program in the language.

2.9 First-order logic extended with transitive-closure operators

In the previous section, we have seen that, at least for linear datasets, the ability to express the transitive closure (TC) suffices to express connectivity. In this section, we describe the extension of first-order logic with various transitive-closure operators more extensively.

We cannot add TC with its standard mathematical semantics, indeed $TC(\{(x, y) \mid y = 2x\})$ is not a semi-algebraic set. We look at the TC-operator as a programming construct with a purely operational semantics and therefore $TC(\{(x, y) \mid y = 2x\})$ is regarded as a non-terminating computation.

Transitive-closure logic More precisely, first-order logic over the reals is extended with expressions of the form

$$[TC_{\vec{x};\vec{y}}\ \psi(\vec{x}, \vec{y})](\vec{s}, \vec{t})$$

where $\vec{x}, \vec{y}$ are k-tuples of real variables bound by the TC-operator and $\vec{s}$ and $\vec{t}$ are k-tuples of variables serving as the parameters of the TC-formula. The evaluation on a input dataset A is then obtained as follows. We set $X_1 := \psi(A)$, and $X_{i+1} := X_i \cup \{(\vec{x}, \vec{y}) \in \mathbb{R}^{2k} \mid (\exists \vec{z})(X_i(\vec{x}, \vec{z}) \wedge X_1(\vec{z}, \vec{y}))\}$, and stop the computation as soon as $X_{i+1} = X_i$. The semantics of $[TC_{\vec{x};\vec{y}}\ \psi(\vec{x}, \vec{y})](\vec{s}, \vec{t})$ is then defined as the $2k$-ary relation $X_i(\vec{s}, \vec{t})$.

For example, $[TC_{x;y} S(x, y)](s, t)$ evaluated on $A = \{(x, y) \mid y = 2x\}$ gives $X_1 = \{(s, t) \mid t = 2s\};\ X_2 = X_1 \cup \{(s, t) \mid t = 4s\} = \{(s, t) \mid t = 2s \vee t = 4s\}; X_3 = X_2 \cup \{(s, t) \mid t = 8s\} = \{(s, t) \mid t = 2s \vee t = 4s \vee t = 8s\}; \ldots$ which is a non-terminating computation (illustrated in Figure 2.15).

On the other hand, the connectivity of linear spatial datasets in the plane can be expressed by the formula

$$\forall \vec{x} \forall \vec{y}(S(\vec{x}) \wedge S(\vec{y}) \rightarrow [TC_{\vec{r},\vec{s}}(Seg(\vec{r}, \vec{s})](\vec{x}, \vec{y}))$$

with $Seg = \{(\vec{r}, \vec{s}) \mid (\exists \lambda)(0 \leqslant \lambda \leqslant 1 \wedge (\forall \vec{t})((\vec{t} = \lambda \cdot \vec{r} + (1 - \lambda) \cdot \vec{s}) \rightarrow S(\vec{t})))\}$.
On linear sets this expression gives rise to a terminating computation. The

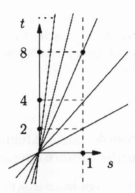

Figure 2.15 A non-terminating transitive-closure computation.

number of iterations is bounded by the number of triangles needed to triangulate the input (or the number of line segments in a piecewise linear line).

Transitive-closure logic with stop conditions   A variant of the above transitive-closure logic is first-order logic with expressions of the form

$$[\text{TC}_{\vec{x};\vec{y}}\ \psi(\vec{x}, \vec{y}) \mid \sigma](\vec{s}, \vec{t})$$

where additionally σ is an first-order definable stop condition. The evaluation of this expression on input database A is again the computation of X_1, X_2, $X_3, \ldots$ as above but with the additional stop condition $(A, X_{i+1}) \models \sigma$. Remark that we do not allow parameters inside TC-expressions, i.e., $\vec{s}$ and $\vec{t}$ are the only free variables of the entire TC-subformula.

For example, if $[\text{TC}_{x;y}\ S(x, y) \mid X(1, 8)](s, t)$ is evaluated on $A = \{(x, y) \mid y = 2x\}$, then $X_1 = \{(s, t) \mid t = 2s\}$; $X_2 = X_1 \cup \{(s, t) \mid t = 4s\} = \{(s, t) \mid t = 2s \vee t = 4s\}$; $X_3 = X_2 \cup \{(s, t) \mid t = 8s\} = \{(s, t) \mid t = 2s \vee t = 4s \vee t = 8s\}$; and the computation terminates because $(1, 8) \in X_3$.

The above two languages were introduced in (Geerts and Kuijpers, 2005).

K-transitive-closure logic   Another variant of transitive-closure logic was proposed in (Kreutzer, 2001) and we call it *K-transitive-closure logic*. In K-transitive-closure logic the transitive-closure operator may be applied to parameterized sets and the evaluation of a transitive-closure expression may be controlled by the termination of particular paths in its computation rather than by the termination of the transitive closure of the complete set.

More formally, it is first-order logic over the reals extended with expressions of the form

$$[\text{TC}_{\vec{x};\vec{y}}\ \psi(\vec{x}, \vec{y}, \vec{u})](\vec{s}, \vec{t}),$$

where $\vec{u}$ is an ℓ-tuple, thus allowing a parameter *inside* the TC-formula. The evaluation on an input dataset A is obtained in stages as follows. First, we set $X_1 := \psi(A) \wedge \bigwedge_{i \in I}(s_i = x_i)$; and then continue $X_{i+1} := X_i \cup \{(\vec{x}, \vec{y}, \vec{u}) \in \mathbb{R}^{2k+\ell} \mid (\exists \vec{z})(X_i(\vec{x}, \vec{z}, \vec{u}) \wedge \psi(\vec{z}, \vec{y}, \vec{u}))\}$; and we stop the computation as soon as $X_i = X_{i+1}$. The semantics of $[\mathrm{TC}_{\vec{x};\vec{y}} \, \psi(\vec{x}, \vec{y}, \vec{u})](\vec{s}, \vec{t})$, is then defined to be X_i. An example follows in Section 2.11.

We remark that the initial transitive-closure logic is a subset of K-transitive-closure logic.

2.10 Expressiveness properties of transitive-closure logics

The results in this section can be found in (Geerts and Kuijpers, 2000; Geerts et al., 2006). For the transitive-closure logic with stop condition we have the following expressiveness result.

Theorem 2.12 *All computable queries on linear spatial datasets, definable by linear polynomials with coefficients in $\mathbb{Z}$, are expressible in the transitive-closure logic with stop conditions.*

The previous result even holds when we disallow the use of multiplication in the query expression.

The proof of Theorem 2.12 can be sketched as follows. Let Q be a computable query on linear spatial datasets. Then we will write Q as a composition $Q_5 \circ Q_4 \circ Q_3 \circ Q_2 \circ Q_1$ of five queries that are expressible in transitive-closure logic with stop conditions. To start with, Q_1 produces on input a spatial dataset S in $\mathbb{R}^n$ a triangulation of S. Since S can described by linear polynomials with coefficients in $\mathbb{Z}$, these corner points of the triangulation will be rational numbers. This encoding (and the corresponding decoding Q_5) can actually be done in first-order logic over the reals.

The queries Q_2 and Q_4 are the encoding/decoding of finite relations over the rational numbers into single natural numbers. Finally Q_3 is the query that simulates Q on the natural number encodings of spatial datasets. The existence of a formula for Q_3 is guaranteed by the following powerful lemma.

Lemma 2.13 *For every partial computable function $f : \mathbb{N}^k \to \mathbb{N}$ there exists a formula $\varphi_f(y)$ in transitive-closure logic with stop conditions over the schema $\mathcal{S} = \{S^{(k)}\}$, such that for any database D over $\mathcal{S}$ with $S^D = \{(n_1, ..., n_k)\}$, we have that $\varphi_f(D)$ is defined if and only if $f(n_1, \ldots, n_k)$ is defined, and in this case $\varphi_f(D) = \{f(n_1, \ldots, n_k)\}$.*

This can be shown by simulating the run of a non-deterministic p-counter machine $M_f = (Q, \delta, q_0, q_f)$ which computes f.

We remark that the above theorem is limited to linear spatial datasets; it is not known whether a finite encoding of an arbitrary semi-algebraic set can be expressed in transitive-closure logic with stop conditions. But for queries on arbitrary spatial datasets we have completeness if we restrict our attention to topological properties.

Theorem 2.14 *All computable* Boolean topological *queries on arbitrary spatial datasets in* $\mathbb{R}^n$ *are expressible in transitive-closure logic with stop conditions.*

This theorem can be proven by showing that we can approximate a spatial dataset by a $\mathbb{Z}$-linear spatial database that is toplogically equivalent to it in transitive-closure logic with stop conditions and by using the first expresiveness (Theorem 2.12) result.

To be more precise about the approximation, we now describe how a rational ε-approximation of a spatial dataset can be expressed in transitive-closure logic with stop conditions. The task here is, given a spatial dataset S and a real number $\varepsilon > 0$, to find a $\mathbb{Z}$-linear spatial dataset that is homeomorphic to S (topological condition) and that ε-approximates S (metric condition).

It is not difficult to show that ε-approximations cannot be expressed in first-order logic over the reals (Geerts and Kuijpers, 2000). We now sketch, for the case of $\mathbb{R}^2$, how a rational ε-approximation of a spatial dataset S can be expressed in transitive-closure logic with stop conditions.

First, we find all points where the boundary of S is not smooth and border points with a vertical tangent line. For these points, we compute their local cone radius (conicity around points was discussed in Section 2.4) and within these radii we locally rectify the database. This is all expressible in first-order logic over the reals (Geerts and Kuijpers, 1999; Geerts et al., 2006) and illustrated in Figure 2.16.

Next, we consider the border of S outside the cone radii determined in the first step. What remains of the border are simple curves. We then compute the maximal cone radius r of all points on these curves in first-order logic. Let $Step$ be the relation of pairs of points (p, q) on these curves such that $d(p, q) = r$. In transitive-closure logic, we can compute TC($Step$) and the termination of this computation is guaranteed. Once this is done, we walk over these curves starting from the endpoints and locally rectify them. This is illustrated in Figure 2.17. We can also do this when the curves are bordering curves of the interior.

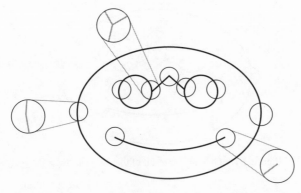

Figure 2.16 Local rectification around non-smooth border points and points with a vertical tangent line.

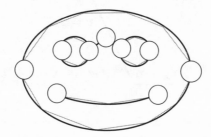

Figure 2.17 Rectification of the border away from non-smooth border points.

Finally, we glue the result of the first step onto the result of the second step. The final result is a $\mathbb{Z}$-linear database homeomorphic to the original semi-algebraic set (illustrated in Figure 2.18).

In dimensions higher than 2, the description of rational ε-approximation in first-order logic with transitive closure with stop condition technically more complicated. It follows a recursive procedure on the dimension and boxes are used instead of spheres to describe the cones of points.

We end this section with a corollary of Theorem 2.14 that concerns connectivity, a property that is important in applications such as geographical information systems.

Corollary 2 *Topological connectivity of (even non-linear!) spatial datasets is expressible in transitive-closure logic with stop conditions.*

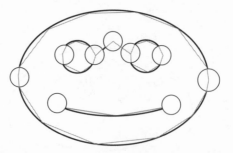

Figure 2.18 Glueing the result of the two steps together.

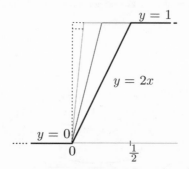

Figure 2.19 A second example of a non-terminating evaluation of transitive closure.

2.11 Deciding termination of transitive-closure logic expressions

The results in this section can be found in (Geerts and Kuijpers, 2005).

From the definition in Section 2.8, it is clear that the evaluation of queries expressed in transitive-closure logic with or without stop conditions may be non-terminating.

Some examples As we have seen before, when we evaluate the transitive closure, that is, the expression $[\text{TC}_{x;y} S(x, y)](s, t)$, on the spatial dataset $A = \{(x, y) \mid y = 2x\}$, we get a growing number of lines (which is illustrated in Figure 2.15). In other words, this is a non-terminating computation.

Even if we modify the function $y = 2x$, as shown by the thick line in Figure 2.19, that is, even if we bound its image, the computation of the transitive closure remains non-terminating (as illustrated in Figure 2.19).

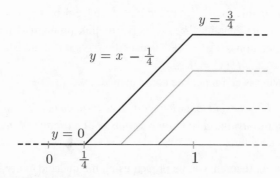

Figure 2.20 An example of a terminating evaluation of transitive closure.

But there are function graphs on which the computation of the transitive closure terminates. The function shown by the thick line in Figure 2.20, is an example. Here we have a terminating computation because $X_4 = X_5 = X_6 = \cdots$ (illustrated by the thinner lines in Figure 2.20).

Obviously, adding stop-conditions may make non-terminating computations terminating, as is illustrated by the example we gave earlier: if we apply the query expressed by $[TC_{x;y} S(x, y) \mid X(1, 8)](s, t)$ to the dataset $A = \{(x, y) \mid y = 2x\}$ given in Figure 2.15, the computation terminates because, $(1, 8) \in X_3$.

Also, when $[TC_{x;y} S(x, y) \mid \exists x \exists y X(x, y) \wedge y = 1 \wedge 10x < 1](s, t)$ is applied to the function graph given in Figure 2.19, we get a terminating computation because the fourth set we compute satisfies the stop condition, that is, $\exists x \exists y X_4(x, y) \wedge y = 1 \wedge 10x < 1$.

Obviously, K-transitive closure logic also has the problem of non-terminating evaluations, since it contains transitive-closure logic. We give an example of a terminating evaluation. Consider the evaluation of $[TC_{x;y} S(x, y)](\frac{1}{4}, t)$ on the graph of the function given in Figure 2.19. This gives $X_1 = \{(\frac{1}{4}, \frac{1}{2})\}$, $X_2 = \{(\frac{1}{4}, \frac{1}{2}), (\frac{1}{2}, 1)\}$ and $X_3 = X_2$.

Deciding termination and undecidability results We can ask whether there is a procedure to decide, for given an expression φ in some transitive-closure logic and a dataset A whether φ has terminating evaluation on A?

It is fairly easy to obtain the following undecidability results (Geerts and Kuijpers, 2005).

Theorem 2.15 *It is undecidable whether a given formula in transitive-closure logic, that uses transitive closure on relations of at most arity 4, terminates on a given input database.*

The proof of this fact is by reduction of undecidability of nilpotency of a piecewise affine function $f : \mathbb{R}^2 \to \mathbb{R}^2$ to this problem (Blondel et al., 2001a,b). A function $f : \mathbb{R}^n \to \mathbb{R}^n$ is called *nilpotent* if there is a $k \geqslant 1$ such that for all $\vec{x} \in \mathbb{R}^n$: $f^k(\vec{x}) = \vec{0}$.

The following is an immediate consequence.

Corollary 3   *It is undecidable whether a given formula in K-transitive-closure logic, that uses transitive closure on relations of at most arity 4, terminates on a given input database.*

The following theorem can be proven using the undecidability of Hilbert's 10th problem.

Theorem 2.16   *It is undecidable whether a given formula in transitive-closure logic with stop conditions, that uses transitive closure on at most binary relations, terminates on a given input database.*

These results are complete for the languages all three types of transitive-closure logics that we have considered, apart from the cases of (K-)transitive-closure restricted to work on binary relations.

We have the following open problem: *Is it decidable whether a given formula in K-transitive-closure logic restricted to binary relations terminates on a given input database?* This problem is related to an open problem in dynamical systems theory, namely the *point-to-fixed-point problem* which asks whehther for a given algebraic number x_0 and a given piecewise affine function $f : \mathbb{R} \to \mathbb{R}$, the sequence $x_0, f(x_0), f^2(x_0), f^3(x_0), \ldots$ reaches a fixed point? This decision problem is open, even for piecewise f consisting of just two line segments.

We have a second open problem: *Is it decidable whether a given formula in transitive-closure logic restricted to binary relations terminates on a given input database?* In this case, this problem is related to deciding nilpotency of functions $f : \mathbb{R}^n \to \mathbb{R}^n$. That is, deciding termination of

$$[\mathrm{TC}_{\vec{x};\vec{y}}\, S(\vec{x}, \vec{y})](\vec{s}, \vec{t})$$

applied to graph(f) adds up to deciding nilpotency of f. But nilpotency of (possibly discontinuous) functions $f : \mathbb{R}^n \to \mathbb{R}^n$ is not known to be undecidable for $n = 1$, whereas it is for $n > 1$.

Terminating functions   We say that a set $A \subseteq \mathbb{R}^2$ has *terminating transitive closure* if the formula $[\mathrm{TC}_{x;y}\, S(x, y)](s, t)$ terminates on input A and we call a function $f : \mathbb{R} \to \mathbb{R}$ *terminating* if graph(f) has a terminating transitive closure.

The function of Figure 2.15 is not terminating, but the one of Figure 2.20 is. We have the following, for what follows, important result.

Theorem 2.17 *There is a procedure that on input a continuous semi-algebraic function $f : \mathbb{R} \to \mathbb{R}$ decides whether it is terminating. Furthermore, this decision procedure is expressible in first-order logic over the reals.*

To get to this result, we need some terminology and lemmas. Let $f : \mathbb{R} \to \mathbb{R}$ be a continuous function. We call $x \in \mathbb{R}$ is a *periodic point of* f if $f^d(x) = x$ for some $d \geqslant 1$ and the smallest such d is the *period of* x. Let $\text{Per}(f)$ denote the set of periodic points of f.

We have the following properties.

Lemma 2.18 *If f is continuous and terminating, then $\text{Per}(f) = f^k(\mathbb{R})$ (for some k) is non-empty, closed and connected. Furthermore, $\text{Per}(f) = \{x \in \mathbb{R} \mid f^2(x) = x\}$.*

A result by Sharkovskiĭ's from 1964 implies that if f is continuous and terminating, then only periods $1, 2, 4, \ldots, 2^d$ can appear for some integer value $d \geqslant 1$. The last part of the lemma is more specific.

So, we can conclude the following.

Corollary 4 *If f is continuous and terminating, then f can only have periodic points with periods 1 and 2.*

We give the following crucial lemma without proof (see (Geerts and Kuijpers, 2005) for details).

Lemma 2.19 *There is an first-order sentence that expresses whether a continuous semi-algebraic function $f : \mathbb{R} \to \mathbb{R}$ is nilpotent.*

We remark that the proof of the correctness of the first-order sentence of Lemma 2.19, given in (Geerts and Kuijpers, 2005), relies on the Bolzano-Weierstrass theorem and therefore does not generalize to arbitrary real closed fields.

Using the above results, we obtain the following procedure to decide termination of a function $f : \mathbb{R} \to \mathbb{R}$.

In the following algorithm, $\tilde{f}$ is obtained from f by contracting the part of f that consists of of points of period 1 or 2, to one point. This is illustrated in Figure 2.21.

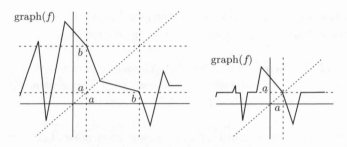

Figure 2.21 The contraction $\tilde{f}$ of a function f.

Algorithm TERMINATE(input f):

Step 1. Compute the sets $C_1 = \{x \mid f(x) = x\}$ and $C_2 = \{x \mid f^2(x) = x\}$. If C_2 is a closed and connected subset of $\mathbb{R}$ and if C_1 is a point with $C_2 \setminus C_1$ around it or if $C_2 \setminus C_1$ is empty, then continue with Step 2, else answer *no*.

Step 2. If C_2 is $\mathbb{R}$, answer *yes*, else compute the function $\tilde{f}$ and decide whether $\tilde{f}$ is NILPOTENT and return the answer.

We remark that $f^k(\mathbb{R}) = C_1 \cup C_2$ if and only if $\tilde{f}^k(\mathbb{R}) = \{0\}$.

We now illustrate the working of the algorithm TERMINATE.

First, consider the a non-terminating function of Figure 2.19. Here, we have $C_1 = \{(0, 0), (1, 1)\}$ and $C_2 = \emptyset$ and the algorithm answers *no* since $C_1 \cup C_2$ is disconnected.

Next, consider the terminating function of Figure 2.20. Here we have $C_1 = \{(0, 0)\}$ and $C_2 = \emptyset$, and thus $C_1 \cup C_2$ is closed and connected and since $\tilde{f} = f$ is nilpotent the algorithm answers *yes*.

Small extensions of first-order logic with transitive closure   The above decidability result concerning termination of functions inspires the following extension of first-order logic with transitive closure restricted to be applied to the graphs of continuous functions $\mathbb{R}^k \to \mathbb{R}^k$. More precisely, let $\psi(\vec{x}, \vec{y})$ be a formula in transitive-closure logic (with or without stop condition). Consider the sentence

$$\gamma_\psi = \gamma_\psi^1 \wedge \gamma_\psi^2,$$

where γ_ψ^1 expresses that $\psi(\vec{x}, \vec{y})$ defines the graph of a function from $\mathbb{R}^k$ to $\mathbb{R}^k$ and γ_ψ^2 expresses that $\psi(\vec{x}, \vec{y})$ defines a continuous function graph. We note that γ_ψ^1 and γ_ψ^2 are first-order related to γ.

Then it is easy to see that $\psi(\vec{x}, \vec{y})$ terminates on input A if and only if γ_ψ terminates on A.

Now, we define *CF-transitive-closure logic*, an extension of first-order logic with a transitive-closure operator that is restricted to graphs of continuous functions $\mathbb{R}^k \to \mathbb{R}^k$.

More formally, CF-transitive-closure logic is the fragment of transitive-closure logic (respectively with stop condition) where expressions of the form

$$[\mathrm{TC}_{\vec{x};\vec{y}}\ \psi(\vec{x},\vec{y}) \wedge \gamma_\psi](\vec{s},\vec{t})$$

(respectively $[\mathrm{TC}_{\vec{x};\vec{y}}\ \psi(\vec{x},\vec{y}) \wedge \gamma_\psi \mid \sigma](\vec{s},\vec{t})$) are allowed, with γ_ψ a sentence that expresses that $\psi(\vec{x},\vec{y})$ defines the graph of a continuous function $\mathbb{R}^k \to \mathbb{R}^k$.

From earlier undecidability results we immediately get

Corollary 5 *It is undecidable whether a given formula in CF-transitive-closure logic with stop condition, where the transitive closure is restricted to work on binary relations, terminates on a given input database.*

Without stop conditions, we have decidability, however.

Theorem 2.20 *It is decidable whether a given formula in CF-transitive-closure logic (without stop conditions), restricted to binary relations, terminates on a given input database. Moreover, this decision procedure is expressible in this language.*

So, for every formula $[\mathrm{TC}_{x;y}\ \psi(x,y)](s,t)$ in CF-transitive-closure logic (without stop conditions), restricted to binary relations, there is a a formula τ_ψ in the same language that expresses that the formula terminates on a given input database (also τ_ψ depends on the input!). We call τ_ψ the *termination guard* of the formula $[\mathrm{TC}_{x;y}\ \psi(x,y)](s,t)$.

Now, we can define *GCF-transitive-closure logic*, the guarded fragment of CF-transitive-closure logic (without stop conditions), restricted to binary relations, in which only transitive-closure expressions of the form

$$[\mathrm{TC}_{x;y}\ \psi(x,y) \wedge \tau_\psi](s,t)$$

are allowed.

By definition, we now have that in GCF-transitive-closure logic every query terminates on all possible input datasets and all terminating queries of CF-transitive-closure logic are expressible in GCF-transitive-closure logic.

We end this section, with an expressibility result concerning these last two languages.

Theorem 2.21 *GCF-transitive-closure logic is more expressive than first-order logic on finite spatial datasets.*

For the last part of the theorem, we remark that the query Q_{int} on 1-dimensional datasets S that expresses "Is S a singleton that contains a natural number?" is expressible in GCF-transitive-closure logic but not in first-order logic.

2.12 Some concluding remarks on transitive-closure logics

One of the motivations to study these different transitive-closure logics, is to compare their expressive power and to establish which languages are computationally complete on linear or arbitrary (semi-algebraic) datasets.

It is not clear whether transitive-closure logic with stop conditions is more expressive than transitive-closure logic without stop conditions. In particular, it is not clear whether transitive-closure logic without stop conditions is also computationally complete on linear spatial datasets.

We also remark that for CF-transitive-closure logic without stop condition, termination is decidable and for CF-transitive-closure logic with stop condition termination is not decidable. This does not separate these languages, however (because equivalence is undecidable).

We also remark that many results on semi-algebraic functions also hold for arbitrary real closed fields. But termination of continuous semi-algebraic functions $f : R \rightarrow R$ for arbitrary real closed fields R is not first-order expressible (for $\mathbb{R}$ the proof relies on Bolzano-Weierstrass).

Bibliography

M. Benedikt, B. Kuijpers, C. Löding, J. Van den Bussche, T. Wilke. A characterization of first-order topological properties of planar spatial data. *Journal of the ACM*, 53(2):273–305, 2006.

V.D. Blondel, O. Bournez, P. Koiran, C.H. Papadimitriou, J.N. Tsitsiklis. Deciding stability and mortality of piecewise affine dynamical systems. *Theoretical Computer Science*, 255(1-2):687–696, 2001.

V.D. Blondel, O. Bournez, P. Koiran, J.N. Tsitsiklis. The stability of saturated linear dynamical systems is undecidable. *Journal of Computer and System Sciences*, 62(3):442–462, 2001.

J. Bochnak, M. Coste, M.-F. Roy. *Real Algebraic Geometry*. Springer, 1998.

M. Coste. Ensembles semi-algébriques. In *Géometrie algébrique réelle et formes quadratiques*, Lectures Notes in Mathematics vol. 959, Springer, 1982.

H.-D. Ebbinghaus, J. Flum. *Finite Model Theory*, second edition. Springer, 1999.

F. Geerts, B. Kuijpers. Expressing topological connectivity of spatial databases. *Proceedings of 7th International Workshop on Database Programming Languages (DBPL'99)*, Lecture Notes in Computer Science 1949, 221-235, Springer-Verlag, 2000.

F. Geerts, B. Kuijpers. Linear approximation of planar spatial databases using transitive-closure logic. *Proceedings of the 19th Symposium on Principles of Database Systems (PODS'00)* 126-135, ACM Press, 2000.

F. Geerts, B. Kuijpers. On the Decidability of Termination of Query Evaluation in Transitive-Closure Logics for Polynomial Constraint Databases. *Theoretical Computer Science*, Vol. 336, Nr. 1, 125-151, 2005.

F. Geerts, B. Kuijpers, J. Van den Bussche. Linearization and completeness results for terminating transitive closure queries on spatial databases. *SIAM Journal on Computing*, Volume 35, Issue 6, 1386-1439, 2006.

107

E. Grädel et al. *Finite Model Theory and Its Applications*. Springer, 2007.

M. Grohe, L. Segoufin. On first-order topological queries. *ACM Transactions on Computational Logic*, 3(3):336–358, 2002.

S. Grumbach, J. Su. Queries with arithmetical constraints. *Theoretical Computer Science*, 173(1):151–181, 1997.

M. Gyssens, J. Van den Bussche, D. Van Gucht. Complete geometric query languages. *Journal of Computer and System Sciences*, 58(1):54–68, 1999.

S. Kreutzer. Operational semantics for fixed-point logics on constraint databases. *Proceedings of the 8th International Conference on Logic for Programming, Artificial Intelligence, and Reasoning (LPAR'01)*, Lecture Notes in Computer Science 2250, 470–484. Springer-Verlag, 2001.

B. Kuijpers, J. Paredaens, J. Van den Bussche. On topological elementary equivalence of closed semi-algebraic sets in the real plane. *Journal of Symbolic Logic*, 65(4):1530–1555, 2000.

B. Kuijpers, J. Paredaens, M. Smits, J. Van den Bussche. Termination properties of spatial Datalog programs *Proceedings of "Logic in Databases" (LID'96)*, Lecture Notes in Computer Science 1154, 101-116, 1996.

L. Libkin. *Elements of Finite Model Theory*. Springer, 2004.

R. McNaughton, S. Papert. *Counter-Free Automata*. MIT Press, 1971.

J. Paredaens, J. Van den Bussche, D. Van Gucht. First-order queries on finite structures over the reals. *SIAM Journal on Computing*, 27(6):1747–1763, 1998.

L. Segoufin, V. Vianu. Querying spatial databases via topological invariants. *Journal of Computer and System Sciences*, 61(2):270–301, 2000.

W. Schwabhäuser, W. Szmielew, A. Tarski. *Metamathematische Methoden in der Geometrie*. Springer-Verlag, 1983.

W. Thomas. Languages, automata, and logic. In G. Rozenberg and A. Salomaa, editors, *Handbook of Formal Language Theory*, volume 3. Springer, 1997.

L. van den Dries, *Tame Topology and O-minimal Structures*, Cambridge University Press, Cambridge, UK, 1998.

M.F. Worboys, M. Duckham, *GIS: A Computing Perspective*, Second Edition, CRC Press, 2004.

3

Some connections between finite and infinite model theory

VERA KOPONEN[a]

3.1 Introduction

Most of the work in model theory has, so far, considered infinite structures and the methods and results that have been worked out in this context cannot usually be transferred to the study of finite structures in an obvious way. In addition, some basic results from infinite model theory fail within the context of finite models. The theory about finite structures has largely developed in connection with theoretical computer science, in particular complexity theory [12]. The question arises whether these two "worlds", the study of infinite structures and the study of finite structures, can be woven together in some way and enrich each other. In particular, one may ask if it is possible to adapt notions and methods which have played an important role in infinite model theory to the context of finite structures, and in this way get a better understanding of fairly large and sufficiently well-behaved classes of finite structures.

If we are to study structures in relation to some formal language, then the question arises which one to choose. Most of infinite model theory considers first-order logic. Within finite model theory various restrictions and extensions of first-order logic have been considered, since first-order logic may be considered as being both too strong and too weak (in different senses) for the study of finite structures. A reasonable candidate for studying finite structures, with a viewpoint from infinite model theory, is the language L^n, first-order logic L restricted to formulas in which at most n variables occur, whether free or bound. Theories consisting of only L^n-formulas, even those which are "complete" within L^n, may have both finite and infinite models, or only finite models, or only infinite models. The language L^n has the nice properties of being closed under subformulas, quantification and negation. Also, there is a

[a] Uppsala Universitet, vera.koponen@math.uu.se

pebble game which distinguishes whether two structures satisfy exactly the same L^n-sentences or not ([23] and implicitly in [29]).

The notion of a type plays an important role in infinite model theory. In finite model theory the notion of an L^n-type, i.e. a type restricted to L^n-formulas, has been used; the number of different L^n-types of an L^n-theory can be seen as a measure of the complexity of the theory. Dawar observed [5] that for every L^n-theory T with finite models there is an upper bound, depending only on the number of L^n-types (in n free variables) of T, of the size of the smallest model of T. Later Grohe proved that this upper bound is not recursive [17]. The language L^n has also been considered in the context of (only) infinite models in the work of Hedman [19] where complete theories (within full first-order logic) which are axiomatizable by L^n-sentences are studied. For a general overview about interactions (and differences) between finite and infinite model theory, see [30]. For a survey about the use of finite variable logics in finite model theory, see [16].

Within infinite model theory the area of stability theory has had great influence. It studies quite a large class of "manageable" (infinite) structures and their complete first-order theories. Work in the direction of developing the basics of a similar theory for finite structures was first carried out by Hyttinen [21]. Then, from a different viewpoint, the author developed some results, inspired by stability theory, aimed at understanding when an L^n-theory with infinite models must also have arbitrarily large finite models [9, 8]. Further developments in this direction were made by Baldwin and Lessmann [2] and by Hyttinen [22]. For an overview, with a historical perspective, of finite and infinite model theory and recent interactions between them, see [1].

Another approach to understanding certain finite and countably infinite structures culminates with the work about smoothly approximable structures in [3]. This line of research started with Lachlan's work on stable finitely homogeneous structures (surveyed in [27]) and Zilber's work on uncountably categorical structures [33]. It continued with joint work by Cherlin, Harrington and Lachlan on ω-categorical ω-stable structures [4] and then with the work of Kantor, Liebeck and Macpherson [24], to reach its current state in [3]. Smoothly approximable structures are infinite but can be approximated by "nicely embedded" finite structures which, intuitively speaking, are quite "homogeneous" or "regular". The theory of smoothly approximable structures can also be seen as a study of finite structures with few types.

More recently, a direction of research initiated by Macpherson and Steinhorn [28] and continued by Elwes [13, 14] and Ryten studies classes of finite structures in which definable sets have a uniform asymptotic behaviour, as the cardinalities of the universes increase. The complete theory T of a non-principal ultraproduct of such a class of finite structures (called an 'asymptotic class') is simple with finite SU-rank and there is a notion of measure on the definable subsets of models of T, but T is not necessarily smoothly approximable. See [15] for a survey of the topic.

In this article an overview is given of a line of research which considers L^n-theories with infinite models and tries to isolate conditions for when these have arbitrarily large finite models and when least upper bounds for the smallest model is recursive in terms of the number of L^n-types in n free variables. Although some results are stated within a more general context, considering some arbitrary fraction of first-order logic which is closed under subformulas, and some results could be stated in a somewhat more general way, we mostly stick to the language L^n for the sake of simplicity. Moreover, when working with L^n we usually consider an L^n-theory T such that T is finitely axiomatizable in L^n and complete (within L^n) in the sense that for every $\varphi \in L^n$, $T \models \varphi$ or $T \models \neg\varphi$. The motivation is that we like to find conditions for T which imply that T has a finite model, and facts 3.2.6 and 3.2.7 below imply the following: if T is an L^n-theory and no complete L^n-theory $T' \supseteq T$ exists such that T' is finitely axiomatizable, then T has no finite model.

The basic idea is to isolate conditions for a finitely axiomatizable complete L^n-theory T which guarantee the existence of a model M of T which is smoothly approximable, since such an M has the property that every sentence which is true in M is true in arbitrarily large finite substructures of M. Moreover, in this situation the theory of smoothly approximable structures implies that a recursive upper bound, in terms of the number of L^n-types in n free variables, of the smallest model exists (in contrary to the general situation, as proved in [17]).

Sections 1 – 7 of this article try to unify, as much as possible, the approaches of [9], [8] and [2]. Hyttinen's paper [22] on canonical finite diagrams and quantifier elimination is highly recommended since it develops, in a more general context, part of the theory and several of the results. Here I have chosen to expose the subject via a more "down-to-earth"-approach focused on L^n-theories, although some generality is lost.

Sections 3.6 and 3.7 discuss infinite structures which have the finite model property but which are not necessarily smoothly approximable (the random bipartite graph is an example [24]). This may be useful for understanding other classes of L^n-theories than those treated in earlier sections. The last section contains a list of questions and problems.

3.2 Preliminaries

3.2.1 Finite variable logic

In this section we introduce the language L^n, the subset of L containing all formulas in which at most n distinct variables occur.

Definition 3.2.1   (i) Let $V = \{v_1, v_2, v_3, \ldots\}$ be the set of variables which are used in formulas of L.

(ii) By $x, y, z, \bar{x}, \bar{y}, \bar{z}$, sometimes with indices, we denote variables and finite sequences of variables. Similarly, $a, b, c, \bar{a}, \bar{b}, \bar{c}$ denote elements and finite sequences of elements from structures. When writing $\bar{a} \in A$ we mean that every element of the sequence $\bar{a}$ belongs to A. If, in addition, we like to stress that $\bar{a}$ has length n, then we may write $\bar{a} \in A^n$.

(iii) L always denotes the set of all first-order formulas over some vocabulary (or signature). We always assume that L is countable.

(iv) If a formula in L is denoted by $\varphi(\bar{x})$ then we mean that every free variable in that formula belongs to the sequence $\bar{x}$.

(v) For any $n < \omega$, L^n denotes the set of all formulas $\varphi(\bar{x}) \in L$ such that at most n distinct variables occur in $\varphi(\bar{x})$ (whether bound or free). We allow $\bar{x}$ to contain "dummy variables" (not occuring in the formula denoted by $\varphi(\bar{x})$). For example, the formula $v_1 = v_2 \vee v_2 = v_3$ may be denoted by $\varphi(v_1, v_2, v_3, v_4)$ and consequently $\varphi(v_1, v_2, v_3, v_4) \in L^3$, because only three variables actually occur in the formula $\varphi(v_1, v_2, v_3, v_4)$.

(vi) An L^n-theory is a set of sentences from L^n.

(vii) An L^n-theory T is called a *complete L^n-theory* if for every sentence $\varphi \in L^n$, $T \models \varphi$ or $T \models \neg\varphi$. Of course, a 'complete L^n-theory' need not be complete with respect to L.

(viii) If M is an L-structure let

$$Th_{L^n}(M) = \{\varphi \in L^n : \varphi \text{ is a sentence and } M \models \varphi\}.$$

So $Th_{L^n}(M)$ is always a complete L^n-theory.

Remark 3.2.2   We have not fixed n special variables to be used in formulas of L^n, but we only say that at most n distinct variables may occur in a formula of L^n. For instance, a formula of L^n may contain variables among $v_1, \ldots, v_n$ or variables among $v_{n+1}, \ldots, v_{2n}$. For example, both $v_1 = v_2 \vee v_2 = v_3$ and $v_2 = v_3 \vee v_3 = v_4$ belong to L^3.

We are interested in finding conditions under which an L^n-theory with infinite models also has (arbitrarily large) finite models. So we first give some easy examples showing that L^n-theories may have only infinite models, only finite models or both infinite and finite models

Examples 3.2.3   (a) Let $M = (\mathbb{N}, S)$, where S is the successor function (or relation). Then $Th_{L^3}(M)$ has no finite model.

(b) If $M = (\mathbb{Q}, <)$ then $Th_{L^3}(M)$ has no finite model.

(c) Let M be a finite graph such that $v_1, \ldots, v_m$ lists all vertices of M and there is an edge between v_i and v_j if and only if $j = i + 1$ or $i = j + 1$. Then every model of $Th_{L^3}(M)$ is isomorphic to M.

(d) Let M be an infinite tree such that for some $m < \omega$, no path in M has length m. Then $Th_{L^n}(M)$ has arbitrarily large finite models, for any n.

(e) Let K be a finite field. Let T be a set of sentences which expresses the axioms of a K-vector space. With scalar multiplication and vector addition represented by function symbols we may assume that T is an L^3-theory. With scalar multiplication and vector addition represented by relation symbols we may assume that T is an L^7-theory. T has arbitrarily large finite models and hence infinite models.

3.2.2 Types

The notion of a 'type' plays an important role in model theory. Here we will in particular be interested in certain types which are restricted to formulas of some sublanguage of the first-order language L. We first give some definitions with associated notation and then state a few well-known results concerning types.

Definition 3.2.4 (i) Let $\Phi \subseteq L$, let M be an L-structure and let $A \subseteq M$.
(ii) Define

$$Th_\Phi(M, A) = \{\varphi(\bar{a}) : \varphi(\bar{x}) \in \Phi, \ \bar{a} \in A, \ M \models \varphi(\bar{a})\},$$

and let $Th_\Phi(M) = Th_\Phi(M, \emptyset)$. So $Th_\Phi(M)$ is the set of sentences in Φ that are true in M.
(iii) For a sequence of variables $\bar{x}$ we define

$$\Phi_{\bar{x}}(A) = \{\varphi(\bar{x}, \bar{a}) : \varphi(\bar{x}, \bar{y}) \in \Phi, \ \bar{a} \in A\},$$

and $\Phi_{\bar{x}} = \Phi_{\bar{x}}(\emptyset)$.
(iv) A Φ-*type* over A (with respect to $Th_\Phi(M, A)$) in the free variables $\bar{x}$ is a set $p(\bar{x}) \subseteq \Phi_{\bar{x}}(A)$ such that $p(\bar{x}) \cup Th_\Phi(M, A)$ is consistent.
(v) A Φ-type $p(\bar{x})$ over A is called a *complete* Φ-type over A if whenever $\varphi(\bar{x}, \bar{a}) \in \Phi_{\bar{x}}(A)$ then $\varphi(\bar{x}, \bar{a}) \in p(\bar{x})$ or $\neg\varphi(\bar{x}, \bar{a}) \in p(\bar{x})$.
(vi) $S_m^\Phi(A, M)$ is the set of all complete Φ-types over A (with respect to $Th_\Phi(M, A)$) in the free variables $v_1, \ldots, v_m$. If $\Phi = L$ then we may omit it.
(vii) We write $S_m^n(A, M)$ instead of $S_m^{L^n}(A, M)$.
(viii) For a complete L^n-theory T we define $S_m^n(T) = S_m^n(\emptyset, M)$, where M is any model of T (so $Th_{L^n}(M, \emptyset) = T$). By Lemma 1.2 in [9] this definition does not depend on the choice of the model M of T.

Below are a few facts about L^n-types.

Fact 3.2.5 *For any complete L^n-theory T, $S_n^n(T)$ is finite if and only if there are only finitely many L^n-formulas in the free variables $v_1, \ldots, v_n$ up to equivalence modulo T.*

The previous fact is a consequence of the "Stone duality theorem for boolean algebras" [20], but it can also be derived in a straightforward way from the definitions.

The next fact can be extracted from the proof of a similar result in [6], and it is also mentioned in [29] (in Exercise 4).

Fact 3.2.6 (Dawar, Lindell, Weinstein; Poizat) *Suppose that the vocabulary of L is finite and contains no function symbols. If T is a complete L^n-theory and $S_n^n(T)$ is finite then there is $\varphi \in L^n$ that axiomatizes T (i.e. $\varphi \models T$ and $T \models \varphi$). Moreover, we can choose φ so that its quantifier rank is at most $|S_n^n(T)| + n$.*

The next fact is easy to prove, but a proof can also be found in [9].

Fact 3.2.7 *(i) If a complete L^n-theory T has a finite model M then $|S_n^n(T)| \leq |M|^n$.*
(ii) If T is a complete L^n-theory and $S_n^n(T)$ is infinite then T has no finite model.

Using Fact 3.2.7, when looking for finite models of a complete L^n-theory T we can rule out the case when $S_n^n(T)$ is infinite.

3.2.3 Closure maps

Definition 3.2.8 Let M be an L-structure.
(i) We call a function cl : $\mathcal{P}(M) \to \mathcal{P}(M)$ a *closure map* (or *closure function*) if whenever $A \subseteq B \subseteq M$ then $A \subseteq \mathrm{cl}(A)$, $\mathrm{cl}(\mathrm{cl}(A)) = \mathrm{cl}(A)$ and $\mathrm{cl}(A) \subseteq \mathrm{cl}(B)$.
(ii) If $\mathrm{cl}(A) = A$ then we say that A is *closed*. A sequence is closed if the set of elements occurring in the sequence is closed.

A few examples of closure maps are given below:

Examples 3.2.9 (a) If $\mathrm{cl}(A) = A$ for every $A \subseteq M$ then cl is a closure map; we say that such a closure map is *trivial*.
(b) If, for every $A \subseteq M$, $\mathrm{cl}(A)$ is the substructure of M which is generated by A, then cl is a closure map.
(c) If, for every $A \subseteq M$, $\mathrm{cl}(A)$ is the algebraic closure of A, in the model theoretic sense, then cl is a closure map. See [20] for a definition of 'algebraic closure' in the model theoretic sense.

(d) If M is an algebraically closed field and cl(A) is the algebraic closure of A, in the sense of field theory, then cl is a closure map.

(e) If M is a vector space and cl(A) is the linear span of A, then cl is a closure map.

3.3 Amalgamation classes

In order to prove that arbitrarily large finite models of a finitely axiomatizable theory T exist we prove that a particularly nice model M of T exists. This M will have the property that every sentence which is true in M is true in a finite substructure of it. Such an M exists if there is an "amalgamation class" of models of T and all models in the amalgamation class are "stable". The definition of an amalgamation class is given in this section (Definition 3.3.5) and the notion of 'stability' is treated in Section 3.4.

Definition 3.3.1 Let $\Phi \subseteq L$ and let M and N be L-structures.
(i) If $a_i \in M$, $b_i \in N$, for $i < \lambda$, then we write

$$(M, (a_i : i < \lambda)) \equiv_\Phi (N, (b_i : i < \lambda))$$

if for every $m < \omega$, every $\varphi(x_1, \ldots x_m) \in \Phi$ and every $\{i_1, \ldots, i_m\} \subseteq \lambda$,

$$M \models \varphi(a_{i_1}, \ldots, a_{i_m}) \text{ if and only if } N \models \varphi(b_{i_1}, \ldots, b_{i_m}).$$

(ii) A function $f : A \to N$, where $A \subseteq M$, is called a Φ-*elementary embedding* if for every $\varphi(\bar{x}) \in \Phi$ and $\bar{a} \in A$ with $|\bar{a}| = |\bar{x}|$, we have

$$M \models \varphi(\bar{a}) \quad \text{if and only if} \quad N \models \varphi(f(\bar{a})).$$

(iii) If M is a substructure of N and for every $\varphi(\bar{x}) \in \Phi$ and every $\bar{a} \in M$ with $|\bar{a}| = |\bar{x}|$, $M \models \varphi(\bar{a})$ if and only if $N \models \varphi(\bar{a})$, then we say that M is a Φ-*elementary substructure* of N and that N is a Φ-*elementary extension* of M, denoted $M \preccurlyeq_\Phi N$. As usual we may write $\preccurlyeq$ instead of $\preccurlyeq_L$.

In the next section we will use the following result which is proved in the same way as the well-known Tarski-Vaught test [20]; for the proof we only need to observe that L^n is closed under subformulas.

Fact 3.3.2 (Tarski-Vaught test for L^n) *Suppose that n is greater than the arity of every function symbol in the vocabulary of L and let M be an L-structure. For any subset N of M, we have $N \preccurlyeq_{L^n} M$ if and only if for every $\varphi(y, \bar{x}) \in L^n$ and every $\bar{a} \in N$ (with $|\bar{a}| = |\bar{x}|$), if $M \models \exists y \varphi(y, \bar{a})$ then there is $b \in N$ such that $M \models \varphi(b, \bar{a})$.*

Assumption 3.3.3 For the rest of this section we assume the following:

(1) $\Phi \subseteq L$ and Φ is closed under subformulas.
(2) $T \subseteq \Phi$ is a set of sentences.
(3) For every $M \models T$ a closure map cl_M is fixed such that if $M, N \models T$, $a_i \in M, b_i \in N$, for $i < \lambda$, and

$$(M, (a_i : i < \lambda)) \equiv_\Phi (N, (b_i : i < \lambda))$$

then $\{a_i : i < \lambda\}$ is closed (with respect to cl_M) if and only if $\{b_i : i < \lambda\}$ is closed (with respect to cl_N). Because of this assumption we can, for simplicity of notation, omit the subscript 'M' in cl_M in the situations where we deal with a closure map.
(4) cl is uniformly locally finite with respect to T; that is, for every $m < \omega$ there is $m' < \omega$ such that if $M \models T$, $A \subseteq M$ and $|A| \le m$, then $|\mathrm{cl}(A)| \le m'$.

Remark 3.3.4 Natural examples of Φ which are closed under subformulas are $\Phi = L^n$ and

$$\Phi = \{\varphi \in L : \text{ the quantifier rank of } \varphi \text{ is at most } n\}.$$

See [20] for a definition of quantifier rank.

Definition 3.3.5 A non-empty class $\mathcal{A}$ of L-structures is called a Φ-*amalgamation class for* T if:

(1) Every structure in $\mathcal{A}$ is a model of T.
(2) $\mathcal{A}$ is closed under isomorphism.
(3) $\mathcal{A}$ is closed under Φ-elementary substructures, i.e. if $N \in \mathcal{A}$ and $M \preccurlyeq_\Phi N$ then $M \in \mathcal{A}$.
(4) Whenever $M_1, M_2 \in \mathcal{A}$ and $\bar{a} \in M_1$, $\bar{b} \in M_2$ are closed sequences of the same length and

$$(M_1, \bar{a}) \equiv_\Phi (M_2, \bar{b}),$$

then there are $N \in \mathcal{A}$ and a Φ-elementary embedding $f : M_2 \to N$ such that $M_1 \preccurlyeq_\Phi N$ and $f(\bar{b}) = \bar{a}$.

Remark 3.3.6 If there is a Φ-amalgamation class for T then there is, by the downward Löwenheim-Skolem theorem, an amalgamation class for T such that all structures in it are countable.

Examples 3.3.7 (a) Let $n \ge 4$, let M be a tree such that for some $m < \omega$ no path in M has length m and let $T = Th_{L^n}(M)$. Moreover, for every $M \models T$

and every $A \subseteq M$ let $\mathrm{cl}(A) = A$. Then the class of all models of T is an L^n-amalgamation class for T.

(b) Let $n \geq 3$, let K be a finite field and let $T \subseteq L^3$ formalize the axioms of K-vector spaces in a language where scalar multiplication and vector addition are represented by function symbols. For every $M \models T$ and every $A \subseteq M$ let $\mathrm{cl}(A)$ be the substructure which is generated by A. Then the class of all models of T is an L^n-amalgamation class for T. If we had defined cl as in (a) then T would not have had any L^n-amalgamation class. This fact is a consequence of results in Section 3.4 and is discussed immediately after Theorem 3.4.7.

Definition 3.3.8 Suppose that $\mathcal{A}$ is a Φ-amalgamation class for T.

(i) We say that $p \subseteq \Phi$ is a *closed* $(\Phi, \mathcal{A})$-*type* if there are $M \in \mathcal{A}$ and a closed finite sequence $\bar{a} \in M$ such that

$$p = \{\varphi(\bar{x}) \in \Phi : M \models \varphi(\bar{a})\}.$$

(ii) We say that an L-structure M is $(\Phi, \omega, \mathcal{A})$-*saturated for closed sets* if whenever $\bar{b} \in N \in \mathcal{A}$, $\bar{a} \in M \cap N$, $\bar{a}$ and $\bar{b}\bar{a}$ are closed finite sequences and

$$(M, \bar{a}) \equiv_\Phi (N, \bar{a}),$$

then there exists $\bar{c} \in M$ such that $(M, \bar{a}\bar{c}) \equiv_\Phi (N, \bar{a}\bar{b})$, and hence, by Assumption 3.3.3 (3), $\bar{c}\bar{a}$ is closed.

Lemma 3.3.9 *If $\mathcal{A}$ is a Φ-amalgamation class for T such that the set of all closed $(\Phi, \mathcal{A})$-types is countable, then there exists a countable L-structure M, such that*

(i) $M \models T$

(ii) M *is* $(\Phi, \omega, \mathcal{A})$-*saturated, and*

(iii) *for every finite $\bar{a} \in M$ there exists $N \in \mathcal{A}$ such that $\bar{a} \in N$ and $(M, \bar{a}) \equiv_\Phi$ $(N, \bar{a})$.*

Proof. By Remark 3.3.6 we may assume that $\mathcal{A}$ is a Φ-amalgamation class for T which consists only of countable structures.

Then we use the idea in the proof of Fraïssé's theorem (see [20] for instance) to construct $M_i \in \mathcal{A}$, for $i < \omega$, such that

- $M_i \preccurlyeq_\Phi M_{i+1}$, for all $i < \omega$, and
- for any $i < \omega$, $\bar{a} \in M_i$ and $N \in \mathcal{A}$, if $\bar{a}, \bar{b} \in N$, $\bar{a}$ and $\bar{a}\bar{b}$ are closed and $(M_i, \bar{a}) \equiv_\Phi (N, \bar{a})$, then there exists $j \geq i$ and $\bar{c} \in M_j$ such that $(M_j, \bar{a}\bar{c}) \equiv_\Phi$ $(N, \bar{a}\bar{b})$.

Let $\pi : \omega^3 \to \omega$ be a bijection such that $\pi(i, j, k) \geq i, j, k$ for all i, j, k and let p_k, $k < \omega$, be an enumeration of all closed $(\Phi, \mathcal{A})$-types. Let $M_0 \in \mathcal{A}$ be arbitrary. Now suppose that M_i is defined for all $i < \ell + 1$, $M_i \preccurlyeq_\Phi M_{i+1}$ for all $i < \ell$ and that $\bar{a}_i^j$, $j < \omega$, is an enumeration of all closed finite sequences of elements from M_i, for $i < \ell + 1$. Suppose that $\ell = \pi(i, j, k)$. If there exists $N \in \mathcal{A}$ and $\bar{b} \in N$ such that $\bar{a}_i^j \in N$, $\bar{a}_i^j \bar{b}$ is closed, $p_k = \{\varphi(\bar{x}, \bar{y}) \in \Phi : N \models \varphi(\bar{a}_i^j, \bar{b})\}$ and $(M_i, \bar{a}_i^j) \equiv_\Phi (N, \bar{a}_i^j)$, then $(M_\ell, \bar{a}_i^j) \equiv_\Phi (N, \bar{a}_i^j)$ so, by condition (4) in the definition of a Φ-amalgamation class, there are $M_{\ell+1} \in \mathcal{A}$ and $\bar{c} \in M_{\ell+1}$ such that $M_\ell \preccurlyeq_\Phi M_{\ell+1}$ and $(M_{\ell+1}, \bar{a}_i^j \bar{c}) \equiv_\Phi (N, \bar{a}_i^j \bar{b})$. Otherwise let $M_{\ell+1} = M_\ell$.

Let $M = \bigcup_{i < \omega} M_i$. Since Φ is closed under subformulas it follows from Lemma 3.3.11 below that $M_i \preccurlyeq_\Phi M$, for all $i < \omega$, and from this we get (i). Conditions (ii) and (iii) follows from the construction of M. □

Definition 3.3.10   If M is a model as in Lemma 3.3.9 then we call M a *limit* of $\mathcal{A}$.

Lemma 3.3.11   *Suppose that Φ is closed under subformulas and that $M_i \preccurlyeq_\Phi M_{i+1}$ for $i < \kappa$. If $M = \bigcup_{i < \kappa} M_i$ then $M_i \preccurlyeq_\Phi M$ for every $i < \kappa$.*

Proof. By induction on the complexity of formulas in Φ. □

See [20] for a definition of an *unnested formula*.

Lemma 3.3.12   *Suppose that every unnested atomic formula of L is equivalent, modulo T, to a formula in Φ. Let $\mathcal{A}$ be a Φ-amalgamation class for T and suppose that M and N are limits of $\mathcal{A}$. Then for all closed finite sequences $\bar{a} \in M$ and $\bar{b} \in N$ with $|\bar{a}| = |\bar{b}|$,*

$$if \quad (M, \bar{a}) \equiv_\Phi (N, \bar{b}) \quad then \quad (M, \bar{a}) \equiv_L (N, \bar{b}),$$

and in fact there is an isomorphism from M onto N which sends $\bar{a}$ to $\bar{b}$.

Proofsketch. Using properties (ii) and (iii) of Lemma 3.3.9 one carries out a back and forth argument which shows that there is an isomorphism from M to N which sends $\bar{a}$ to $\bar{b}$. □

Corollary 3.3.13   *Suppose that every unnested atomic formula of L is equivalent, modulo T, to a formula in Φ. Then a limit of a Φ-amalgamation class $\mathcal{A}$ for T is unique up to isomorphism.*

Note that if, for example, $\Phi = L^n$, $n \geq 2$, every relation symbol has arity at most n and every function symbol has arity less than n, then every unnested atomic L-formula belongs Φ, so the condition about Φ in Lemma 3.3.12 and in Corollary 3.3.13 is satisfied.

Definition 3.3.14   An L-structure M is Φ-*determined* if for any closed finite sequences $\bar{a}, \bar{b} \in M$,

$$\text{if} \quad (M, \bar{a}) \equiv_\Phi (M, \bar{b}) \quad \text{then} \quad (M, \bar{a}) \equiv_L (M, \bar{b}).$$

Hence, if Φ and T satisfies the assumptions of Lemma 3.3.12 and $\mathcal{A}$ is a Φ-amalgamation class for T, then a limit of $\mathcal{A}$ exists, is unique up to isomorphism and is Φ-determined. We also have a partial converse. To state it we need the following definition:

Definition 3.3.15   An L-structure M is ω-*homogeneous* if for all finite sequences $\bar{a}, \bar{b} \in M$ such that $(M, \bar{a}) \equiv_L (M, \bar{b})$ and for every $c \in M$ there is $d \in M$ such that $(M, \bar{a}c) \equiv_L (M, \bar{b}d)$.

Lemma 3.3.16   *If M a model of T which is Φ-determined and either finite, or infinite and ω-homogeneous, then there is a Φ-amalgamation class for T.*

Proofsketch. Take as the Φ-amalgamation class all N which are isomorphic to some countable $N' \preccurlyeq_\Phi M$. $\qquad\qquad\qquad\qquad\qquad\qquad\qquad\qquad\square$

We say that an infinite L-structure M is ω-*categorical* if $Th_L(M)$ is ω-categorical.

Remark 3.3.17   A basic fact is that every ω-categorical structure is ω-homogeneous. Also, a complete theory is ω-categorical if and only if for every model M of the theory and every $n < \omega$, $S_n^L(\emptyset, M)$ is finite and there are only finitely many formulas with at most n free variables, up to equivalence modulo the theory. (This is the well-known "Ryll-Nardzewski theorem" [20]).

For the next proposition, recall that $\Phi_{\bar{x}} = \{\varphi(\bar{x}) : \varphi(\bar{x}) \in \Phi\}$. By combining the previous lemmas we get the following.

Proposition 3.3.18   *Suppose that every unnested atomic formula in L is equivalent to a formula in Φ and that for any $\bar{x}$, $\Phi_{\bar{x}}$ is finite up to equivalence modulo T. Then the following are equivalent:*

 (i)   *There exists a Φ-amalgamation class for T.*
 (ii) *There exists $M \models T$ which is Φ-determined, and if M is infinite, then M is ω-categorical (because, by Assumption 3.3.3 (4), cl is uniformly locally finite on M).*

3.4 Stability

Now we will consider 'stability' and see how imposing a stability condition on an L^n-theory T makes the limit M of every amalgamation class for T ω-stable. This together with the ω-categoricity of M ensures that there are arbitrarily large finite substructures of M which are models of T.

Assumption 3.4.1 In this section we assume the following:

(1) For every theory T that we speak about there is a closure map cl on the models of T which is uniformly locally finite with respect to T. (See Assumption 3.3.3 (4).)
(2) If T is an L^n-theory, $M, N \models T$, $a_i \in M$, $b_i \in N$, for $i < \lambda$ and

$$(M, (a_i : i < \lambda)) \equiv_{L^n} (N, (b_i : i < \lambda))$$

then $\{a_i : i < \lambda\}$ is closed if and only if $\{b_i : i < \lambda\}$ is closed.

Definition 3.4.2

(i) Suppose that $\mathcal{A}$ is an L^n-amalgamation class (for T). We say that $\mathcal{A}$ is *stable in L^n* if for every $\varphi(\bar{x}, \bar{y}) \in L^n$ there exists $k_\varphi < \omega$ such that there does *not* exist $M \in \mathcal{A}$ and $\bar{a}_i, \bar{b}_i \in M$, for $i < k_\varphi$, satisfying $M \models \varphi(\bar{a}_i, \bar{b}_j) \iff i \leq j$.
(ii) We adopt the convention that every finite structure is *stable*.
(iii) An infinite L-structure M is *stable* if for every $\varphi(\bar{x}, \bar{y}) \in L$, there exists $k_\varphi < \omega$ such that there do *not* exist $\bar{a}_i, \bar{b}_i \in M$, for $i < k_\varphi$, satisfying $M \models \varphi(\bar{a}_i, \bar{b}_j) \iff i \leq j$.
(iv) A formula $\varphi(\bar{x}, \bar{y})$ is *unstable* with respect to a theory T if there exist $M \models T$ and $\bar{a}_i, \bar{b}_i \in M$, $i < \omega$, such that $M \models \varphi(\bar{a}_i, \bar{b}_j) \iff i \leq j$; otherwise $\varphi(\bar{x}, \bar{y})$ is *stable* with respect to T.

Proposition 3.4.3 *Suppose that n is greater than the arity of every function symbol in the vocabulary of L and that n is at least as great as the arity of every relation symbol in the vocabulary of L. If T is a complete L^n-theory such that $S_n^n(T)$ is finite then the following are equivalent:*

(i) There is a stable L^n-amalgamation class for T.
(ii) T has a stable model which is L^n-determined (and hence ω-categorical if it is infinite).

Proofsketch. By Fact 3.2.5, the assumption that $S_n^n(T)$ is finite implies that there are only finitely many L^n-formulas up to equivalence modulo T and hence for any sequence of variables $\bar{x}$ (of any finite length) $L_{\bar{x}}^n$ is finite up to equivalence modulo T.

Hence Proposition 3.3.18 gives all except the statement about stability. But one direction of this follows from the fact that

- in (ii) we take the model of T to be the limit of a stable L^n-amalgamation class for T, and
- if the formulas $\varphi_i(\bar{x}, \bar{y})$, $i = 1, \ldots, m$, are stable with respect to a complete L-theory, then every boolean combination of the φ_i's is stable with respect to the same complete L-theory. (This can be proved directly by using Ramsey's theorem, but it also follows from the basic work on stable formulas by Shelah [31].)

And conversely, given a model M satisfying the conditions in (ii), a stable L^n-amalgamation class is obtained by taking all L^n-elementary substructures of M (and structures isomorphic to these) as in Lemma 3.3.16. $\qquad\square$

Definition 3.4.4 An L-structure M is ω-*stable* if whenever $M' \equiv_L M$, $A \subseteq M'$ and $|A| \le \omega$ then $|S_1^L(A, M)| \le \omega$.

A basic fact from stability theory is that if M is ω-stable then M is stable. The next lemma, which shows that under certain circumstances the converse also holds, will be essential here.

Lemma 3.4.5 *Suppose that M is an infinite L-structure such that $\mathrm{cl}(A) = A$ for every $A \subseteq M$ and $S_n^n(\emptyset, M)$ is finite. If M is stable and L^n-determined then M is ω-stable.*

Proofsketch. Under the premises of the lemma it follows that any L-formula is equivalent, modulo $Th_L(M)$, to a boolean combination of L^n-formulas and there are only finitely many L^n-formulas up to equivalence modulo $Th_{L^n}(M)$. Thus for every complete L-type $p(\bar{x})$ over a set A, $p(\bar{x})$ is determined by $p(\bar{x}) \cap L^n(A)$. Now the lemma follows from the fact that if $0 < m < \aleph_0$ then $(\aleph_0)^m = \aleph_0$ and from Shelah's "unstable formula theorem" ([31], Theorem II.2.2), which tells us that if $\varphi(v_1, \bar{y})$ is stable and $A \subseteq M$ is countable then $S_1^{\{\varphi(v_1, \bar{y})\}}(A)$ is countable. $\qquad\square$

Suppose that n is greater than the arity of every function symbol in the vocabulary of L and greater than or equal to the arity of every relation symbol in the vocabulary of L. From Proposition 3.4.3 and Lemma 3.4.5 it follows that if T is a complete L^n-theory such that $S_n^n(T)$ is finite and there is a stable L^n-amalgamation class for T, then T has a model which is ω-stable and, if it is infinite, ω-categorical.

Next we state the crucial result which will give us finite models for every complete L^n-theory T such that $S_n^n(T)$ is finite and there is a stable L^n-amalgamation class for T.

Theorem 3.4.6 (Cherlin, Harrington, Lachlan [4]) *If M is ω-categorical and ω-stable, $M \models \varphi$ and A is a finite subset of M, then there exists a finite substructure $N \subseteq M$ such that $N \models \varphi$ and $A \subseteq N$.*

Recall that, by Fact 3.2.6, if T is a complete L^n-theory such that $S_n^n(T)$ is finite then T is axiomatized by an L^n-sentence. Thus, applying Theorem 3.4.6 and previous results we get:

Theorem 3.4.7 *Suppose that n is greater than the arity of every function symbol in the vocabulary of L and greater or equal to the arity of every relation symbol in the vocabulary of L. If T is a complete L^n-theory such that $S_n^n(T)$ is finite then the following hold:*

(i) *If M is a model of T which is ω-categorical and ω-stable, then for any finite $A \subset M$ there is a finite $N \preccurlyeq_{L^n} M$ such that $A \subseteq N$, and consequently $N \models T$.*

(ii) *Suppose that for every $M \models T$ and $A \subseteq M$, $\mathrm{cl}(A) = A$. If there is a stable L^n-amalgamation class $\mathcal{A}$ for T such that $\mathcal{A}$ contains at least one infinite structure then T has arbitrarily large finite models.*

Observe the assumption in part (ii) of the above theorem that the closure operation is trivial. We now turn to the case when 'cl' is not trivial (i.e. for some A, $\mathrm{cl}(A) \neq A$). An example of such a situation is if M is an infinite vector space over a finite field and $T = Th_{L^n}(M)$, for sufficiently large n. Then there cannot exist a trivial closure operation cl and a stable L^n-amalgamation class for T, with respect to this closure operation, which contains an infinite structure. The reason is that it would imply the existence of an infinite vector space, over the same field, which is L^n-determined and this is impossible. For in every infinite model of T we can choose m greater than n and on the one hand a linearly independent sequence $\bar{a} = (a_1, \ldots, a_m)$ and on the other hand a sequence $\bar{b} = (b_1, \ldots, b_m)$ such that $\bar{b}$ is not linearly independent but every proper subtuple of $\bar{b}$ is linearly independent. With this choice, $\bar{a}$ and $\bar{b}$ have the same L^n-type but not the same L-type, so the structure is not L^n-determined.

Motivated by this example we would like to find some amalgamation property for complete L^n-theories T which holds also for the example of vector spaces and which implies the existence of an ω-categorical and ω-stable model of T, so that we are in position to apply Theorem 3.4.6.

Definition 3.4.8 Let T be an L^n-theory. T has the *strong L^n-amalgamation property over countable models* if, whenever

$M_0 \models T$, $M_0 \preccurlyeq_{L^n} M_i$, where M_i is countable for $i = 1, 2, \bar{a}_1 \in M_1, \bar{a}_2 \in M_2$ are finite sequences and $(M_1, \mathrm{cl}(\bar{a}_1)M_0) \equiv_{L^n} (M_2, \mathrm{cl}(\bar{a}_2)M_0)$

then

there are M and L^n-elementary embeddings $f_i : M_i \to M$, $i = 1, 2$, such that $f_1(\bar{a}_1) = f_2(\bar{a}_2)$ and f_i is the identity on M_0 for $i = 1, 2$.

Remark 3.4.9 Suppose that M is a vector space over a finite field and $T = Th_{L^n}(M)$, for n larger than the number of elements in the field. Using the elementary theory of vector spaces it is now easy to verify that T has the strong L^n-amalgamation property over countable models. Also one can easily verify that if cl is taken to be linear closure then there is a stable L^n-amalgamation class for T with respect to this closure operation.

Theorem 3.4.10 (Baldwin, Lessmann [2]) *Suppose that T is a complete L^n-theory such that $S_n^n(T)$ is finite and T has the strong L^n-amalgamation property over countable models.*

(i) If $M \models T$ and M is stable and L^n-determined then M is ω-stable.

(ii) If there is a stable L^n-amalgamation class $\mathcal{A}$ for T such that $\mathcal{A}$ contains at least one infinite structure then T has arbitrarily large finite models; these can be taken as L^n-elementary substructures of the limit of $\mathcal{A}$.

Proofsketch. (ii) follows from (i) and earlier results. Concerning (i): The assumptions that $S_n^n(T)$ is finite, M is stable and L^n-determined (so M is ω-categorical) imply, via Shelah's "unstable formula theorem" [31], that for any countable $A \subseteq M' \equiv_L M$, $S_1^{L^n}(A, M')$ is countable. The useful consequence of the strong L^n-amalgamation property over countable models is that if $N \preccurlyeq N' \models T$, where N is countable, and $\bar{a}, \bar{b} \in N'$ are finite sequences then

$$(N', \mathrm{cl}(\bar{a})N) \equiv_{L^n} (N', \mathrm{cl}(\bar{b})N)$$

implies

$$(N', \mathrm{cl}(\bar{a}N)) \equiv_{L^n} (N', \mathrm{cl}(\bar{b}N)).$$

This property together with the assumption that M is L^n-determined (which by the ω-categoricity of M implies that any $N \equiv_L M$ is L^n-determined) shows that M is ω-stable, by a counting types argument. $\square$

3.5 Recursive bounds

In this section we derive results about recursive upper bounds of the size of the least model of L^n-theories.

We will use the theory of smoothly approximable structures which is presented in detail in [3]. Every structure which is ω-categorical and ω-stable is smoothly approximable which essentially follows from [4], but also see [24]. One of several equivalent ways of defining 'smoothly approximable' is the following:

Definition 3.5.1   An L-structure M is *smoothly approximable* if M is ω-categorical and if every L-sentence which is true in M is true in a finite substructure $N \subseteq M$ such that

(1)   for every $\theta(\bar{x}) \in L$ there is $\chi(\bar{x}) \in L$ such that

$$\{\bar{a} \in N : M \models \theta(\bar{a})\} = \{\bar{a} \in N : N \models \chi(\bar{a})\}, \text{ and}$$

(2)   for all $\bar{a}, \bar{b} \in N$ of the same finite length

$$(N, \bar{a}) \equiv_L (N, \bar{b}) \iff (M, \bar{a}) \equiv_L (M, \bar{b}).$$

We derive our results from the following theorem which does not directly speak about smoothly approximable structures.

Theorem 3.5.2   (Cherlin, Hrushovski [3]) *We can effectively decide for a given sentence and $k < \omega$ if that sentence has a finite model M such that $|S_4^L(\emptyset, M)| = k$.*

For our purposes we now define a recursive function $\mathfrak{f} : \omega^2 \to \omega$ as follows:

Let $\mathfrak{f}(n, k) = k$ if $n < 2$ or $k = 0$.
Now suppose that $n \geq 2$ and $k \geq 1$.

- Let $\varphi_1, \ldots, \varphi_m$ be an enumeration of all sentences (up to equivalence) of quantifier rank at most $k + n$ in a language $\mathcal{L}_k$ such that for every $1 \leq i \leq n$ the vocabulary of $\mathcal{L}_k$ contains exactly k constant symbols and exactly k i-ary relation symbols, but no function symbols, and we assume that '$=$' is one of the binary relation symbols.
- For $1 \leq i \leq m$, let L_i be the language built up from the constant symbols and the relation symbols that occur in φ_i and the identity symbol '$=$' (but no other symbols from the vocabulary of $\mathcal{L}_k$). Then use Theorem 3.5.2 to decide if φ_i has a finite model M_i such that $|S_4^{L_i}(\emptyset, M_i)| \leq k$;
- if such a model of φ_i exists then search until we find such M_i and let $\ell_i = |M_i|$; otherwise let $\ell_i = 0$.
- Then let $\mathfrak{f}(n, k) = \max\{\ell_1, \ldots, \ell_m\}$.

If we had allowed function symbols in the language $\mathcal{L}_k$ appearing in the definition of $\mathfrak{f}$, then there would have been infinitely many formulas of rank at most $n \leq k + n$ (even quantifier free formulas) that are non-equivalent. When computing $\mathfrak{f}$ we depend on the fact that (with the stated definition) there are only finitely many non-equivalent $\mathcal{L}_k$-formulas with quantifier rank at most $k + n$.

Definition 3.5.3 Let T be a complete L^n-theory. As in the previous sections we associate a uniformly locally finite closure operation cl with models of T. We now define a function $\text{cl}^* : \omega \to \omega$ as follows:

$$\text{cl}^*(n) = \max\{|\text{cl}(A)| : A \subseteq M \models T, \ |A| \leq n\}.$$

Corollary 3.5.4 *Let T be a complete L^n-theory such that $S_n^n(T)$ is finite and let cl be the closure operation associated with models of T. Also assume that $n \geq \text{cl}^*(4)$ and that the vocabulary of L contains no function symbols and that the arity of every relation symbol is at most n. If $M \models T$, where M is smoothly approximable and $|S_{\text{cl}^*(4)}^n(\emptyset, M)| = |S_{\text{cl}^*(4)}(\emptyset, M)|$, then φ has a finite model of cardinality at most $\mathfrak{f}(n, |S_n^n(\emptyset, M)|)$ (where $S_n^n(\emptyset, M) = S_n^n(T)$).*

Proof. Suppose that T, L, n and M satisfies the premises of the corollary. By renaming symbols if necessary we may assume, without loss of generality, that $L \subseteq \mathcal{L}_k$ where $\mathcal{L}_k$ is the language that occurs in the definition of $\mathfrak{f}$, with $k = |S_n^n(T)|$. By Fact 3.2.6, T is axiomatized by an L^n-sentence with quantifier rank at most $|S_n^n(T)| + n = k + n$. Since M is smoothly approximable it follows that M has a finite substructure N such that $N \models T$ and $|S_4(\emptyset, N)| \leq |S_4(\emptyset, M)|$. From the assumptions that $n \geq \text{cl}^*(4)$ and $|S_{\text{cl}^*(4)}^n(\emptyset, M)| = |S_{\text{cl}^*(4)}(\emptyset, M)|$ it follows that

$$|S_4(\emptyset, N)| \leq |S_4(\emptyset, M)| \leq |S_{\text{cl}^*(4)}(\emptyset, M)| = |S_{\text{cl}^*(4)}^n(\emptyset, M)| \leq |S_n^n(\emptyset, M)| = k.$$

By the definition of $\mathfrak{f}$, there is a model of T with cardinality at most $\mathfrak{f}(n, |S_n^n(\emptyset, M)|)$. $\qquad\square$

Corollary 3.5.5 *Let $n \geq 4$ and let L be a language with finite vocabulary which contains no function symbols and in which all relation symbols have arity at most n. If T is a complete L^n-theory such that*

- *$S_n^n(T)$ is finite,*
- *$n \geq \text{cl}^*(4)$,*
- *for every $M \models T$ and every $A \subseteq M$, $\text{cl}(A) = A$, or T has the strong L^n-amalgamation property over countable models, and*
- *there is a stable L^n-amalgamation class for T (with respect to cl),*

then T has a model of cardinality at most $\mathfrak{f}(n, |S_n^n(T)|)$.

Proof. Suppose that T satisfies the above conditions. First note that (by Fact 3.2.6) T is axiomatized by an L^n-sentence with quantifier rank $\leq$ $|S_n^n(T)| + n$. By results in the previous section, T has an ω-categorical and ω-stable (hence smoothly approximable) model M which is L^n-determined (with respect to the given closure operator). Then $|S_{\mathrm{cl}^*(4)}^n(\emptyset, M)| = |S_{\mathrm{cl}^*(4)}(\emptyset, M)|$, so by Corollary 3.5.4, T has a model N with cardinality at most $\mathfrak{f}(n, |S_n^n(\emptyset, M)|) = \mathfrak{f}(n, |S_n^n(T)|)$. $\square$

Remark 3.5.6   Grohe [17] has shown that if $n \geq 3$ then there does *not* exist a recursive function $f_n : \omega \to \omega$ such that for every complete L^n-theory T with finite models,

$$\min\{|M| : \ M \models T\} \leq f_n(|S_n^n(T)|).$$

Except for the results presented here, an existence result about recursive upper bounds has also been obtained by Dawar in [5]. The hypothesis of Dawar's result is that the class C of finite structures considered (where C could be the class of all finite models of an L^n-theory, for example) has the 'weak n-Ehrenfeucht-Mostowski property'. Roughly speaking, this property says that every sufficiently long "L^n-indiscernible" sequence in a structure in C can both be extended in some L^n-elementary extension which belongs to C and reduced (as long as it does not become too short) in some L^n-elementary substructure that belongs to C.

The question arises: How general can a class, $\mathcal{T}$, of complete L^n-theories with finite models be if we require that there exists a recursive function f such that $\min\{|M| : \ M \models T\} \leq f(|S_n^n(T)|)$ for all $T \in \mathcal{T}$? Another problem is to determine such a function f more precisely (polynomial, exponential, etc.), perhaps starting with some smaller class of theories over which we have more control.

3.6 Simple, possibly not smoothly approximable structures

This paper has focused on obtaining finite models for a complete L^n-theory T by showing that T has an infinite model M which has the *finite submodel property*, by which we mean that every sentence which is true in M is true in a finite substructure of M. As stated in Theorem 3.4.6, every ω-categorical ω-stable structure has the finite submodel property. The same holds for the more inclusive class of smoothly approximable structures which also contains unstable examples (see [3]).

There are natural examples of structures which have the finite submodel property but are not smoothly approximable, such as the random (bipartite) graph G_{rg} [24]. G_{rg} can be defined as the Fraïssé limit of the class of all finite graphs, or alternatively one can give an explicit axiomatization of the complete theory of G_{rg}; see for instance [12, 20] for more about the random graph. (The random bipartite graph is obtained similarly by considering the class of all finite graphs expanded with an equivalence relation with exactly two classes subject to the condition that edges may only occur between elements in different classes.) The random (bipartite) graph has the following model theoretical properties: it is ω-categorical with elimination of quantifiers, (super)simple (but unstable) with SU-rank 1 and has trivial forking; see for instance [20] and [32] for these model theoretic and stability/simplicity theoretic notions. The fact that G_{rg} has SU-rank 1 implies that the algebraic closure operation 'acl' forms a pregeometry on G_{rg} (see [20]).

Before continuing we note that there is a line of research [28, 13, 14, 15], not discussed here, which studies the connection between classes of finite structures in which definable sets have uniform behaviour, asymptotically, and (infinite) simple structures with finite SU-rank and with a measure on the definable subsets, but which are not necessarily smoothly approximable. A question not answered here is whether the approach in this article has anything in common with the work about 'asymptotic classes' and 'measurable structures' (the random graph fits within both frameworks).

Work in two different directions has been carried out by the author to prove the finite submodel property for classes of structures which contain the random (bipartite) graph. One direction of research [10] studies ω-categorical structures on which the algebraic closure operation forms a pregeometry. The other direction of research [11, 26] studies structures which are ω-categorical, simple with finite SU-rank and have trivial forking. In both directions a probabilistic argument is involved in proving the finite submodel property. In order to carry out this argument we need to assume that definable relations are "sufficiently independent" from each other in senses that are made precise in [10] and [26]. It seems that without any assumption about "sufficient independence" we are in a difficult situation with respect to proving or disproving the finite submodel property. We say more about this in the last paragraph of Section 3.7.

The notion of "sufficient independence" which is considered in [26] is called the 'n-embedding of types property' (for a natural number $n \geq 2$), with respect to certain kinds of "generators". Before stating the main result of [26] we introduce some notation from stability/simplicity theory and explain, roughly, the involved notions. We assume familiarity with imaginary elements (see [20] or [32] for example). By $A \underset{C}{\downarrow} B$ we mean that A *is independent from* B *over* C (see for example [32] for a definition of 'independence'). The negation of $A \underset{C}{\downarrow} B$ is denoted by $A \underset{C}{\not\downarrow} B$. A complete first-order theory T has *trivial dependence* *(or trivial forking)* if whenever A, B_1, B_2, C are subsets of M^{eq} where $M \models T$ and $A \underset{C}{\not\downarrow} (B_1 \cup B_2)$, then $A \underset{C}{\not\downarrow} B_i$ for $i = 1$ or $i = 2$.

Here is a rough description of the *n-embedding of types property* (with respect to all/simple generators). Suppose that T is a complete first-order theory which is simple (see [32] for a definition of 'simple') and assume that $M \models T$. As usual, M^{eq} denotes the extension of M by imaginary elements and 'algebraic closure' is taken in the structure M^{eq} (see [20] for a definition of algebraic closure). We identify every natural number n with the set $\{0, \ldots, n-1\}$ and let $\mathcal{P}(n)$ be the set of all subsets of n and $\mathcal{P}^-(n) = \mathcal{P}(n) - \{n\}$. Suppose that A_i^0 and B_i^0, for $i \in n$, are subsets of M^{eq} and that, for every $w \in \mathcal{P}^-(n)$, A_w is the algebraic closure of $\bigcup_{i \in w} A_i^0$ and B_w is the algebraic closure of $\bigcup_{i \in w} B_i^0$. Also assume that

- for all $w, w' \in \mathcal{P}^-(n)$, $A_w \underset{A_{w \cap w'}}{\downarrow} A_{w'}$ and $B_w \underset{B_{w \cap w'}}{\downarrow} B_{w'}$, and
- for every $w \in \mathcal{P}^-(n)$ there is an elementary map (see [20]) f_w from A_w onto B_w such that $f_w(A_i^0) = B_i^0$ for every $i \in w$, and if $w \subseteq w'$ then $f_{w'}$ extends $f_w \restriction \bigcup_{i \in w} A_i^0$.

The *n-embedding of types property* says (omitting some details) that if $\bar{a} = (a_1, \ldots, a_r)$ is a sequence of elements from M^{eq} which does not contain any element from the algebraic closure of $\bigcup_{w \in \mathcal{P}^-(n)} A_w$, then there are $\bar{b} = (b_1, \ldots, b_r)$ in M^{eq} (we can assume that M is sufficiently saturated) and, for $w \in \mathcal{P}^-(n)$, elementary maps $g_w : A_w \cup \{a_1, \ldots, a_r\} \rightarrow B_w \cup \{b_1, \ldots, b_r\}$ such that $g_w \restriction \bigcup_{i \in w} A_i^0 = f_w \restriction \bigcup_{i \in w} A_i^0$ and if $w \subseteq w'$ then $g_{w'}$ extends $g_w \restriction \bigcup_{i \in w} A_i^0$.

The phrase 'with respect to all/simple generators' when stating the condition '*n-embedding of types property with respect to all/simple generators*' in the next theorem refers to the conditions (if any) that we impose on the sets A_i^0, B_i^0, for $i \in n$. In [26] the sets A_i^0, B_i^0, $i \in n$, are called the "generators" of the sets A_m, B_m, $m \in \mathcal{P}^-(n)$. Every stable theory (see [20, 31, 32] for the notion 'stable') has the *n-embedding of types property with respect to simple* generators for every $1 < n < \omega$ [26]. The complete theory of the random graph [20] has the *n-embedding of types property with respect to all* generators for every $1 < n < \omega$ [26].

Theorem 3.6.1 [26] *Suppose that there is $m < \omega$ such that every function symbol of the language of M has arity at most m. If $Th(M)$ is ω-categorical,* simple with finite SU-rank, has trivial dependence and, for every $1 < k < \omega$, has the k-embedding of types property with respect to all generators, then M has the finite submodel property. If the SU-rank of $Th(M)$ is 1, then the phrase 'with respect to all generators' can be replaced by the phrase 'with respect to simple generators', a weaker hypothesis.

It follows from Lemma 3.6.3, below, that Theorem 3.6.1 holds also if we replace 'trivial dependence' with 'n-degenerate dependence for some $n < \omega$', where n-degenerate dependence is defined as follows.

Definition 3.6.2   Let T be a complete simple (L-) theory. We say that T has *n-degenerate dependence* if the following holds: Whenever $M \models T$ and $A, B, C \subseteq M$ and $A \underset{C}{\not\!\smile} B$ then there is $B' \subseteq B$ such that $|B'| \leq n$ and $A \underset{C}{\not\!\smile} B'$.

Observe that trivial dependence implies 1-degenerate dependence.

Lemma 3.6.3   *Suppose that T is ω-categorical, simple with finite SU-rank and with n-degenerate dependence for some $n < \omega$. Then T has trivial dependence.*

Proof sketch. Suppose that T satisfies the premises of the lemma. By Corollary 4.7 in [18] and Lemma 3.22 in [7], it is sufficient to show that every type with SU-rank 1 is trivial, i.e. if D is the set of realizations of the type, in M^{eq} where M is a sufficiently saturated model of T, then the restriction to D of the algebraic closure operator forms a trivial pregeometry. For a contradiction, suppose that there is a nontrivial type of SU-rank 1. By Corollary 3.17 in [7] there is a definable subset of $\mathcal{M}^{eq}$ on which the algebraic closure is a projective geometry over a finite field. Now a contradiction can be derived in the same way as in the last two paragraphs of the proof of Proposition 8.7 in [8]. $\square$

Now we can derive a corollary which applies to L^n-theories and amalgamation classes. Suppose that Φ is a subset of L and T is a set of sentences from Φ such that Φ is closed under subformulas, every unnested atomic formula of L is equivalent, modulo T, to a formula in Φ, and for every $\bar{x}$, $\Phi_{\bar{x}}$ is finite up to equivalence modulo T. Also suppose that $\mathcal{A}$ is a Φ-amalgamation class for T. Note that the assumptions about Φ hold if $\Phi = L^n$ and n is greater than the arity of every symbol in the vocabulary. Then the assumptions of Lemma 3.3.9 and of Corollary 3.3.13 are satisfied, so by these results a unique limit of $\mathcal{A}$ exists. By Lemma 3.3.12 and Proposition 3.3.18 the limit of $\mathcal{A}$ is finite or ω-categorical.

Corollary 3.6.4   *Assume that there is $m < \omega$ such that every function symbol has arity at most m. Let T be a set of sentences from $\Phi \subseteq L$ where Φ is closed* *under subformulas. Suppose that, for every $\bar{x}$, $\Phi_{\bar{x}}$ is finite up to equivalence modulo T and that every unnested atomic formula is equivalent to a formula in Φ, modulo T. Moreover, suppose that T has a Φ-amalgamation class $\mathcal{A}$ with a limit M such that M is simple with finite SU-rank and has both n-degenerate forking for some $n < \omega$ and the k-embedding of types property with respect to all generators for every $1 < k < \omega$. Then T has arbitrarily large finite models (which can be taken as substructures of M).*

The following example illustrates the notions and assumptions of the previous corollary.

Example 3.6.5 Let the vocabulary of the language L be $\{=, E\}$ where E is a binary relation symbol. Let χ be the sentence $\forall x_1, x_2\big(\neg E(x_1, x_1) \wedge (E(x_1, x_2) \rightarrow E(x_2, x_1))\big)$. For every $n \geq 2$ and every $w \subseteq \{1, \ldots, n-1\}$ let $\theta_w^n(x_1, \ldots, x_n)$ be the formula

$$\bigwedge_{i \in w} E(x_i, x_n) \wedge \bigwedge_{i \notin w} \neg E(x_i, x_n)$$

and let φ_w^n be the sentence

$$\forall x_1, \ldots, x_{n-1}\Big(\bigwedge_{i \neq j} x_i \neq x_j \rightarrow \exists x_n \theta_w^n(x_1, \ldots, x_n)\Big).$$

Then let φ^n be the conjunction of χ and every φ_w^n as w ranges over subsets of $\{1, \ldots, n-1\}$, so $\varphi^n \in L^n$. Also, every model of φ^n is an undirected graph, or just 'graph' for brevity.

Fix an arbitrary natural number $n \geq 2$. Recall that, by Definition 3.3.1, $M \equiv_{L^n} N$ means that M and N are L^n-elementarily equivalent, i.e. satisfy exactly the same L^n-sentences, From the n-pebble game characterization of L^n-elementary equivalence [23, 29] it follows that if M and N are models of φ^n, then Duplicator (or "player II" or "$\exists$") has a winning strategy in the n-pebble game on M and N in ω rounds, and therefore $M \equiv_{L^n} N$. It follows that $T^n = \{\varphi^n\}$ is a complete L^n-theory (in the sense of Definition 3.2.1 (vii)).

Infinite models of φ^n exist since for every sequence $a_1, \ldots, a_{n-1}$ from a graph and every $w \subseteq \{1, \ldots, n-1\}$ we can add a new element b to the graph and extend the interpretetation of E so that $E(a_i, b)$ holds if and only if $i \in w$. By repeating this process systematically in ω steps (if we start with a finite structure) we can make sure that the union of the graphs created in the process is a model of φ^n. In fact, by reasoning similarly as has been outlined, we can show the following.

(I) For every graph G (finite or infinite) there is a graph $M \models \varphi^n$ such that G is a substructure of M.

From the n-pebble game characterization of '$M \equiv_{L^n} N$' we can also derive the the following.

(II) If M and N are models of φ^n, $m < \omega$, $a_1, \ldots, a_m \in M$ are different elements and $b_1, \ldots, b_m \in N$ are different elements, then $(M, a_1, \ldots, a_m) \equiv_{L^n} (N, b_1, \ldots, b_m)$ if and only if for all $i, j \in \{1, \ldots, m\}$, $M \models E(a_i, a_j) \iff N \models E(b_i, b_j)$.

Consequently, if $M, N \models \varphi^n$ then every embedding $f : M \to N$ is an L^n-elementary embedding. In particular, if $M, N \models \varphi^n$ then $M \preccurlyeq_{L^n} N$ if and only if M is a substructure of N.

For every $M \models \varphi^n$ and every $A \subseteq M$, define $\mathrm{cl}(A) = A$, so every subset of every model of φ^n is closed. Let $\mathcal{A}$ be the class of all models of $T^n = \{\varphi^n\}$. From the definition of $\mathcal{A}$ it immediately follows that (1)–(3) in the definition of an L^n-amalgamation class for T^n (Definition 3.3.5) are satisfied. We verify that also (4) in the same definition holds. Suppose that $M_1, M_2 \in \mathcal{A}$, $a_1, \ldots, a_m \in M_1$, $b_1, \ldots, b_m \in M_2$ and that

$$(M_1, a_1, \ldots, a_m) \equiv_{L^n} (M_2, b_1, \ldots, b_m).$$

To simplify the argument, without loss of generality, we may assume that $a_i = b_i$ for $i = 1, \ldots, m$, and that $M_1 \cap M_2 = \{a_1, \ldots, a_m\}$. By (I), the graph $M_1 \cup M_1$ (where $M_1 \cup M_2 \models E(a, b) \iff M_1 \models E(a, b)$ or $M_2 \models E(a, b)$) is a substructure of some model N of $T^n = \{\varphi^n\}$. Since $M_1, M_2 \models \varphi^n$, it follows from (II) that $M_i \preccurlyeq_{L^n} N$ for $i = 1, 2$. Hence $\mathcal{A}$ is an L^n-amalgamation class for T^n.

By Lemma 3.3.9 and Corollary 3.3.13, $\mathcal{A}$ has a unique limit M which is countable (by the definition of limit). By Lemma 3.3.12, M is L^n-determined which together with (II) implies that

(III) if $m < \omega$, $a_1, \ldots, a_m \in M$ are different elements and $b_1, \ldots, b_m \in M$ are different elements, then $(M, a_1, \ldots, a_m) \equiv_L (M, b_1, \ldots, b_m)$ if and only if for all $i, j \in \{1, \ldots, m\}$, $M \models E(a_i, a_j) \iff M \models E(b_i, b_j)$.

Hence $Th_L(M)$ has elimination of quantifiers and is ω-categorical. Since M is the limit of $\mathcal{A}$, conditions (ii) and (iii) in Lemma 3.3.9 are satisfied. This together with (I) implies that for every $m < \omega$, every choice of distinct $a_1, \ldots, a_m \in M$ and every $w \subseteq \{1, \ldots, m\}$, there is $b \in M$ such that $M \models E(a_i, b) \iff i \in w$. Hence $M \models \varphi^m$ for *every* $1 < m < \omega$. This implies that M is the *random graph* [12, 20]. It is well-known that $Th_L(M)$ is simple with SU-rank 1 and has trivial dependence [32]. In [26] it is shown that $Th_L(M)$ has the k-embedding of types property with respect to all generators, for every $1 < k < \omega$. Note that since $M \models \varphi^m$ for every $1 < m < \omega$, we can, for every $m < \omega$, find $a_i, b_i \in M$ for

$i < m$ such that $M \models E(a_i, b_j)$ if and only if $i \leq j$. Hence M and $\mathcal{A}$ are not stable in L^n.

Corollary 3.6.4 implies that $T^n = \{\varphi^n\}$ has arbitrarily large finite models (all of which are isomorphic to substructures of M). This is nothing new, since the proof of the so-called 0-1 law for the random graph shows that, for every $1 < n < \omega$, the number of graphs with universe $m = \{0, \dots, m-1\}$ which satisfy T^n divided by the number of all graphs with universe m approaches 1 as $m \to \infty$ [12, 20]. At the core of the proof is a probabilistic argument ("What is the probability that φ^n holds in a graph with universe m?"). The method in this example of showing that φ^n has arbitrarily large finite models does not avoid the main idea, the probabilistic argument, in the proof of the 0-1 law. On the contrary, our approach has utilized this idea in a more general setting, but this is not evident in this paper since we don't discuss the proof of Theorem 3.6.1 or of Theorem 3.7.7, on which the former theorem relies.

Remark 3.6.6 It would be nice if we could specify some properties of complete L^n-theories (without speaking about limits of amalgamation classes) which, if they hold for such a theory, would allow us to derive the existence of a structure M as in Theorem 3.6.1. However, while the notion of stability straightforwardly transfers from the context of complete L-theories to complete L^n-theories (in Section 3.4) the notions 'n-degenerate forking', 'SU-rank' and 'n-embedding of types property' involve the stability/simplicity theoretic notion of forking (or (in)dependence) and the author does not currently see a straightforward, or "natural", way of defining forking with respect to a complete L^n-theory (which, according to our definition, need not be complete in the usual sense). The notion of simplicity may, on the other hand, be straightforwardly transferred to the context of complete L^n-theories by saying that a complete L^n-theory is simple if no L^n-formula has the tree property (see [32]) in any model of the theory. However, the question remains whether simplicity, defined in this way, has any interesting consequences for L^n-theories.

3.7 Structures on which algebraic closure forms a pregeometry

In this section we give a brief overview of the main results in [10], about the finite submodel property, which are stated as Theorems 3.7.6 and 3.7.7 below. Theorem 3.7.7 is used to prove the main result in [26] (stated as Theorem 3.6.1 in this survey). We will assume throughout this section that M is an

ω-categorical L-structure such that the algebraic closure in M, denoted acl_M, forms a pregeometry on M; see for instance [10] or [20] for a definition of a pregeometry. A consequence of M being ω-categorical is that for every finite $A \subseteq M$, $\mathrm{acl}_M(A)$ is finite. Moreover, in this context every subset $A \subseteq M$ has a *dimension* defined by

$$\dim_M(A) = \inf\{|B| : B \subseteq A \text{ and } A \subseteq \mathrm{acl}_M(B)\}.$$

A type is called *algebraic* if it has only finitely many realizations.

Definition 3.7.1 Let $0 < k < \omega$. We say that M is *polynomially k-saturated* if there is a polynomial $P(x)$ such that for every $n_0 < \omega$ there is a natural number $n \geq n_0$ and a finite substructure $N \subseteq M$ such that:

(1) $n \leq |N| \leq P(n)$.
(2) N is algebraically closed (in M).
(3) Whenever $A \subseteq N$, $\dim_M(A) < k$ and $q(x) \in S_1^L(A, M)$ is non-algebraic, then there are distinct $b_1, \ldots, b_n \in N$ such that $M \models q(b_i)$ (i.e. b_i realizes q in M) for each $1 \leq i \leq n$.

Examples of structures on which the algebraic closure forms a pregeometry and which are polynomially k-saturated for every $0 < k < \omega$ include the "infinite empty structure" (having only the relation '='), the random (bipartite) graph, infinite vector spaces, projective spaces and affine spaces over any finite field [10]. Another example is obtained by "independently" expanding a vector space (for instance) with the random graph [10].

We also have the following result from [10] which relates polynomial k-saturation to the finite submodel property.

Lemma 3.7.2 *If M is polynomially k-saturated for every $0 < k < \omega$, then M has the finite submodel property.*

Assumption 3.7.3 From now on $\mathcal{L}$ is a first-order language such that $\mathcal{L}$'s vocabulary is included in L's vocabulary, so $\mathcal{L} \subseteq L$. We suppose that acl_M coincides with $\mathrm{acl}_{M \upharpoonright \mathcal{L}}$ (i.e. $\mathrm{acl}_M(A) = \mathrm{acl}_{M \upharpoonright \mathcal{L}}(A)$ for every $A \subseteq M$). Moreover, we assume that both M and $M \upharpoonright \mathcal{L}$ have elimination of quantifiers, where $M \upharpoonright \mathcal{L}$ denotes the reduct of M to $\mathcal{L}$. If these conditions are not fulfilled in the beginning, then we can just add new relation symbols to L and $\mathcal{L}$ so that the resulting expansions satisfy these conditions and all previous assumptions about M.

Before going to the next definition we note that if acl_M and $\mathrm{acl}_{M \upharpoonright \mathcal{L}}$ coincide and $\bar{a}, \bar{b} \in M$ satisfy exactly the same $\mathcal{L}$-formulas, then $\bar{a}$ is algebraically closed if and only if $\bar{b}$ is.

Definition 3.7.4   We say that M *satisfies the k-independence hypothesis over* $\mathcal{L}$ if the following holds:
Whenever A and B are algebraically closed substructures of M and

(1) $\dim_M(B) \le k$,
(2) $\dim_M(A) < k$,
(3) $f : A{\restriction}\mathcal{L} \to B{\restriction}\mathcal{L}$ is an $\mathcal{L}$-embedding (i.e. it preserves all atomic $\mathcal{L}$-formulas and negations of atomic $\mathcal{L}$-formulas), and
(4) if $A' \subset A$ (proper inclusion) is an algebraically closed substructure then the restriction $f : A' \to B$ is an L-embedding,

then there are an algebraically closed substructure $C \subseteq M$ and an $\mathcal{L}$-ismorphism $g : B{\restriction}\mathcal{L} \to C{\restriction}\mathcal{L}$ such that

(5) $gf : A \to C$ is an L-embedding, and
(6) for every algebraically closed substructure $B' \subseteq B$ such that $f(A) \nsubseteq B'$, $g : B' \to C$ is an L-embedding.

The definition of the k-independence hypothesis given above looks a bit different from the definition of it given in [10], but the two ways of defining the k-independence hypothesis are equivalent under Assumption 3.7.3. Below follow some examples which illustrate the k-independence hypothesis.

Examples 3.7.5   (a) *The random graph*: Let the vocabulary of $\mathcal{L}$ be $\{=\}$ and let the vocabulary of L be $\{=, E\}$ where E is a binary relation symbol. Let M be the random graph in the language L where E is interpreted as the edge relation. Then M and $M{\restriction}\mathcal{L}$ have elimination of quantifiers and $\mathrm{acl}_M(A) = A$ and $\dim_M(A) = |A|$ for any $A \subseteq M$. Since any finite graph embeds into M it follows that M satisfies the k-independence hypothesis over $\mathcal{L}$ for every $k < \omega$. With the notation of the definition, the case when $\dim_M(A) = 2$ is the most interesting. The reason is that M has elimination of quantifiers in a language with only binary relations symbols and that dimension coincides with cardinality.

(b) *The random structure*: Let $\mathcal{L}$ be as in (a) and let the vocabulary of L be $\{=, R_1, \ldots, R_m\}$ where R_i are relation symbols of any arity. Let M be the random structure in the language L, i.e. M is the Fraïssé limit of the class of all finite L-structures. For the same reasons as in (a), M satisfies the k-independence hypothesis over $\mathcal{L}$ for every $k < \omega$. However the verification becomes a little bit more interesting for A of dimension > 2 if L contains relation symbols of arity greater than 2.

(c) *A vector space expanded with the bipartite random graph*: Let K be the class of all finite structures
$N = (V, P, E, +, f_0, f_1, 0)$ such that:
1. V, the universe of N, is a vector space over the field $F = \{0, 1\}$.
2. P is a unary relation.
3. E is a binary relation symbol interpreted as an irreflexive and symmetric relation.
4. $+$ is a binary function symbol interpreted as vector addition and the constant symbol 0 is interpreted as the zero vector.
5. $f_i(v) = i \cdot v$, for $i = 0, 1$ and any $v \in V$ (so f_i represents scalar multiplication by i).
6. $N \models \forall xy \big(E(x, y) \to \big[\big(P(x) \wedge \neg P(y) \big) \vee \big(\neg P(x) \wedge P(y) \big) \big] \big)$.
7. $N \models P(0)$.
It is easy to verify that K is nonempty and has the hereditary property, the joint embedding property and the amalgamation property and is uniformly locally finite (see [20]). Hence the Fraïssé limit of K, which we call M, exists and is ω-categorical with elimination of quantifiers. Since the reduct of M to the language with vocabulary $\{=, P, E\}$ is the random bipartite graph, M is not smoothly approximable [3].

Let $\mathcal{L} \subseteq L$ be the sublanguage which contains all symbols of L except P and E. Then $M \!\restriction\! \mathcal{L}$ is a vector space over a finite field, so $M \!\restriction\! \mathcal{L}$ has elimination of quantifiers. It is not hard to see, using quantifier elimination of M and the fact that any structure in K can be embedded into M (since M is the Fraïssé limit of K), that $\mathrm{acl}_M(A)$ is linear span of A. Hence acl_M and $\mathrm{acl}_{M \restriction \mathcal{L}}$ coincide. Again using the fact that M is the Fraïssé limit of K it follows that M satisfies the k-independence hypothesis over $\mathcal{L}$, for every $k < \omega$.

(d) *The random pyramid-free (3)-hypergraph*: As shown in [10], the random pyramid-free (3)-hypergraph does not satisfy the 4-independence hypothesis over the language $\mathcal{L}$ with vocabulary $\{=\}$ (as opposed to the case of the random graph).

Having Assumption 3.7.3 in mind, we now state the two main results of [10].

Theorem 3.7.6 *Suppose that $M \!\restriction\! \mathcal{L}$ is polynomially k-saturated and that M satisfies the k-independence hypothesis over $\mathcal{L}$. If $\varphi \in L$ is an unnested sentence, in which at most k distinct variables occur, and $M \models \varphi$, then φ has arbitrarily large finite models.*

Note that Theorem 3.7.6 only speaks about arbitrarily large finite models, but does not claim that these can be taken as substructures of M.

Theorem 3.7.7   *Suppose that, for every* $0 < k < \omega$, $M \upharpoonright \mathcal{L}$ *is polynomially k-saturated and that* M *satisfies the k-independence hypothesis over* $\mathcal{L}$. *Then* M *is polynomially k-saturated, for every* $0 < k < \omega$, *and thus* M *has the finite submodel property.*

We say that M has *trivial* (also called *degenerate*) algebraic closure if for every $A \subseteq M$, $\mathrm{acl}_M(A) = \bigcup_{a \in A} \mathrm{acl}_M(a)$. Examples of ($\omega$-categorical) M which are simple with SU-rank 1 and trivial algebraic closure include the random (bipartite) graph, the random structure and the random pyramid-free (3)-hypergraph. The following is a consequence of the first theorem:

Corollary 3.7.8   *Suppose that* M *is simple with SU-rank 1 and has trivial algebraic closure. If* $\varphi \in L^3$ *is unnested and* $M \models \varphi$ *then* φ *has arbitrarily large finite models.*

The assumption that M satisfies the k-independence hypothesis (over $\mathcal{L}$) in the previous two theorems is used in the probabilistic argument at the core of the proofs. It generalizes the argument used when showing that the random graph (or random structure) satisfies a 0-1 law with the uniform probability measure. The proofs of Theorems 3.7.6 and 3.7.7 do not however lead to 0-1 laws in general, with the uniform probability measure.

It seems that without assuming any kind of independence we get into a difficult situation with respect to proving or disproving the finite submodel property, as witnessed by the complete theory of the random pyramid-free (3)-hypergraph (example (d) above). It is ω-categorical with elimination of quantifiers, simple with SU-rank 1, has trivial dependence and trivial algebraic closure. However, for all $k \geq 4$, it does not satisfy the k-independence hypothesis over (the only proper sublanguage) $\mathcal{L}$ with vocabulary $\{=\}$ [10]. It has neither the n-embedding of types property for any $n \geq 4$, nor the n-amalgamation property for any $n \geq 4$ [25, 26]. It is an open problem whether or not the random pyramid-free (3)-hypergraph has the finite submodel property.

3.8 Questions and problems

In connection with the approach expounded in this paper one may of course ask many questions, some of which are stated below.

(1) Can we find "natural" amalgamation properties and stability/simplicity theoretic properties for L^n-theories T (or other fragments of first-order logic) which imply the existence of an infinite model M of T with the finite model

property (i.e. every sentence which is true in M is true in a finite model), for other classes of theories T than those that fit into the framework presented here (in sections 1–3)?

(2) In particular, can we find "natural" amalgamation properties and stability/simplicity theoretic properties for "simple" L^n-theories T (without a stable amalgamation class) which guarantee that T has a model such as M in Theorem 3.6.1?

(3) Can stronger upper bounds than recursive (exponential, polynomial etc.) on the size of the smallest model be obtained for some interesting classes of theories?

(4) Are there other approaches, than the one presented here, towards understanding when (arbitrarily large) finite models exist and when a recursive (or better) upper bound of the smallest model exists, in terms of the number of L^n-types, for instance?

(5) Can one derive the conclusions of Theorem 3.6.1 from a weaker assumption than that forking is trivial?

(6) Can the approach in Section 9, about structures on which algebraic closure forms a pregeometry, be helpful for understanding L^n-theories (or, say, theories in a language with a finite bound on the quantifier rank)?

(7) The random graph fits within the framework presented in sections 3.6 and 3.7 as well as within the framework of 'asymptotic classes' and 'measurable structures' [28, 13, 14, 15]. Do the two approaches have anything in common? If 'yes', can both approaches together enrich our knowledge about relationships between infinite structures and classes of finite structures.

(8) Does the random pyramid-free (3)-hypergraph (Example 3.7.5 (d)) have the finite submodel property?

References

[1] J. Baldwin, Finite and infinite model theory - a historical perspective, Logic Journal of the IGPL, Vol. 8 (2000) 605-628.

[2] J. Baldwin, O. Lessmann, Amalgamation properties and finite models in L^n-theories, Archive for Mathematical Logic, Vol. 41 (2002) 155–167.

[3] G. Cherlin, E. Hrushovski, Finite structures with few types, Annals of Mathematics Studies 152, Princeton University Press 2003.

[4] G. Cherlin, L. Harrington, A. H. Lachlan, $\aleph_0$-categorical, $\aleph_0$-stable structures, Annals of Pure and Applied Logic, 28 (1985) 103–135.

[5] A. Dawar, Types and indiscernibles in finite models, in Logic Colloquium '95, Lecture Notes in Logic, Vol. 11, Springer.

[6] A. Dawar, S. Lindell, S. Weinstein, Infinitary logic and inductive definability over finite structures, Information and Computation, Vol. 119 (1995) 160–175.

[7] T. De Piro, B. Kim, The geometry of 1-based minal types, Transactions of the Americal Mathematical Society, Vol. 355 (2003) 4241–4263.

[8] M. Djordjević, Stability theory in finite variable logic, Ph.D. thesis, Uppsala University 2000.

[9] M. Djordjević, Finite variable logic, stability and finite models, The Journal of Symbolic Logic, Vol. 66 (2001) 837–858.

[10] M. Djordjević, The finite submodel property and ω-categorical expansions of pregeometries, Annals of Pure and Applied Logic Vol. 139 (2006) 201–229.

[11] M. Djordjević, Finite satisfiability and $\aleph_0$-categorical structures with trivial dependence, The Journal of Symbolic Logic, Vol. 71 (2006) 810–830.

[12] H-D. Ebbinghaus, J. Flum, Finite Model Theory, Second Edition, Springer Verlag, 1999.

[13] R. Elwes, Dimension and measure in finite first order structures, PhD thesis, University of Leeds, 2005.

[14] R. Elwes, Asymptotic classes of finite structures, The Journal of Symbolic Logic, Vol. 72 (2007) 418–438.

[15] R. Elwes, H.D. Macpherson, A survey of asymptotic classes and measurable structures, in Model theory and applications to algebra and analysis. Eds. Z. Chatzidakis, H.D. Macpherson, A. Pillay, A.J. Wilkie, Cambridge University Press, to appear.

[16] M. Grohe, Finite variable logic in descriptive complexity theory, The Bulletin of Symbolic Logic, Vol. 4 (1998) 345–398.

[17] M. Grohe, Large finite structures with few L^k-types, Information and Computation, Vol. 179 (2002) 250–278.

[18] B. Hart, B. Kim, A. Pillay, Coordinatisation and canonical bases in simple theories, The Journal of Symbolic Logic, Vol. 65 (2000) 293–309.

[19] S. Hedman, Finitary axiomatizations of strongly minimal theories, Ph.D. thesis, University of Illinois at Chicago, 1999.

[20] W. Hodges, Model theory, Cambridge University Press 1993.

[21] T. Hyttinen, On stability in finite models, Archive for Mathematical Logic, Vol. 39 (2000) 89–102.

[22] T. Hyttinen, Canonical finite diagrams and quantifier elimination, Mathmatical Logic Quarterly, Vol. 48 (2002) 533–554.

[23] N. Immerman, Upper and lower for first order expressibility, Journal och Computer and Systems Sciences, Vol. 25 (1982) 76–98.

[24] W. M. Kantor, M. W. Liebeck, H. D. Macpherson, $\aleph_0$-categorical structures smoothly approximated by finite substructures, Proceedings of the London Mathematical Society, Vol. 59 (1989) 439–463.

[25] B. Kim, A.S. Kolesnikov, A. Tsuboi, Generalized amalgamation and n-simplicity, Annals of Pure and Applied Logic, Vol. 155 (2008) 97–114.

[26] V. Koponen, Independence and the finite submodel property, Annals of Pure and Applied Logic, Vol. 158 (2009) 58–79.

[27] A. H. Lachlan, Stable finitely homogeneous structures: a survey, in B. T. Hart, A. H. Lachlan, M. A. Valeriote, editors, Algebraic model theory, Kluwer Academic Publishers (1997) 145–159.

[28] H.D. Macpherson, C. Steinhorn, One-dimensional asymptotic classes of finite structures, Transactions of the American Mathematical Society, Vol. 360 (2008) 411–448.

[29] B. Poizat, Deux ou trois choses que je sais de L_n, The Journal of Symbolic Logic, Vol. 47 (1982) 641–658.

[30] E. Rosen, Some aspects of model theory and finite structures, The Bulletin of Symbolic Logic, Vol. 8 (2002) 380–403.

[31] S. Shelah, Classification Theory, North–Holland 1990.

[32] F. O. Wagner, Simple theories, Kluwer Academic Publishers 2000.

[33] B. Zilber, Uncountably categorical structures, AMS translations of mathematical monographs 117 (1993).

4

Definability in classes of finite structures

DUGALD MACPHERSON[a] AND CHARLES STEINHORN[b]

4.1 Introduction

This paper provides an overview of recent work by the authors and others on two topics in the model theory of finite structures. The point of view here differs from that usually associated with the term 'finite model theory', as presented for example in [21] or [46], in which the emphasis and motivation come primarily from computer science. Instead, the inspiration for this work has its origins in contemporary (infinite) model theoretic themes such as dimension, independence, and various measures of the complexity of definable sets. Each of the topics deals with classes of finite structures for first-order logic that are isolated by conditions that are drawn from these model-theoretic considerations. Moreover, in both cases, connections exist to areas in infinite model theory such as stability and simplicity theory, and o-minimality. This survey is intended for both mathematical logicians and computer scientists whose work focuses on logical aspects of the subject.

The first theme concerns *asymptotic classes* of finite structures. This subject has its origins in the model theory of finite fields, via the work of Chatzidakis, van den Dries and Macintyre [13] (see Theorem 4.2.1) and the earlier model theory of finite fields developed by Ax [4], and ultimately rests on the Lang-Weil bounds for the number of points in a finite field of an irreducible variety defined over that field. Given a first-order formula φ in the language of rings, the analysis in [13] provides estimates for the cardinality of the set defined by this formula in all finite fields in terms of two parameters, dimension and measure. Denoting the universe of a finite field by F, the cardinality estimate has the

[a] School of Mathematics, University of Leeds,Leeds LS2 9JT England
   h.d.macpherson@leeds.ac.uk
[b] Vassar College, Poughkeepsie, New York 12604, USA steinhorn@vassar.edu; partially
   supported by U.S. National Science Foundation Grant DMS-0070743.

form $\mu|F|^k$, where k represents the dimension and μ the measure of the size of the set defined by φ. Asymptotic classes are, roughly speaking, classes of finite structures with a strong uniformity condition on the cardinality of definable sets that mirrors precisely that for finite fields (see [22], [25], or [48]). Indeed, finite fields, as developed in [13], provides a key example, but there are many others. Ryten [52] has shown that certain difference fields, and hence any family of finite simple groups of fixed Lie rank, fit into this framework, and Elwes [23] has established that every smoothly approximable structure admits a class of (finite) envelopes which forms an asymptotic class that witnesses smooth approximability. The uniformity properties of asymptotic classes feed through to ultraproducts of the members of any such class: they are supersimple of finite rank, with an additional ingredient, still rather mysterious, called *measure*.

The second topic, *robust classes* of finite structures, has its origins in an attempt to bring an appropriate version of o-minimality to classes of finite structures. Obstacles quickly present themselves: for example, no class of finite totally ordered structures can be an asymptotic class – see [25] or [48]. In fact, every o-minimal structure elementarily equivalent to an ultraproduct of totally ordered finite structures must be discretely ordered, and thus carry very limited structure [51]. A robust class is a directed system of finite structures *with embeddings*, such that any formula, interpreted in a structure in the class that is sufficiently large relative to the parameters appearing in the formula, assumes a constant truth value. In this setting, a non-trivial notion of o-minimality may be defined. For example, the group $(\mathbb{Q}, +, <)$, with $+$ interpreted as a ternary relation, is the direct limit of an o-minimal robust class of finite structures. The initial theory of robust classes has been developed by the authors, and in the Ph.D. thesis of Macpherson's student, R. Marshall [49]. It is our hope is that connections eventually will emerge between the two topics of this survey and current concerns of finite model theory. This seems particularly possible for robust classes; here Ehrenfeucht-Fraïssé games intervene, locality can play a role, and there are very natural, if still rudimentary, notions of complexity.

The organization of this paper is as follows. Asymptotic classes are introduced in Section 4.2, and examples are the focus of Sections 4.3, 4.4 and 4.6. Smoothly approximable structures, which provide important examples for both asymptotic classes and robust classes, are given a brief overview in Section 4.4. Section 4.5 links asymptotic classes with contemporary infinite model theory, in particular simple theories, and introduces the notion of a measurable structure. Asymptotic classes of groups, with connections to simple theories, are treated in Section 4.6. Robust classes are introduced in Section 4.7 and Section 4.8 is devoted to examples. In the final section of the paper, Section 4.9, an 'o-minimal' robust approximation of the ordered group of rational numbers is presented. Notation throughout is standard; any uncommon terminology or notation is defined where it arises.

4.2 Asymptotic classes

The starting point here is the following theorem of Chatzidakis, van den Dries, and Macintyre. One considers, uniformly across finite fields, families of definable sets determined by formulas $\varphi(\bar{x}, \bar{y})$, where the $\bar{y}$ are parameter-variables.

Theorem 4.2.1 ([13]) *Let $\varphi(\bar{x}, \bar{y})$ be a formula in the language $L_{\text{rings}} = (+, \times, -, 0, 1)$ for rings, with $\bar{x} = (x_1, \ldots, x_n)$ and $\bar{y} = (y_1, \ldots, y_m)$. Then there is a positive constant C, and a finite set D of pairs (d, μ) with $d \in \{0, \ldots, n\}$ and μ a non-negative rational number, such that for each finite field $\mathbb{F}_q$ and $\bar{a} \in \mathbb{F}_q^m$,*

$$\left| |\varphi(\mathbb{F}_q^n, \bar{a})| - \mu q^d \right| \leq C q^{d - (1/2)} \qquad (*)$$

for some $(d, \mu) \in D$.

Furthermore, for each $(d, \mu) \in D$, there is a formula $\varphi_{(d,\mu)}(\bar{x})$ which defines in each finite field $\mathbb{F}_q$ the set of tuples $\bar{a}$ such that $(*)$ holds.

Each such pair (d, μ) may be understood as providing a finite combinatorial version of the *dimension d* and measure μ of those definable sets to which the pair corresponds. For formulas which define absolutely irreducible varieties – without the μ – this is the result of Lang-Weil [45]. A 'near model completeness' result of Kiefe [40], coming rapidly out of Ax's work [4], asserts that every formula $\varphi(\bar{x})$ is equivalent, uniformly across finite fields, to a boolean combination of formulas $\exists y g(\bar{x}, y) = 0$ where $g(\bar{X}, Y) \in \mathbb{Z}[\bar{X}, Y]$. This suggests why the above theorem should hold: one can reduce definable sets to sets built from finite-to-one projections of varieties. The details are intricate.

Theorem 4.2.1 suggests that one might consider arbitrary classes of finite structures satisfying asymptotic uniformities in the spirit of the theorem. Note that there are possible natural weakenings of the conditions. First, one could weaken the error term $C q^{d - (1/2)}$. There also is perhaps no reason to require μ to be rational. Additionally, and most importantly, the class of finite fields is in a sense 1-dimensional: any formula uniformly picking out an arbitrarily large subset of the field (in affine 1-space) picks out a positive fraction of the field. Thus, for example, Theorem 4.2.1 answers a question of Felgner (that in fact inspired the paper), showing that $\mathbb{F}_q$ is not uniformly definable in $\mathbb{F}_{q^2}$. One could easily consider the universe F of a structure to be N-dimensional if all definable subsets of F are roughly of size $\mu |F|^{d/N}$ for $d \in \{0, 1, \ldots, N\}$

and μ a constant. These considerations lead to the following definition of
Elwes [22] of an *N-dimensional asymptotic class* of finite structures. For the
initially considered concept, 1-dimensional asymptotic classes, see [48]. A
more extensive survey of asymptotic classes than provided here, with more
emphasis on the infinite limits, may be found in [25].

Definition 4.2.2 (Elwes, [23]) Let $N \in \mathbb{N}$, and let C be a class of finite L-
structures, where L is a finite language. Then we say that C is an *N-dimensional
asymptotic class* if the following hold.

(i) For every L-formula $\varphi(\bar{x}, \bar{y})$ where $l(\bar{x}) = n$ and $l(\bar{y}) = m$, there is a
finite set of pairs $D \subseteq (\{0, \ldots, Nn\} \times \mathbb{R}^{>0}) \cup \{(0, 0)\}$ and for each $(d, \mu) \in D$
a collection $\Phi_{(d,\mu)}$ of pairs of the form $(M, \bar{a})$ where $M \in C$ and $\bar{a} \in M^m$, so
that $\{\Phi_{(d,\mu)} : (d, \mu) \in D\}$ is a partition of $\{(M, \bar{a}) : M \in C, \bar{a} \in M^m\}$, and

$$\left| |\varphi(M^n, \bar{a})| - \mu |M|^{\frac{d}{N}} \right| = o(|M|^{\frac{d}{N}})$$

as $|M| \longrightarrow \infty$ and $(M, \bar{a}) \in \Phi_{(d,\mu)}$.

(ii) Each $\Phi_{(d,\mu)}$ is $\emptyset$-definable, that is to say $\{\bar{a} \in M^m : (M, \bar{a}) \in \Phi_{(d,\mu)}\}$ is
uniformly $\emptyset$-definable across C.

We may write D_φ for D in the definition above, and call $\{\Phi_{(d,\mu)} : (d, \mu) \in D\}$
a *(definable) asymptotic partition*. We define $h(\varphi(M^n, \bar{a}))$ to be the pair

$$(\mathrm{Dim}(\varphi(M^n, \bar{a})), \mathrm{Meas}(\varphi(M^n, \bar{a}))),$$

which equals (d, μ) if $(M, \bar{a}) \in \Phi_{(d,\mu)}$, except that if $d = \mu = 0$ we work with
the convention that $\mathrm{Dim}(\varphi(M^n, \bar{a})) = -1$.

The o-notation in (i) here means that for every $\varepsilon > 0$ there is $Q \in \mathbb{N}$ such
that for all $M \in C$ with $|M| > Q$ and all $\bar{a} \in M^m$, where $(M, \bar{a}) \in \Phi_{d,\mu}$, we
have

$$\left| |\varphi(M^n, \bar{a})| - \mu |M|^{\frac{d}{N}} \right| < \varepsilon |M|^{\frac{d}{N}}.$$

We call C a *weak asymptotic class* if C satisfies the asymptotic criterion (i)
for all φ, but the $\Phi_{(d,\mu)}$ are not assumed to be definable. We do not discuss
the intermediary condition that the $\Phi_{(d,\mu)}$ are definable but not necessarily
$\emptyset$-definable.

It is clear that clause (i) is preserved by reducts; that is, by the process of
restricting the *class* of structures to the reducts in a sublanguage. In fact, by
Lemma 3.7 of Elwes [22], it is preserved by parameter-interpretations which
are uniform in the sense that the interpreting formulas range through a finite set.
However, clause (ii) may by lost under interpretations, though it is preserved
under parameter-free bi-interpretations ([22], Lemma 3.7).

In infinite model theory, the definition of o-minimality (see e.g., [20]) places a restriction on the definable sets in one variable: a totally ordered structure is *o-minimal* if every definable set is a finite union of open intervals and points. The Cell Decomposition Theorem in this context yields topological and logical finiteness properties for n-variable definable sets. By the following theorem, which might be viewed as a combinatorial cell decomposition theorem, the story is essentially the same for asymptotic classes, though we have chosen here to have the *definition* focus on the n-variable condition. The proof makes heavy use of clause (ii).

Theorem 4.2.3 (Lemma 2.1.2 of [23]; Theorem 2.1 of [48]) *Suppose that C is a class of finite structures which satisfies Definition 4.2.2 (clauses (i) and (ii)) for $n = 1$, i.e. for definable sets in one variable. Then C is an N-dimensional asymptotic class.*

Its proof, an induction on n, is analogous to that of the Cell Decomposition Theorem for o-minimal structures, with asymptotic calculations replacing topological arguments. For a definable subset X of M^{n+1} let $\pi : M^{n+1} \to M$ denote projection to the first coordinate. Then apply the definition to $\pi(X)$ and the inductive hypothesis to the fibers X_a for $a \in \pi(X)$. Note for every pair (d, μ) that $\{a \in M : h(X_a) = (d, \mu)\}$ is an $\emptyset$-definable subset of M, by clause (ii), so itself has a specified dimension and measure.

The next section provides many examples of asymptotic classes. As a paradigmatic *non-example* – see Remark 4.5.2 (d) – observe that the collection of all finite total orders is not an asymptotic class: if $\varphi(x, y)$ is the formula $x < y$, then as a ranges through a finite totally ordered structure M, $\varphi(M, a)$ is a subset of M of arbitrary size.

4.3 Examples of asymptotic classes

The most interesting of the examples below are associated with finite fields. We see it as an area of significant interest to find new classes of examples with no connection to finite fields.

Example 4.3.1 The class of all finite fields forms a 1-dimensional asymptotic class, by the main theorem of [13].

Example 4.3.2 A *difference field* is a pair (F, σ) where F is a field and σ is an automorphism of F. Fix a prime p, and positive integers m, n with $m \geq 1, n > 1$, and $(m, n) = 1$. Let $C_{(m,n,p)}$ be the collection of difference fields $(\mathbb{F}_{p^{kn+m}}, \text{Frob}^k)$ where $k > 0$ and Frob denotes the Frobenius automorphism

$x \mapsto x^p$; so m, n, p are fixed in the class, but k is varying. Then, by Theorem 3.5.8 of [52], $C_{(m,n,p)}$ is a 1-dimensional asymptotic class. Note that the fixed field of $\sigma : x \mapsto x^{p^k}$ on $\mathbb{F}_{p^{kn+m}}$ is $\mathbb{F}_{p^t}$ where $t = (k, m)$, so has bounded size. Ryten's result rests upon the main results of Hrushovski [37]: the asymptotic results for difference varieties, and the identification of the ultraproduct theory of $(\bar{\mathbb{F}}_p, x \mapsto x^{p^k})$ with the theory $ACFA_p$, the model companion of the theory of characteristic p difference fields; here $\bar{\mathbb{F}}_p$ denotes the algebraic closure of $\mathbb{F}_p$.

Example 4.3.3 By further results of Ryten [52, Chapter 5], every family of finite simple groups of *fixed Lie type* is an asymptotic class. The Lie type here in particular determines the Lie rank, that is the number of nodes of the associated Dynkin diagram (or orbits on nodes under the corresponding graph automorphism, in the case of twisted groups). For example, the groups $PSL_3(q)$, with q varying, form a family of finite simple groups of fixed Lie type and Lie rank 2, corresponding to the Dynkin diagram with two nodes joined by a single edge, so form an asymptotic class. The groups $PSU_3(q)$, which are subgroups of $PSL_3(q^2)$, also form an asymptotic class. Here we view the alternating group Alt_n as having Lie rank n, and these do not form an asymptotic class as n grows with the size of the group; indeed, the model theory of the finite alternating groups, like the model theory of groups $PSL_n(q)$ for fixed q and increasing n, seems to be completely wild. For undecidability of the theory of all finite symmetric groups, or for various families of $n \times n$ matrix groups with unbounded n over a fixed field, see [27], or [12, Section 6.3] for a survey. Also, see [2] for a treatment of non-standard alternating groups as objects in Peano Arithmetic. For separate reasons, the (simple) cyclic groups of prime order also form an asymptotic class – see Example 4.3.5 below.

Most of the families of finite simple groups are *uniformly parameter bi-interpretable* (even bi-definable), in a natural sense, with finite fields (see Chapter 4 of [52]). Using results of Elwes and Ryten, it follows that the property of being an asymptotic class transfers from the fields to the groups, though care is needed with clause (ii) in Definition 4.2.2, due to the role of parameters in the interpretations.

For example, for any fixed n the groups $PSL_n(q)$ and $PSU_n(q)$ are both (uniformly in q) bi-interpretable with the field $\mathbb{F}_q$; details can be found in Chapter 5 of [52]. The uniform interpretation of the $PSL_n(q)$ in $\mathbb{F}_q$ is almost immediate from the definition of $PSL_n(q)$. More generally, for a given family of Chevalley groups of Lie type $\mathbb{L}$, such as the family of symplectic groups $PSp_{2m}(q)$, we can uniformly in $\mathbb{F}_q$ interpret the Lie algebra $\mathbb{L}(\mathbb{F}_q)$, hence the general linear group $GL(\mathbb{L}(\mathbb{F}_q))$, and inside this the family of root subgroups which generate $\mathrm{Sp}_{2m}(q)$. Since the symplectic group is a product of a bounded number, dependent only on m, of these root subgroups, it is itself uniformly interpretable in $\mathbb{F}_q$, and hence so is $\mathrm{PSp}_{2m}(q)$. In the other direction, to construct the field $\mathbb{F}_q$ inside $\mathrm{PSL}_n(q)$, the additive structure is given by a root group. The multiplicative structure of the field arises from a torus – which is conjugate to the image in $\mathrm{PSL}_n(q)$ of an appropriate diagonal subgroup of $\mathrm{SL}_n(q)$ – acting on the root group. See also [59] and [42].

For the families of Suzuki and Ree twisted simple groups, the situation is rather more complicated. The construction involves an automorphism of the Dynkin diagram which does not preserve lengths of roots. As a result, these groups are uniformly parameter bi-interpretable not with pure fields, but with difference fields. The class of Suzuki groups $^2B_2(2^{2k+1})$ is uniformly parameter bi-interpretable with the clas $\mathcal{C}_{(1,2,2)}$, as is the class of Ree groups $^2F_4(2^{2k+1})$. The Ree groups $^2G_2(3^{2k+1})$ are uniformly parameter bi-interpretable with the members of $\mathcal{C}_{(1,2,3)}$. In these cases we apply Example 4.3.2 above.

Example 4.3.4 The families of simple groups of Lie type all arise as automorphism groups of Tits buildings. The building blocks for these are the so-called 'rank 2 residues', which are generalized polygons. Here, a generalized polygon is an incidence structure of points and lines such that the associated bipartite incidence graph – which has the points and lines as vertices with incidence for adjacency – has diameter n and girth $2n$. A generalized n-gon is said to be *thin* if it is an ordinary n-gon, and is *thick* if every point (respectively line) is incident with at least three lines (respectively points). A thick generalized 3-gon is just a projective plane. The generalized polygons involved in finite simple groups satisfy an additional symmetry condition, the 'Moufang' property. Moufang generalized polygons have been classified by Tits and Weiss [61]. In particular, there are seven families of finite Moufang generalized polygons, each such polygon associated with its corresponding 'little projective group.' Dello Stritto [19] shows that each of these seven families forms an asymptotic class by proving that the polygons are uniformly parameter bi-interpretable with their corresponding little projective groups, as each corresponding class of groups forms an asymptotic class, by Example 4.3.3 above.

Example 4.3.5 By [48, Theorem 3.14], the collection of all finite cyclic groups is a 1-dimensional asymptotic class. This is hardly surprising, as the multiplicative groups of finite fields are cyclic. In general, the result follows from Szmielew's Theorem (see for example [31, Theorem A.2.2]), which says that in every abelian group every formula $\varphi(x, \bar{y})$ is equivalent to a boolean combination of formulas of the form $p^m | t(x, \bar{y})$ or $t(x, \bar{y}) = 0$, where p is prime, t is a term, and $p^m|t(x, \bar{y})$ abbreviates $(\exists z)\, p^m z = t(x, \bar{y})$. By the Compactness Theorem, there is a finite family of such boolean combinations, one of which will be equivalent to φ in each abelian group. The argument then reduces to examining a conjunction of such conditions and their negations. Observe, by Theorem 4.2.3, that it suffices to consider formulas $\varphi(x, \bar{y})$, i.e., families of definable sets in *one* variable x.

Example 4.3.6 Recall that the *random graph* is the unique countably infinite graph that satisfies, for each $n > 0$, the following sentence σ_n, where Rxy denotes that vertices x and y are adjacent:

$$\forall x_1 \ldots x_n, \forall y_1, \ldots y_n \left[\bigwedge_{1 \le i, j \le n} x_i \ne y_j \right.$$
$$\left. \to \exists z \Big(\bigwedge_{1=1}^{n} Rzx_i \wedge z \ne x_i \wedge \bigwedge_{i=1}^{n} \neg Rzy_i \wedge z \ne y_i \Big) \right].$$

The *Paley graph* P_q, where q is a prime power with $q \equiv 1 \pmod 4$, has vertex set $\mathbb{F}_q$ and edge relation given by $x, y \in \mathbb{F}_q$ are adjacent if and only if $x - y$ is a square in $\mathbb{F}_q$. The collection $\mathcal{C}$ of all Paley graphs forms a 1-dimensional asymptotic class – see [48, Example 3.4]. The essential point here is due to Bollobás and Thomason [9] (see also [10, Ch. XIII.2]). If U and W are disjoint sets of vertices in the Paley graph P_q with $|U \cup W| = m$, and $v(U, W)$ is the number of vertices of P_q not in $U \cup W$ joined to each vertex of U and none of W, then

$$|v(U, W) - 2^{-m}q| \le \frac{1}{2}(m - 2 + 2^{-m+1})q^{\frac{1}{2}} + m/2.$$

It follows that every non-principal ultraproduct satisfies each σ_n, so is elementarily equivalent to the random graph. As the latter has quantifier elimination, to check that $\mathcal{C}$ is an asymptotic class it suffices to consider quantifier-free formulas, which are handled by the above asymptotic estimates.

Thus the Paley graphs form a class of finite graphs whose theory approximates the random graph and witness that the random graph has the finite model property, that is, every sentence true of the random graph, and in particular the above axioms σ_n for $n \ge 1$, has a finite model and in fact hold in almost all finite graphs.

If one works instead with primes congruent to 3 mod 4, then -1 is a non-square, so the relation R defined above is antisymmetric, and one obtains a 1-dimensional asymptotic class of *Paley tournaments*, whose theory approximates the *random tournament*. Here a *tournament* is a directed graph such that any two vertices are connected by an arc. The analogue of the result of Bollobás and Thomason can be found in [29].

It would be interesting to find other asymptotic classes of graphs approximating the random graph, e.g. not corresponding to edge probability 1/2. Work of Szönyi [57] may be relevant here.

Example 4.3.7   The random graph has, for each $k > 2$, an arity k analogue, the *countable universal homogeneous k-uniform hypergraph*. Here, a k-uniform hypergraph is just a set equipped with a collection of k-element subsets, the hyperedges. Like the random graph, its theory is axiomatized by 'extension axioms', which hold with probability tending to one in finite k-uniform hypergraphs. There is no naïve arity k analogue of the Paley graphs, but Beyarslan [7] has shown that the random k-uniform hypergraph is interpretable in a pseudofinite field, that is, an infinite model of the theory of finite fields. Hence, there is a family of finite k-uniform hypergraphs, uniformly interpretable in finite fields, with an ultraproduct elementarily equivalent to the random k-uniform hypergraph. This certainly yields a weak asymptotic class of finite k-uniform hypergraphs whose theory approximates the random one. So far as we know, it has not been checked whether it is an asymptotic class, i.e., satisfies Definition 4.2.2(ii).

Example 4.3.8   Recall that a theory T is *strongly minimal* if, in all models of T, every definable subset of the domain is finite or cofinite. Suppose that $\mathcal{C}$ is a class of finite structures such that every non-principal ultraproduct is strongly minimal. It follows rather easily, by Theorem 4.2.3, that $\mathcal{C}$ is a 1-dimensional asymptotic class – see Example 3.9 of [48]. In particular, for every positive integer $d > 2$, the collection of all finite vertex transitive graphs of valency d is a 1-dimensional asymptotic class. Indeed, an ultraproduct is a vertex transitive graph of valency d, and all such are well-known to be strongly minimal; see for example [8], Lemma 2.2.11.

Example 4.3.9   Lastly, let M be a *smoothly approximable structure* – see Section 4.4 directly following this example for a fuller discussion of smooth approximability. Then M is a union of a chain of finite so-called 'envelopes'. It was shown by Elwes [22] that these envelopes can be chosen to form an asymptotic class. This rests on the asymptotic information on the sizes of definable sets in [18, Proposition 5.2.2]. In particular, suppose that M is *unidimensional*, that is, up to non-orthogonality has a unique family of definable Lie geometries. Then the asymptotic bounds are much tighter than in Definition 4.2.2. Namely, one finds that if the class $\mathcal{C}$ is N-dimensional, $E \in \mathcal{C}$, and D is a d-dimensional definable subset of measure μ of E, then for some fixed constant C depending just on M, it follows that

$$\left| |D| - \mu |E|^{\frac{d}{N}} \right| < C |E|^{\frac{d}{N} - \frac{1}{N}}.$$

In the non-unidimensional case, the asymptotic behavior is not so clear – consider for example a countably infinite structure consisting of the disjoint union of two $\aleph_0$-dimensional vector spaces, one over $\mathbb{F}_2$ and the other over $\mathbb{F}_3$. It is not clear how to approximate these by a class of finite substructures so that a conclusion like that in Theorem 4.2.1 holds. This is the reason for the weaker error term, in the form $o(|M|^{\frac{d}{N}})$, in Definition 4.2.2. Smoothly approximable structures are relevant also to robust classes, and we discuss them further in the next section.

4.4 Smoothly approximable structures

The notion of smooth approximation appears to be due originally to Lachlan; it plays a role already in [15] and [16]. The first systematic investigation of smooth approximation is [39], and a deep theory was developed by Cherlin and Hrushovski in the monograph [18]. There also are two excellent survey accounts, [35] and [17].

A finite substructure N of a structure M is a k-*homogeneous substructure* of M if all $\emptyset$-definable relations on M induce $\emptyset$-definable relations on N, and for every pair $\bar{a}, \bar{b}$ of k-tuples from N, they have the same type in N if and only if they have the same type in M. An $\aleph_0$-categorical structure M is *smoothly approximated* if it is the union of a chain $(M_i : i \in \mathbb{N})$ of finite substructures, where for each i, M_i is an $|M_i|$-homogeneous substructure of M.

As a very basic example, let M be an $\aleph_0$-dimensional vector space over the finite field $\mathbb{F}_p$, and let the M_i for $i \in \mathbb{N}$ form a sequence of finite subspaces with $M_i \leq M_{i+1}$ and with $\bigcup_{i \in \mathbb{N}} M_i = M$. These structures are parsed in the language of $\mathbb{F}_p$-modules, that is the language for M as a group under addition, and a unary function symbol for multiplication by each element of $\mathbb{F}_p$. The point here is that if $\mathrm{Dim}(M_i) = n_i$ then we have a natural sequence of embeddings $\mathrm{GL}_{n_0}(p) \leq \mathrm{GL}_{n_1}(p) \leq \mathrm{GL}_{n_2}(p) \leq \cdots$, and the union of this sequence of groups has the same orbits on n-tuples from M as $\mathrm{Aut}(M)$, which is $\mathrm{GL}_{\aleph_0}(p)$.

For a slightly more complicated example, suppose that M is endowed with a symplectic form, that is, a non-degenerate bilinear form $\beta : M \times M \to \mathbb{F}_p$ such that $\beta(v, v) = 0$ for all $v \in M$, and let the M_i form a sequence of finite (even dimensional) subspaces with union M, on each of which β induces a non-degenerate form. Here, β can be given by a family of binary relations, one for each element of $\mathbb{F}_p$. The fact that each M_i is an $|M_i|$-homogeneous substructure of M is a consequence of Witt's Lemma – see for example Section 20 of [3].

These two smoothly approximable structures are both examples of *Lie geometries* (see [39] or [18]), about which we say more below. The first is totally categorical, even strongly minimal. The second is not stable (it has the independence property; see, e.g., [31]) but is supersimple of rank 1 (see Remark 4.5.2(c) for more about simple theories). We remark that the random graph is *not* smoothly approximable, even though it is also supersimple of rank 1, $\aleph_0$-categorical, and arises in a natural way as a union of a chain of Paley graphs (see Example 4.8.3).

Let M be smoothly approximable. As M is $\aleph_0$-categorical, the Ryll-Nardzewski Theorem yields a function $g : \mathbb{N} \to \mathbb{N}$ such that for each $k \in \mathbb{N}$, the automorphism group of M, Aut(M), has at most $g(k)$ orbits on M^k. It follows that if $(M_i : i \in \omega)$ is a sequence of finite substructures witnessing smooth approximability, then Aut(M_i) has at most $g(k)$ orbits on M_i^k, for each $i, k \in \omega$. This is a very strong condition on a family of arbitrarily large finite permutation groups. In fact, *a posteriori*, the condition holds for all k if one just knows the condition for $k \leq 4$. Using finite permutation group theory – the classification of finite simple groups, the O'Nan-Scott Theorem, and Aschbacher's structure theory for subgroups of classical groups – the authors in [39] were able to classify all smoothly approximable structures with *primitive* automorphism groups, that is, automorphism groups which preserve no proper non-trivial equivalence relation. In particular, certain building blocks, the *Lie geometries*, were identified. These include pure sets, examples like those described in the preceding paragraphs (possibly with orthogonal or unitary bilinear forms), and their projective and affine versions. There is also the 'self-dual geometry', which is really a pair of infinite dimensional vector spaces V, V' over a finite field $\mathbb{F}_q$ equipped with a non-degenerate bilinear map $V \times V' \to \mathbb{F}_q$, and the 'quadratic geometry'. The somewhat more mysterious latter is essentially the collection of all quadratic forms associated with a given symplectic form on an $\aleph_0$-dimensional vector space over a finite field of characteristic 2.

A very beautiful structure theory of smoothly approximable structures is developed in [18]. We provide a brief overview.

As shown in [15], every $\aleph_0$-categorical ω-stable structure – in particular every totally categorical structure, and every stable structure homogeneous over a finite relational language – is smoothly approximable. This is already quite a deep result, and includes rather complicated structures built, by a sequence of finite and affine covers, from pure sets and projective or affine spaces over finite fields. See for example [15], [1] and [32].

Smoothly approximable structures which are not ω-stable are all unstable. Indeed, any stable ω-categorical structure which is not ω-stable interprets a pseudoplane [43], and by [18, Corollary 5.5.5] a pseudoplane cannot be interpreted in any smoothly approximable stucture. However, all smoothly approximable structures have a simple theory–see Remark 4.5.2(c) for basic facts about simple theories, and [62] for a general source – and in fact they are supersimple of finite rank (see Remark 4.5.2). Indeed, some crucial ideas in simplicity theory, such as the Independence Theorem, first appeared in [18] (see e.g. Section 5.1 and Proposition 8.4.3 of [18]). They are 1-based (see Section 4.5 for the definition; that they are 1-based may be found in [18]), and thus in particular, no infinite field is definable. Although finite fields play a key role in any family of finite approximating structures, any field involved remains fixed throughout the family.

All groups definable in a smoothly approximable structure are finite-by-abelian-by-finite (see [18]). As noted in Section 4.6 (see Proposition 4.6.2 ff.), extraspecial p-groups of exponent p are smoothly approximable and finite-by-abelian, but not abelian-by-finite.

Cherlin and Hrushovski define the notion of a *Lie coordinatizable structure*. This is a structure bi-interpretable with a *Lie coordinatized* structure, which is, roughly speaking, one coordinatized by a tree of finite height of Lie geometries. A key fact proved in [18] is that the Lie coordinatizable structures are exactly the smoothly approximable structures. Arguments by induction on the height of the coordinatizing tree thus are often used.

Smoothly approximable structures also are *quasi-finitely axiomatizable*. This means that the theory is axiomatized by a single sentence together with a schema of axioms saying that each 'non-orthogonality class' of Lie geometries (again, see [18]) is infinite dimensional. In particular, if the smoothly approximable structure is unidimensional, that is, any two interpretable Lie geometries are non-orthogonal, then the theory is axiomatized by a single sentence together with, for each n, a sentence saying that the structure has size at least n. This generalizes an earlier result of Hrushovski [32], itself extending work of Ahlbrandt and Ziegler [1], that all totally categorical structures are quasi-finitely axiomatized in this last sense.

Built into Lie coordinatizability is a theory of the *envelopes* – that is, the finite approximating substructures – of smoothly approximable structures. There are precise results in [18] on the cardinalities of definable sets in envelopes, given by certain polynomials. This is exploited in [22] to show the envelopes can be chosen to form an asymptotic class, so yields Example 4.3.9 above. It also suggests that there may be interesting strengthenings of the notion of asymptotic class, where one considers classes of finite structures in which the sizes of definable sets, with respect to a fixed formula $\varphi(\bar{x}, \bar{y})$, are given not by asymptotic conditions, but by one of finitely many polynomials. So far as we know, this has not been explored. In addition, Lachlan's theory [44] of 'shrinking and stretching', developed for the class of finite structures homogeneous over a fixed finite relational language, holds in the smoothly approximable context too.

It also follows from the theory that, over a fixed finite language L, if C is a class of finite structures for which there is some k such that Aut(M) has at most k orbits on M^4 for all $M \in C$, then there is $g : \mathbb{N} \to \mathbb{N}$ such that for all $M \in C$ and all $k \in \mathbb{N}$, Aut(M) has at most $g(k)$ orbits on M^k. See for example Theorem 6 of [18].

Smooth approximability is not preserved by reducts, due to problems with the quadratic geometries mentioned above – see the example due to Evans in [18, p. 149]. Yet the class of reducts, namely the *weakly Lie coordinatizable structures*, also is fairly well understood. In particular, the class of weakly Lie coordinatizable structures is characterized by the conjunction of nine model-theoretic properties (Theorem 7 in [18]), among which are $\aleph_0$-categoricity, pseudofiniteness (or the finite model property), finiteness of a certain rank, the Independence Theorem, and some more technical conditions. Intriguingly, whereas the bulk of the structure theory of [18] rests ultimately on the classification of finite simple groups, this last result does not, even though it characterizes a class of structures intimately connected to finite simple groups.

4.5 Asymptotic classes and their ultraproducts

We have already seen that investigations of asymptotic classes can be assisted by working with ultraproducts. This is already explicit in [13] on finite fields. Indeed, recall that a *pseudofinite field* is an infinite model of the theory of finite fields. Equivalently, by Ax [4] it is a field F which is perfect, quasifinite (has a unique extension of degree n for each $n > 1$), and is *pseudo-algebraically closed* (PAC) (that is, every absolutely irreducible variety defined over F has an F-rational point). It is shown in [13] that, by Theorem 4.2.1, if F is a pseudofinite field then it is possible to associate with each definable subset D of each power F^l a pair (d, μ), where d is a nonnegative integer and $\mu \in \mathbb{Q}^{>0}$, such that d is the algebraic-geometric dimension of the Zariski closure of D, and the pairs (d, μ) satisfy certain counting conditions. This led the authors in [48] to introduce the following notion of *measurable structure*. The definition below, taken from [25], is slightly different from Definition 5.1 of [48], but equivalent.

Definition 4.5.1 An infinite L-structure M is *measurable* if there is a function $h : \mathrm{Def}(M) \to \mathbb{N} \times \mathbb{R} \cup \{(0, 0)\}$ (we write

$$h(X) = (\mathrm{Dim}(X), \mathrm{Meas}(X)) = (\mathrm{Dim}, \mathrm{Meas})(X))$$

such that the following hold.

1. For each L-formula $\varphi(\bar{x}, \bar{y})$ there is a finite set $D \subset \mathbb{N} \times \mathbb{R}^{>0} \cup \{(0, 0)\}$, so that for all $\bar{a} \in M^m$ we have $h(\varphi(M^n, \bar{a})) \in D$.
2. If $\varphi(M^n, \bar{a})$ is finite then $h(\varphi(M^n, \bar{a})) = (0, |\varphi(M^n, \bar{a})|)$.
3. For every L-formula $\varphi(\bar{x}, \bar{y})$ and all $(d, \mu) \in D_\varphi$, the set $\{\bar{a} \in M^m : h(\varphi(M^n, \bar{a})) = (d, \mu)\}$ is $\emptyset$-definable.
4. (Fubini) Let $X, Y \in \mathrm{Def}(M)$ and $f : X \to Y$ be a definable surjection. Then there are $r \in \mathbb{N}$ and $(d_1, \mu_1), \ldots, (d_r, \mu_r) \in (\mathbb{N} \times \mathbb{R}^{>0}) \cup \{(0, 0)\}$ so that if $Y_i := \{\bar{y} \in Y : h(f^{-1}(\bar{y})) = (d_i, \mu_i)\}$, then $Y = Y_1 \cup \ldots \cup Y_r$ is a partition of Y into non-empty disjoint definable sets. Let $h(Y_i) = (e_i, \nu_i)$ for $i \in \{1, \ldots, r\}$. Also let $c := \mathrm{Max}\{d_1 + e_1, \ldots, d_r + e_r\}$, and suppose (without loss) that this maximum is attained by $d_1 + e_1, \ldots, d_s + e_s$. Then $h(X) = (c, \mu_1\nu_1 + \cdots + \mu_s\nu_s)$.

If $X \in \mathrm{Def}(M)$ and $h(X) = (d, \mu)$, we call d the *dimension* of X and μ the *measure* of X, and h the *measuring function*.

We do not emphasize measurable structures in this paper. For more information, see [48] or [25]. We do note the following observations.

Remark 4.5.2 a. If M is measurable and $N \equiv M$ then N is measurable; hence one may speak of a *measurable theory*.

b. If $\mathcal{C}$ is an N-dimensional asymptotic class, then every non-principal ultraproduct of $\mathcal{C}$ is measurable – essentially, the pairs (d, μ) transfer through to the ultraproduct.

c. Shelah introduced in [55] the concept of a *simple* theory (see also [41]). Simplicity is a generalization of stability in which model-theoretic non-forking still provides a good notion of independence. Indeed, it satisfies all the main properties of non-forking in stable theories except stationarity, which controls the number of non-forking extensions of a complete type. Stationarity (i.e., the Finite Equivalence Relation Theorem) is replaced by the 'Independence Theorem' in simple theories. The analogue of a superstable theory in stability is, in the context of simplicity, the notion of a *supersimple theory*. In a supersimple theory every definable set has an ordinal-valued 'D-rank'; in fact, there are several notions of rank on a definable set in a supersimple theory – D-rank, SU-rank, and S_1-rank – all of which coincide if any of them is finite. It is shown in [25, Corollary 3.4] – but was noted earlier by Ryten – that if M is measurable then it is supersimple, and for any definable set X in M, the D-rank of X is at most its dimension and hence is finite.

d. It follows from (b) and (c) that an ultraproduct of an asymptotic class cannot have the *strict order property*: there cannot be a definable partial order (even on a power of the structure) with an infinite chain. This generalizes the observation made at the end of Section 4.2 that the collection of finite linear orders does not form an asymptotic class.

e. There are measurable structures that are not elementarily equivalent to any ultraproduct of an asymptotic class. Vector spaces over an infinite field, in the language of modules over the field, provide one example. A more interesting example of Elwes [22, Section 3.4] consists of a structure with two different pseudofinite field structures (in disjoint languages) of different prime characteristics. It arises by taking the fixed point set of a generic automorphism of the 'Hrushovski fusion' [37] of two different algebraically closed fields in distinct positive characteristics. Such a structure cannot be an ultraproduct of finite structures, since no positive integer can be a power of two distinct primes.

We next discuss how some stability-theoretic notions interact with asymptotic classes. The following result enables us to detect in the 1-dimensional case, just from asymptotic information, whether or not every ultraproduct of an asymptotic class is stable.

Proposition 4.5.3 ([48]) *Let $\mathcal{C}$ be a 1-dimensional asymptotic class. Then some ultraproduct of $\mathcal{C}$ is unstable if and only if there is a formula $\varphi(x, \bar{y})$, and for each $k \in \mathbb{N}$ some $M \in \mathcal{C}$ and $\bar{a}_1, \ldots, \bar{a}_k \in M^{\ell(\bar{y})}$ with*

(a) $|\varphi(M, \bar{a}_i)| \geq k$ for each $i = 1, \ldots, k$
(b) $|\varphi(M, \bar{a}_i) \triangle \varphi(M, \bar{a}_j)| \geq k$ for all distinct $i, j \in \{1, \ldots, k\}$.

Of the examples of asymptotic classes considered in Section 4.3, only the following have all ultraproducts stable: the class of finite cyclic groups; the asymptotic classes of Example 4.3.8 with all ultraproducts strongly minimal; in Example 4.3.9, if M is a smoothly approximable structure which is ω-categorical and ω-stable (or in particular, which is totally categorical), then every asymptotic class consisting of its envelopes.

We now recall the construction, from a complete theory T, of T^{eq}. For each $n > 0$, and each $\emptyset$-definable equivalence relation E on n-tuples, one adjoins a new sort interpreted, for $M \models T$, by M^n / E and a function taking each n-tuple to its corresponding E-equivalence class in the new sort. There is a corresponding language L^{eq}, and all models M of T have corresponding expansions M^{eq} in the language L^{eq} with theory T^{eq}.

Recall also that if A is a subset of a structure M, then $b \in M$ is *algebraic* over A if there is a finite A-definable subset of M containing b. The algebraic closure of A, namely the set of elements algebraic over A, is denoted $\mathrm{acl}(A)$.

A supersimple theory T is said to be *1-based* if, for every $M \models T$ and all subsets A and B of M^{eq}, we have that A and B are independent, in the sense of non-forking, over $\mathrm{acl}^{\mathrm{eq}}(A) \cap \mathrm{acl}^{\mathrm{eq}}(B)$. Using the main theorem of [33] Elwes [22] derived the result below, which *a fortiori* gives structural restrictions on asymptotic classes all of whose ultraproducts are stable. Note that there is an error in [22] stemming from a misunderstanding of [33] – an invalid use of compactness in [22, Lemma 6.4]. A valid argument in its place has been given by Kestner and Pillay (personal communication).

Theorem 4.5.4 *Every measurable stable theory is 1-based.*

As a very special case, note that an algebraically closed field cannot be measurable. To illustrate, for the complex field $\mathbb{C}$, the map $x \mapsto x^2$ is a surjection $\mathbb{C} \setminus \{0\} \to \mathbb{C} \setminus \{0\}$ which is 2-to-1, contrary to Definition 4.5.1(iv). In fact, there is evidence that measurable fields must be pseudofinite. Scanlon has shown that every infinite measurable field is quasifinite, and easily, every measurable field is perfect – see [48, Theorems 5.18 and 6.1], and also [54]. It is not known if the PAC pseudo-algebraically closed property (see the first paragraph in this section) holds for all measurable fields; this would yield pseudofiniteness.

If M is a 1-dimensional measurable structure – e.g., an ultraproduct of a 1-dimensional asymptotic class – then the algebraic closure operator defines a pre-geometry on subsets of M (as it is supersimple of rank 1). In particular it satisfies the *exchange property*: if $b \in \mathrm{acl}(A \cup \{c\}) \setminus \mathrm{acl}(A)$, then $c \in \mathrm{acl}(A \cup \{b\})$. The exchange property can be formalized fairly concretely for 1-dimensional asymptotic classes. The definitions can be finitized, and the formula making c algebraic over $A \cup \{b\}$ can be identified up to finitely many possibilities over the given data. Indeed, suppose that $\mathcal{C}$ is a 1-dimensional asymptotic class, $\varphi(x, \bar{y})$ is a formula, and D is the corresponding subset of $\{0, 1\} \times \mathbb{R}^{>0}$ of dimension-measure pairs as provided in Definition 4.2.2(i). Let $E := \{\mu : (1, \mu) \in D\}$ and for $\mu \in E$ let $\varphi_\mu(\bar{y})$ be a formula defining $\Phi_{(1,\mu)}$ as in Definition 4.2.2(ii). If $M \in \mathcal{C}$, $A \subset M$, and $b \in M$, we say b is in the *φ-closure* of A, written $b \in \mathrm{cl}_\varphi(A)$, if there is some $\bar{a}$ from A such that $M \models \varphi(b, \bar{a})$ and $M \not\models \varphi_\mu(\bar{a})$ for each $\mu \in E$. For a set of formulas Σ, we say that b is in the Σ-closure of A, written $b \in \mathrm{cl}_\Sigma(A)$, if $b \in \mathrm{cl}_\varphi(A)$ for some $\varphi \in \Sigma$. We have:

Proposition 4.5.5 *[48, Proposition 4.4] Let C be a 1-dimensional asymptotic class, $M \in C$, and $A \subset M$. There are finite sets $\Delta(\varphi)$ and $\Gamma(\varphi)$ of formulas (depending on φ) such that for $a, b \in M$ with $a \in \mathrm{cl}_\varphi(Ab) \setminus \mathrm{cl}_{\Delta(\varphi)}(A)$, if M is 'large enough' relative to A and φ, then $b \in \mathrm{cl}_{\Gamma(\varphi)}(Aa)$.*

The notion of a 1-based theory is related to the trichotomy conjecture of Zilber. One version of this conjecture, now known to be false, asserts that if M is a strongly minimal structure, then: either M is *disintegrated*, that is, $\mathrm{acl}(A) = \bigcup(\mathrm{acl}(a) : a \in A)$ for every $A \subset M$; or M is locally modular, equivalently, 1-based; or M interprets an infinite field. Counterexamples to this conjecture were found by Hrushovski [34], but the conjecture has remained extremely influential, and versions hold in key contexts. Furthermore, disintegrated and locally modular strongly minimal sets now are fairly well understood.

It makes sense to investigate the conjecture for 1-dimensional asymptotic classes. The following result, Proposition 4.5 of [48], gives a clear notion of a 'disintegrated 1-dimensional asymptotic class', identified by the asymptotic condition (iii).

Proposition 4.5.6 *Let C be a 1-dimensional asymptotic class. Then the following are equivalent.*

(i) for every formula $\varphi(x, \bar{y})$, there is a formula $\psi(x, \bar{z})$ and some $K \in \mathbb{N}$ such that if $M \in C$ with $|M| > K$ and $A \subset M$, then $\mathrm{cl}_\varphi(A) \subset \bigcup_{a \in A}(\mathrm{cl}_\psi(\{a\}))$;

(ii) in every infinite ultraproduct M of members of C, if $A \subset M$ then $\mathrm{acl}(A) = \bigcup_{a \in A}(\mathrm{acl}(\{a\}))$;

(iii) for every $\varphi(x, \bar{y})$ there is some $K_\varphi \in \mathbb{N}$ such that for all $M \in C$ with $|M| > K_\varphi$, if $A \subset M$ then $|\mathrm{cl}_\varphi(A)| \le K_\varphi |A|$.

It would be of interest to investigate the Zilber trichotomy further for asymptotic classes. Certainly, by one of the main results of [18], all smoothly approximable structures are 1-based. This includes in particular the Lie geometries. We have no idea if, in a 1-dimensional asymptotic class that is not 1-based – that is, ultraproducts of which are not 1-based, and so by Theorem 4.5.4 are unstable – arbitrarily large finite fields must be uniformly interpretable. It would be intriguing to investigate this already for classes of structures interpretable in finite fields, or even for reducts of finite fields.

4.6 Asymptotic classes of groups

As mentioned in Example 4.3.3, the work of Ryten on difference fields yields the following theorem; we emphasize again that the bound on Lie rank is essential.

Theorem 4.6.1 (Ryten) *If C is any family of finite simple groups of fixed Lie type, then C is an asymptotic class.*

The structure of asymptotic classes of groups is an attractive area of study. For groups in which definability is governed by definability in finite fields, or by definable subgroups of cartesian powers (as in one-based groups), one expects good control of definability. It is not clear whether 'asymptotic class' is the optimal model theoretic assumption on a family of finite groups; often the same conclusions can be drawn just assuming that all ultraproducts have supersimple finite rank theory. This is developed in [24].

In developing a structure theory one wants, as far as possible, not to assume the classification of finite simple groups (CFSG). It would be wonderful to recover parts of the classification just from model theoretic hypotheses. As one step in this direction, Hrushovski has shown that any family of finite simple groups uniformly definable in finite fields is a family of (possibly twisted) Lie type [36, Theorem 9.2], and the same holds for groups uniformly definable in a family of the difference fields $C_{(m,n,p)}$ (see [37, Theorem 1.8] – the proof is unpublished). In the same spirit, Theorem 7.5.6 of [18] identifies, by model-theoretic hypotheses, a class of structures closely associated with finite simple groups.

As mentioned in Example 4.3.5, the class of finite cyclic groups is a 1-dimensional asymptotic class. It should be feasible to describe all asymptotic classes of abelian groups. At higher levels of complexity – but within the class of soluble groups – very little is known, though one expects unipotent and Borel subgroups of finite Chevalley groups of fixed Lie type to fall into asymptotic classes. We also mention the following ([48, Proposition 3.11]). A p-group is *extraspecial* if $G' = Z(G) = \Phi(G) \cong C_p$. In particular, extraspecial groups are nilpotent of class 2.

Proposition 4.6.2 *If p is an odd prime, then the class of finite extraspecial groups of exponent p is a 1-dimensional asymptotic class.*

In fact, finite extraspecial groups of odd exponent p are envelopes of a smoothly approximable structure, the unique countably infinite extraspecial group of exponent p. This group has cyclic center, and the quotient by the center is an elementary abelian p-group, equipped with an alternating bilinear form given by the commutator map to the center. Its theory is supersimple but unstable.

The remaining results in this section dealing with families of finite groups are obtained under weaker hypotheses than that of being an 'asymptotic class' – see our remarks following Theorem 4.6.1. We say that a family C of finite structures is *supersimple of finite rank* (respectively, *supersimple of rank n*) if all non-principal ultraproducts have these properties. The results below are all analogues of theorems about groups of finite Morley rank. Typically, the proofs use ultraproducts, and facts about measurable groups, or, more generally, groups with a supersimple finite rank theory. In some cases – Propositions 4.6.3(i), 4.6.5, 4.6.6, 4.6.8 – the results really belong in some such setting, and there is no use of finiteness or pseudofiniteness.

Proposition 4.6.3 *(i) [48, Theorem 3.12] Let C be a supersimple rank 1 family of finite groups. Then there is $d \in \mathbb{N}$ such that each group $G \in C$ has normal subgroups H and N, where $|G : N| \leq d$, $|H| \leq d$, $H \leq Z(N)$, and N/H is abelian.*

(ii) [24] Let C be a supersimple rank 2 family of finite groups. Then there is $d \in \mathbb{N}$ such that all groups in C have a normal subgroup of index at most d which is soluble of derived length at most 4.

Neither result requires CFSG. Part (ii) was proved by Elwes and Ryten in [26] under the extra assumption that C is a (2-dimensional) asymptotic class, using CFSG. The bound 4 on the derived length is probably not optimal; it should perhaps be 2, arising from the class of 1-dimensional affine groups $\mathrm{AGL}_1(\mathbb{F}_q)$.

One key ingredient in the proof is the body of results on so-called BFC groups: that is, groups with a finite bound d on the size of all conjugacy classes. If G is such a group with bound d on the size of conjugacy classes, then there is a bound $B(d) \in \mathbb{N}$ such that $|G'| \leq B(d)$; see Wiegold [63], for example, where $B(d) = d^{\frac{1}{2}d^4(\log_2 d)^3}$ is obtained. Another element is a theorem of Schlichting, strengthened by Bergman and Lenstra [6]. It asserts that if G is a group and $\mathcal{F}$ is a family of subgroups of G which is (setwise) invariant under a group K of automorphisms of G such that for some d and every $F_1, F_2 \in \mathcal{F}$ we have $|F_1 : F_1 \cap F_2| \leq d$, then for some d' dependent only on d, there is a K-invariant subgroup N of G so that $|F : F \cap N| \leq d'$ and $|N : F \cap N| \leq d'$ for all $F \in \mathcal{F}$.

To establish further results, one often needs a version for supersimple theories of the Zilber Indecomposability Theorem, applied to ultraproducts. The version below follows from [62, Theorem 5.5.4].

Theorem 4.6.4 *Let G be a group definable in a supersimple structure of finite rank, and let $\{X_i : i \in I\}$ be a collection of definable subsets of G. Then there exists a definable subgroup H of G such that:*

(i) $H \leq X_{i_1}^{\pm 1} \ldots X_{i_m}^{\pm 1}$ for some $i_1, \ldots, i_m \in I$;
(ii) X_i/H is finite for each $i \in I$.

Moreover, if the collection $\{X_i : i \in I\}$ *is invariant under the group* K *of definable automorphisms of* G, *then* H *can be chosen to be* K-*invariant.*

Consequences of Theorem 4.6.4 include the next results. If C is a family of groups, we say that the subsets X of members G of C are *uniformly definable* if just finitely many formulas $\varphi(x, \bar{y})$ are required to define the sets X as G ranges through C.

Proposition 4.6.5 *(from [36, Corollary 7.4]) If* C *is a supersimple finite rank family of finite groups with no uniformly definable proper non-trivial normal subgroups, then all but finitely many of the groups in* C *are simple.*

Proposition 4.6.6 *(from [36, Corollary 7.1]) If* C *is a supersimple finite rank family of finite groups, then the derived subgroups of members of* C *are uniformly definable.*

Recall that the *soluble radical* $R(G)$ of a finite group G is its largest soluble normal subgroup.

Proposition 4.6.7 *[24, Theorem 1.1] If* C *is a supersimple finite rank family of finite groups, then the soluble radicals* $R(G)$ *of the groups* $G \in C$ *are uniformly definable.*

Proposition 4.6.7 has further structural consequences. For a group G we denote its socle, the direct product of its minimal normal subgroups, by $\mathrm{Soc}(G)$. It follows fairly easily from Proposition 4.6.7 that if C is a supersimple finite rank class, then the groups $\mathrm{Soc}(G/R(G))$ are uniformly interpretable in G as G ranges through C, and are a direct product of a *bounded number* of finite simple groups of bounded Lie rank.

There are also the beginnings of a model theory for families of finite *permutation* groups. We view a permutation group model-theoretically as a pair (X, G), with a definable group structure on G and a definable action of G on X. By the Orbit-Stabilizer Theorem, if G is transitive on X, we may parse (X, G) as a pair (G, H), where H is a subgroup of G, the stabilizer of some $x \in X$.

Recall that a permutation group (X, G) is *primitive* if there is no proper non-trivial G-invariant equivalence relation – that is, G-congruence – on X, equivalently, if all point stabilizers G_x for $x \in X$ are maximal subgroups of G. We say that the family C of finite permutation groups (X, G) is *definably primitive* if, for each non-principal ultraproduct (X^*, G^*), there is no definable proper non-trivial G^*-congruence on X^*.

Proposition 4.6.8 ([26])   *If* C *is a supersimple finite rank class of definably primitive finite permutation groups, then all but finitely many of the permutation groups in* C *are primitive.*

Theorem 4.6.9 ([24])   *Let* C *be a supersimple finite rank family of finite primitive permutation groups, and suppose that for every ultraproduct* (X^*, G^*), $\mathrm{rk}(X^*) = 1$. *Then one of the following holds for ultraproducts* (X^*, G^*).

(i) $\mathrm{rk}(G^*) = 1$, G^* *acts regularly on* X^*, *and* G^* *is elementary abelian or torsion-free divisible abelian.*

(ii) $\mathrm{rk}(G^*) = 2$, *and there is an interpretable pseudofinite field* F *such that* $G \leq \mathrm{AGL}_1(F)$ *(the one-dimensional affine group* $(F, +).(F, \times)$*) in the natural action on* F.

(iii) $\mathrm{rk}(G^*) = 3$, *there is an interpretable pseudofinite field* F, *and* $\mathrm{PSL}_2(F) \leq G \leq \mathrm{P\Gamma L}_2(F)$ *in the natural action on the projective line* $\mathrm{PG}_1(F)$.

In [47] a structure theory is given for families of finite permutation groups all of whose non-principal ultraproducts are primitive. We do not give details here. One feature concerns families of primitive permutation groups (X, G) such that $\mathrm{Soc}(G)$ is a non-abelian simple group of fixed Lie rank. Except in very specific cases (essentially where the point stabilizers are bounded, or are 'subfield subgroups' associated with subfields for which the field extension degree is unbounded) families of this type have primitive ultraproducts. The proof uses much of the above work of Elwes and Ryten, Theorem 4.6.4, and also knowledge of maximal subgroups of finite simple groups. In particular, we have the following result. It generalizes [38, Proposition 8.1], which is over prime fields, but unlike the latter, it makes heavy use of the classification of finite simple groups.

Theorem 4.6.10 ([47])   *Let* Chev *be a fixed Lie type (possibly twisted) of finite simple groups, and* d *a positive integer. Let* C *be a family of pairs* (G, H) *where* $G = \mathrm{Chev}(q)$ *and* H *is a maximal subgroup of* G, *and suppose that if* $H = \mathrm{Chev}(q_0)$ *then the degree* $[\mathbb{F}_q : \mathbb{F}_{q_0}] \leq d$. *Then the groups* H *are uniformly definable in the groups* G, *and* C *is an asymptotic class.*

4.7 Robust classes

Robust classes consist of chains of finite structures in which the truth value of every formula, with parameters from some structure in the chain, eventually stabilizes when the formula is interpreted in a sufficiently larger structure. That is, one must "look ahead" in the chain to determine satisfaction of a formula with parameters in some structure in the chain. This framework has provided a setting in which to investigate notions of stability and o-minimality, as well as some provisional versions of complexity of such a chain. In addition to investigating these topics, the interest has mainly been in finding examples.

We begin with the definition of a robust class, and some initial observations, working over an arbitrary – and, unless otherwise specified, finite – first-order language L. Examples are discussed in the next section, and a robust approximation to the ordered additive group of rational numbers is established in Section 4.9.

Definition 4.7.1 A sequence $\mathcal{C} = (M_i : i \in \mathbb{N})$ of finite L-structures forms a *chain* of structures if $M_i \subseteq M_{i+1}$ (as a substructure) for each $i \in \mathbb{N}$. A chain $\mathcal{C}$ is said to be *robust* if for each $n \in \mathbb{N}$ and L-formula $\varphi(x_1, \ldots, x_n)$ there is a function $f = f_\varphi : \mathbb{N} \to \mathbb{N}$ such that for each $i \in \mathbb{N}$ and $a_1, \ldots, a_n \in M_i$, and for every $j \geq f(i)$, we have

$$M_{f(i)} \models \varphi(\bar{a}) \text{ if and only if } M_j \models \varphi(\bar{a}).$$

If $\mathcal{C}$ is robust and $\bar{a}$ and M_i are as above, we write $\mathcal{C} \models_{\text{ev}} \varphi(\bar{a})$ if $M_j \models \varphi(\bar{a})$ for all sufficiently large j.

Remark 4.7.2 (a) In what follows, we always assume that f_φ grows as slowly as possible subject to witnessing robustness. Thus, f_φ is uniquely determined by $\mathcal{C}$.

(b) There is a natural generalization of Definition 4.7.1, where we replace the chain $\mathcal{C}$ by a directed system of finite structures, equipped with specified embeddings. This is our original context, but as nothing has yet been done with the greater generality, we here use the more concrete version in Definition 4.7.1.

Initial results on robust classes have been obtained over the last few years by the authors (first published here) and in the Ph.D. thesis of the first author's student, Richard Marshall [49]. The study of robust classes is still in its early stages and thus not yet fully systematic. For example, one could envisage a theory of 'asymptotic robust classes', in which the asymptotic estimates described in Definition 4.2.2 are required to hold only when a formula is interpreted in a sufficiently larger structure, but this has not been considered.

Observe that Definition 4.7.1 applies in particular to *sentences*. Thus, if $\mathcal{C} = (M_i : i \in \mathbb{N})$ is a robust class, then for every L-sentence σ, precisely one of σ or $\neg\sigma$ holds in cofinitely many M_i. If $\mathcal{C}$ is an arbitrary chain of finite structures, we define the *asymptotic theory* $T_\mathcal{C}^{\text{as}}$ to be the collection of L-sentences true in cofinitely many $M_i \in \mathcal{C}$. If $\mathcal{C}$ is robust then this is a complete theory. Let M be the (countable) direct limit structure of the chain $\mathcal{C}$. The *limit theory* $T_{\mathcal{C}}^{\lim}$ of $\mathcal{C}$ is defined to be Th(M). In general, even assuming robustness, we do not expect $T_{\mathcal{C}}^{as} = T_{\mathcal{C}}^{\lim}$ (see Propositions 4.7.4 and 4.8.2, and the comment following 4.8.2).

Recall that a first-order structure M is *locally finite* if every finite subset of M is contained in a finite substructure of M. Given a countably infinite locally finite structure M, it is natural to ask if it is possible to construct a robust chain with direct limit M. The next result demonstrates that we must refine this initial question to investigate conditions under which we can obtain an *explicitly described* robust chain. This question can take several forms, a theme that we explore throughout the rest of this paper. If a chain of finite structures $\mathcal{C}'$ is a subsequence of a chain $\mathcal{C}$, we call it a *coarsening* of $\mathcal{C}$. Obviously, any coarsening of a robust chain is robust.

Proposition 4.7.3 *[49, Theorem 2.4.5]*

(i) *Let* $\mathcal{C} = (M_i : i \in \mathbb{N})$ *be a chain of finite L-structures. Then there is a coarsening of $\mathcal{C}$ which is robust.*

(ii) *Every countably infinite locally finite structure is the direct limit of a robust chain.*

Proof. (i) One systematically, for each structure in the chain and choice of parameters in the structure, replaces the sequence of larger structures by an infinite subsequence in which the formula takes an eventually constant truth value.

(ii) Immediate from (i).   □

Recall that a theory T is *near model complete* if every formula is equivalent modulo T to a boolean combination of existential formulas. Many familiar theories are near model complete; in particular, every model complete theory is near model complete. The theory of pseudofinite fields is near model complete by [40] (see also [14, Section 3]), and so is any complete theory of abelian groups, or, more generally, of modules. Several 'Hrushovski constructions' have near model complete but not model complete theories; see for example Baldwin and Shelah [5].

Proposition 4.7.4 *Let* $\mathcal{C} = (M_i : i \in \mathbb{N})$ *be a chain of finite structures.*

(i) *If* $T_{\mathcal{C}}^{as}$ *is near model complete then $\mathcal{C}$ is robust.*

(ii) *If* $T_{\mathcal{C}}^{as}$ *is* ∀∃*-axiomatized then* $T_{\mathcal{C}}^{\lim} \models T_{\mathcal{C}}^{as}$.

(iii) *If* $T_{\mathcal{C}}^{as}$ *is* ∀∃*-axiomatized and complete, then* $T_{\mathcal{C}}^{\lim} = T_{\mathcal{C}}^{as}$.

Proof. (i) Let M be the direct limit of $\mathcal{C}$. For every tuple $\bar{a}$ from M and existential formula $\varphi(\bar{x})$, we have that $M \models \varphi(\bar{a})$ if and only if $M_i \models \varphi(\bar{a})$ for sufficiently large $i \in \mathbb{N}$. For every formula $\psi(\bar{x})$ there is, modulo $T_{\mathcal{C}}^{as}$, a formula

$\theta(\bar{x})$, which is a boolean combination of existential formulas, and a sentence $\sigma \in T_{\mathcal{C}}^{\text{as}}$, such that $\sigma \models (\forall \bar{x}) \, \psi(\bar{x}) \leftrightarrow \theta(\bar{x})$. As there is some $N \in \mathbb{N}$ such that $M_i \models \sigma$ for all $i \geq N$, it follows for sufficiently large i that $M_i \models \psi(\bar{x})$ if and only if $M \models \theta(\bar{x})$. Hence $\mathcal{C}$ is robust.

(ii) Each of the $\forall\exists$-axioms of $T_{\mathcal{C}}^{\text{as}}$ holds in sufficiently large M_i, thus hold in the union M, and hence so do their consequences.

(iii) This is immediate from (ii). $\qquad\qquad\qquad\qquad\qquad\qquad\qquad\qquad\square$

Recall that a theory T has the *finite submodel property* if for every $M \models T$ and $\sigma \in T$, there is a finite substructure of M satisfying σ. An extension of the arguments above yields the following.

Proposition 4.7.5 *[49, Theorem 2.4.13] Let L be a finite language and let T be a complete L-theory that is near model complete, $\forall\exists$-axiomatized, and has the finite submodel property. Then there is a robust chain $\mathcal{C}$ such that $T_{\mathcal{C}}^{\text{as}} = T_{\mathcal{C}}^{\text{lim}} = T$.*

One of the original aims behind the introduction of robust classes is to develop a framework in which a (countable) stable or o-minimal structure might be approximated by a chain of finite structures that reflects these properties. To this end, we propose the following analogues of stability and o-minimality.

Definition 4.7.6 Let $\mathcal{C} = (M_i : i \in \mathbb{N})$ be a robust chain of L-structures with limit M.

(i) The L-formula $\varphi(x_1, \ldots, x_m, y_1, \ldots, y_n)$ is *unstable in $\mathcal{C}$* if for all $t \in \mathbb{N}$ there are $\bar{a}_1, \ldots \bar{a}_t \in M^m$ and $\bar{b}_1, \ldots, \bar{b}_t \in M^n$ such that for all $i, j \leq t$ we have

$$\mathcal{C} \models_{\text{ev}} \varphi(\bar{a}_i, \bar{b}_j) \Leftrightarrow i \leq j.$$

(ii) The chain $\mathcal{C}$ is *unstable* if and only if some formula is unstable in $\mathcal{C}$.

(iii) The chain $\mathcal{C}$ is *strongly minimal* if for every formula $\varphi(x, y_1, \ldots, y_n)$ there is $n_\varphi \in \mathbb{N}$ such that for all $\bar{a} \in M^n$ either $|\{x \in M : \mathcal{C} \models_{\text{ev}} \varphi(x, \bar{a})\}| \leq n_\varphi$, or $|\{x \in M : \mathcal{C} \models_{\text{ev}} \neg\varphi(x, \bar{a})\}| \leq n_\varphi$.

(iv) Assume that L contains a binary relation $<$ that totally orders M. Then $\mathcal{C}$ is said to be *o-minimal* if for every formula $\varphi(x, y_1, \ldots, y_n)$ there is an $n_\varphi \in \mathbb{N}$ such that for all $\bar{a} \in M^n$ we have $\{x \in M : \mathcal{C} \models_{\text{ev}} \varphi(x, \bar{a})\}$ is the union of at most n_φ singletons and open intervals of $(M, <)$.

Proposition 4.7.7 *Let $\mathcal{C} = (M_i : i \in \mathbb{N})$ be a robust chain with limit M.*

(i) *[49, 4.2.12] For a formula $\varphi(\bar{x}, \bar{y})$, if φ is unstable in $\mathcal{C}$ then φ is unstable in $T_{\mathcal{C}}^{\text{as}}$.*

(ii) *If $T_{\mathcal{C}}^{\text{as}}$ is near model complete, then:*

(a) [49, 4.2.4] a formula φ is stable in $\mathcal{C}$ if φ is stable in $T_{\mathcal{C}}^{\lim}$;

(b) $\mathcal{C}$ is strongly minimal if $T_{\mathcal{C}}^{\lim}$ is strongly minimal;

(c) if M is totally ordered by $<$, then $\mathcal{C}$ is o-minimal if M is o-minimal.

(iii) If $T_{\mathcal{C}}^{\lim}$ is near model complete, then the converses to (ii)(a)–(c) hold.

Proof. (i) If φ is unstable, then for all t the following sentence holds eventually in $\mathcal{C}$, and hence belongs to $T_{\mathcal{C}}^{as}$:

$$\exists \bar{x}_1, \ldots \bar{x}_t \exists \bar{y}_1 \ldots \bar{y}_t [\bigwedge_{i<j} \varphi(\bar{x}_i, \bar{y}_j) \wedge \bigwedge_{i \geq j} \neg \varphi(\bar{x}_i, \bar{y}_j)].$$

(ii) In each case, this follows from the observation that every formula is equivalent, in sufficiently large members of $\mathcal{C}$, to a boolean combination of existential formulas, and the latter holds eventually in $\mathcal{C}$ if and only if it holds in M. (It is important to note that we *do not claim* that $\mathcal{C} \models_{ev} \varphi \Leftrightarrow M \models \varphi$ for every formula φ; indeed, this already may fail for sentences.)

(iii) The argument is similar to that for (ii).   □

The beginnings of a stability theory for robust chains are explored by Marshall in [49, Chapter 4]. For example, a version of Shelah's φ-rank (for a formula ψ) is defined, and it is shown that for a robust class $\mathcal{C}$, some formula $\varphi(\bar{x}, \bar{y})$ is unstable in $\mathcal{C}$ if and only if the φ-rank of $\bar{x} = \bar{x}$ is infinite in this sense. An interesting feature here is that the infinitary arguments of Shelah [55, II.2] seem to be unavailable, and the combinatorial result of [30] is used instead. Marshall shows [49, 4.5.19] that if $\mathcal{C}$ is a strongly minimal robust class with $T_{\mathcal{C}}^{as}$ model complete, then $\mathcal{C}$ is stable, but this has not been proved without the model completeness assumption. An initial theory of Morley rank is also developed. Versions of the independence property and the strict order property for robust classes are defined, and each implies that $\mathcal{C}$ is unstable. Conversely, under the assumption that $T_{\mathcal{C}}^{as}$ is near model complete and $\forall\exists$-axiomatized, Marshall proves that $\mathcal{C}$ has either the independence property or the strict order property.

The "look ahead" aspect of satisfaction in a robust chain suggests the introduction of various notions of complexity. Partly adapting [49, Definition 6.2.1], we make the following provisional definitions.

Definition 4.7.8 Let $\mathcal{C} = (M_i : i \in \mathbb{N})$ be a robust chain.

a. We say $\mathcal{C}$ has *chain complexity* 0 if for every formula $\varphi(\bar{x})$ there is n_φ such that $f_\varphi(i) \leq \text{Max}\{i, n_\varphi\}$. The chain $\mathcal{C}$ has *linear* (respectively, *polynomial*) chain complexity, if, for each formula φ, the function f_φ is bounded above by a linear (respectively, polynomial) function.

b. We say that C has linear (respectively, polynomial, exponential) *model growth* if the function $i \mapsto |M_i|$ is bounded above by a linear (respectively, polynomial, exponential) function.
c. The chain C has polynomial *satisfaction complexity* if for every formula $\varphi(\bar{x})$ there are constants $C > 0$ and $d \in \mathbb{N}$ so that $|M_{f_\varphi(i)}| \leq C|M_i|^d$ for all for all $M_i \in C$.

Note that chain complexity and satisfaction complexity can be defined for each formula separately. Observe also that Proposition 4.7.3 suggests that chain complexity and model growth play off against each other. Chain complexity and model growth have been explored in [49]; satisfaction complexity is newer and not yet well-explored. Note also that chain complexity 0 implies even linear satisfaction complexity. Satisfaction complexity can be refined by specifying how the constants C and d depend on φ; for example one could demand that C be a recursive or even polynomial function of φ or $|\varphi|$.

As a corollary to Proposition 4.7.5, we have

Corollary 4.7.9 *[49, 6.2.3] Let T be a complete theory with quantifier elimination and the finite submodel property. Then there is a robust chain C with $T_C^{\text{as}} = T_C^{\text{lim}} = T$, and every such chain has chain complexity 0.*

Proof. The first assertion is just Proposition 4.7.5. For the second, let $C = (M_i : i \in \mathbb{N})$ be such a chain. For every formula $\varphi(\bar{x})$ there is a quantifier-free formula $\psi(\bar{x})$ such that T contains the sentence $\sigma := (\forall \bar{x}) \varphi(\bar{x}) \leftrightarrow \psi(\bar{x})$. Choose n_φ least such that $M_i \models \sigma$ for all $i \geq n_\varphi$. Since for all $i > j \geq n_\varphi$ and $\bar{a}$ in M_j we have $M_j \models \psi(\bar{a}) \Leftrightarrow M_i \models \psi(\bar{a})$, it follows that $M_j \models \varphi(\bar{a}) \Leftrightarrow M_i \models \varphi(\bar{a})$. $\square$

As an extension of 4.7.9, we give a syntactic characterization of robust classes of chain complexity 0. If $L \subset L^+$ are languages and $C = (M_i : i \in \mathbb{N})$ is a chain of finite L-structures, we say that $C^+ = (M_i^+ : i \in \mathbb{N})$ is an L^+-*expansion of* C if each M_i^+ is an L^+ expansion of M_i and C^+ is a chain, that is, M_i^+ is an L^+-substructure of M_{i+1}^+ for each $i \in \mathbb{N}$.

Proposition 4.7.10 *Let $C = (M_i : i \in \omega)$ be a chain of finite L-structures. The following are equivalent.*
(i) C is robust with chain complexity 0.
(ii) There is a language $L^+ \supset L$ and an expansion C^+ of C to L^+ such that $T_{C^+}^{\text{as}}$ is model-complete.

Proof. (ii) $\Rightarrow$ (i). Suppose that C and C^+ are as in (ii). Let $\varphi(\bar{x})$ be an L-formula. Then there are formulas $\exists \bar{y} \, \psi(\bar{x}, \bar{y})$ and $\forall \bar{z} \, \chi(\bar{x}, \bar{z})$, where ψ and χ are quantifier-free L^+-formulas, such that $T_{C^+}^{\text{as}}$ contains both $(\forall \bar{x}) \varphi(\bar{x}) \leftrightarrow$

$\exists \bar{y} \psi(\bar{x}, \bar{y})$ and $(\forall \bar{x}) \, \varphi(\bar{x}) \leftrightarrow \forall \bar{z} \chi(\bar{x}, \bar{z})$. Hence there is $n_\varphi \in \mathbb{N}$ such that each of these sentences holds in M_i^+ for $i \geq n_\varphi$. Let $j > i \geq n_\varphi$ and $\bar{a} \in M_i^{l(\bar{x})}$. If $M_i \models \varphi(\bar{a})$, then there is $\bar{b} \in M_i^{l(\bar{y})}$ such that $M_i \models \psi(\bar{a}, \bar{b})$, and thus, as ψ is quantifier-free, $M_j^+ \models \psi(\bar{a}, \bar{b})$. Hence $M_j \models \varphi(\bar{a})$. If $M_i \models \neg\varphi(\bar{a})$ we argue similarly, using χ.

(i) $\Rightarrow$ (ii). We 'Morleyize', uniformly. That is, we expand L to L^+ by introducing, for each L-formula $\varphi(\bar{x})$, a relation symbol $R_\varphi(\bar{x})$, and let

$$T^+ := T_\mathcal{C}^{\text{as}} \cup \{(\forall \bar{x}) \, \varphi(\bar{x}) \leftrightarrow R_\varphi(\bar{x}) : \varphi \text{ an } L\text{-formula}\}.$$

Then it follows from robustness of $\mathcal{C}$ that T^+ is a complete L^+-theory.

We expand $\mathcal{C}$ to L^+ as follows. For each new relation symbol R_φ, let n_φ be chosen least so that for all $j > i \geq n_\varphi$ and $\bar{a} \in M_i^{l(\bar{x})}$, we have that $M_i \models \varphi(\bar{a})$ if and only if $M_j \models \varphi(\bar{a})$; such an n_φ exists as $\mathcal{C}$ is robust with chain complexity 0. For $i \geq n_\varphi$, interpret R_φ in M_i by φ and for $i < n_\varphi$ interpret R_φ as the relation induced by the interpretation of R_φ in M_{n_φ}. Then $\mathcal{C}^+$ is a chain of L^+-structures, and its asymptotic theory contains T^+, so is model-complete. □

4.8 Examples of robust classes

We present several examples of robust classes here and in the next section. The emphasis here is two-fold: to provide examples that illustrate the properties introduced in Section 4.7, and to produce classes with various properties that have prescribed limit structures.

Proposition 4.8.1 *Let M be a smoothly approximable structure, approximated by a chain $\mathcal{C} = (M_i : i \in \omega)$ of finite substructures, where M_i is an $|M_i|$-homogeneous substructure of M. Then*

(i) $\mathcal{C}$ is robust with chain complexity 0.

(ii) The chain $\mathcal{C}$ approximating M can be chosen to have exponential model growth.

Proof. (i) We may suppose that $\text{Th}(M)$ admits quantifier elimination by adding a new relation symbol for each formula. Now $\text{Th}(M)$ is model complete, and every sentence in $\text{Th}(M)$ holds in M_i for all sufficiently large i. Since every expansion of M by finitely many constants also is smoothly approximable, the same holds for formulas, that is, $\mathcal{C}$ is robust. By Corollary 4.7.9, $\mathcal{C}$ has chain complexity 0.

(ii) The fact that $\mathcal{C}$ can be chosen to have exponential model growth follows from [18, 5.2.2]. □

Example 4.8.2 Let L be the language of rings, and in this language let $\mathcal{C} = (M_i : i \in \mathbb{N})$ be any chain of finite fields of characteristic p with union $\bar{\mathbb{F}}_p$, the algebraic closure of $\mathbb{F}_p$. The theory $T_{\mathcal{C}}^{as}$ includes the theory of pseudofinite fields, so is near model complete. By Propositions 4.7.4 and 4.7.7 it follows that $\mathcal{C}$ is a strongly minimal robust chain (and is stable).

Note here that $T_{\mathcal{C}}^{as} \neq T_{\mathcal{C}}^{lim}$, as $T_{\mathcal{C}}^{as}$ has the independence property and is thus unstable. The point essentially is that formulas – or sentences – which are boolean combinations of existential formulas hold in the direct limit of $\mathcal{C}$ if and only if they hold eventually in $\mathcal{C}$, but this is not true for arbitrary sentences, in particular the axioms of the theory of pseudofinite fields.

Example 4.8.3 Let $p \equiv 1$ (mod 4) be prime and M_i be the field $\mathbb{F}_{p^{2^i}}$ for $i \in \mathbb{N}$. Let P_i be the Paley graph on M_i (see Example 4.3.6 for the definition), and put $\mathcal{C} := (P_i : i \in \mathbb{N})$. Then $T_{\mathcal{C}}^{as}$ is the theory of the random graph, so has quantifier elimination and the finite submodel property. Thus, $\mathcal{C}$ is robust of chain complexity 0, with the random graph as its direct limit.

By probabilistic arguments, it is possible to realize the random graph, or its arity k analogue, as the direct limit of a chain complexity 0 robust chain $(M_i : i \in \mathbb{N})$, with model growth given by the identity function – i.e., $|M_i| = i$ for all i. Indeed, by [60, Theorem 3.2], if Γ_k denotes the universal homogeneous k-uniform hypergraph, then we may write Γ_k as the union of a chain $(M_i : i \in \mathbb{N})$ of finite substructures such that $|M_i| = i$ for each i, and for every sentence σ, if $\Gamma_k \models \sigma$ then $M_i \models \sigma$ for all but finitely many i. This suffices, by quantifier elimination.

Example 4.8.4 *[49, Section 3.5]* Using the invariants for the elementary theories of Boolean algebras (in the language $(\vee, \wedge, -, 0, 1)$), due to Tarski [58], as described in [11, p.288], it can be shown that the theory of pseudofinite Boolean algebras is complete and near model complete, although not model complete. Hence, any chain $(M_i : i \in \mathbb{N})$ of finite Boolean algebras is robust by 4.7.4. We may obtain the countable atomless Boolean algebra as a direct limit by choosing the embeddings $M_i \to M_{i+1}$ carefully, for example, putting $M_{i+1} := M_i \times M_i$ for each i, with the diagonal embedding $M_i \to M_{i+1}$ given by $a \mapsto (a, a)$.

Example 4.8.5 *[49, Section 3.4.1]* The analysis of theories of abelian groups based on the Szmielew invariants yields that any complete theory of abelian groups is near model complete (see e.g., [31, p.663]). Hence, if $\mathcal{C}$ is a chain of finite abelian groups such that $T_{\mathcal{C}}^{as}$ is complete, then $\mathcal{C}$ is robust. Since theories of abelian groups are stable, it follows from Proposition 4.7.7 that in this case $\mathcal{C}$ is stable.

Example 4.8.6 [49, Section 3.2] The theory of pseudofinite total orders is complete and near model complete. This can be proved, for example, by an

Ehrenfeucht-Fraïssé game argument. It follows that any infinite chain of finite total orders is robust.

It is easy to build such a chain so that it is o-minimal and has direct limit $(\mathbb{Q}, <)$. Let M_n be the natural total order on $\{1, 2, \ldots, 2n + 1\}$, and embed M_n into M_{n+1} via the map $i \mapsto 2i$. The model growth of $\mathcal{C}$ is clearly linear.

Of course, any other countable total order can be realized as the direct limit of a robust chain by Proposition 4.7.3, and $(\mathbb{N}, <)$ and $(\mathbb{Z}, <)$ each is the direct limit of an o-minimal robust chain, by Lemma 4.7.7.

We conclude this section with an example which is closer in spirit to finite model theory, in that it uses Gaifman's Locality Theorem and locality arguments.

Example 4.8.7   Let L be a finite relational language, and let M be a countably infinite L-structure of *finite valency*, in the sense that every $a \in M$ lies in just finitely many tuples satisfying relations of L. Suppose in addition that the automorphism group of M, $\mathrm{Aut}(M)$, is transitive on M. Thus, there is $d \in \mathbb{N}$ such that all elements of M have valency d in the sense above. This assumption is for convenience, and can surely be weakened. At any rate, Cayley graphs of finitely generated groups provide a rich source of examples.

There is a natural notion of distance in M: for $a, b \in M$, we write $d(a, b) = r$ if r is least such that there is a sequence $a = a_0, a_1, \ldots, a_r = b$ such that each pair a_i, a_{i+1} lie in a tuple satisfying a relation in M. We further suppose that M is connected, in the sense that $d(a, b)$ is finite for all $a, b \in M$. Lastly, for each $n \in \mathbb{N}$ and $a \in M$, let $S_n(a)$ be $\{x \in M : d(a, x) \le n\}$ with the L-structure induced from M.

Fix $a \in M$. We recursively construct finite substructures M_n of M as follows. Let $M_1 := S_1(a)$. If M_n has been defined, let M_{n+1} be the union of $M_n \cup S_{n+1}(a)$ and a disjoint isomorphic copy of M_n that is not connected to $M_n \cup S_{n+1}(a)$. Note that each M_n has at least n connected components. Let $\mathcal{C} := (M_n : n > 0)$. Then the direct limit of $\mathcal{C}$ is isomorphic to M.

We claim that $T_{\mathcal{C}}^{\mathrm{as}}$ is complete and near model complete. From this it follows by Proposition 4.7.4 that $\mathcal{C}$ is robust.

We use the presentation of Gaifman's Locality Theorem from [21, Section 2.5]. For an L-structure N and $\bar{a} = (a_1, \ldots, a_n)$ from N, let $S_k(\bar{a}) := S_k(a_1) \cup \ldots \cup S_k(a_n)$. We first recall that for every L-formula $\varphi(\bar{x})$ and $k \in \mathbb{N}$ one can associate a formula $\varphi^{S_k(\bar{x})}$, called a *local formula*, such that for every L-structure N and $\bar{a}$ in N,

$$N \models \varphi^{S_k(\bar{x})}(\bar{a}) \text{ if and only if } S_k(\bar{a}) \models \varphi(\bar{a}).$$

A *basic local sentence* has the form

$$\exists x_1 \ldots \exists x_n \bigwedge_{1 \le i < j \le n} d(x_i, x_j) > 2r \, \wedge \, \varphi^{S_r(x_i)}(x_i).$$

Gaifman's Theorem asserts that every first-order L-sentence is logically equivalent to a boolean combination of basic local sentences.

Since M_{n+1} includes some connected components whose union U_n is isomorphic to M_n, it is clear that if σ is a basic local sentence and $M_n \models \sigma$, then $M_{n+1} \models \sigma$; the witnesses for the existential quantifiers of σ in M_n have copies in U_n that witness σ in M_{n+1}. Thus, every basic local sentence is eventually true or eventually false in $\mathcal{C}$. The completeness of $T_{\mathcal{C}}^{as}$ follows.

It remains to verify near model completeness. An extension of Gaifman's Theorem (see [28]) asserts that every *formula* $\varphi(\bar{x})$ is logically equivalent to a boolean combination of local formulas and basic local sentences. A local formula $\varphi^{S_r(\bar{x})}$ is a boolean combination of formulas which describe possible atomic diagrams of $S_r(\bar{x})$. Since there is a fixed upper bound on valency in substructures of M, the formula $\varphi^{S_r(\bar{x})}$ can itself be assumed to be a boolean combination of existential formulas. This yields near model completeness.

Note that if M has valency d, then it can be checked that $|M_{n+1}| \le (d + 2)|M_n|$ for all n. It follows that $|M_n| \le (d + 2)^n$ for each n. In particular, $\mathcal{C}$ has exponential model growth. We have not attempted to minimize growth and chain complexity for a chain with limit M.

4.9 A robust approximation of $(\mathbb{Q}, <, +)$

Our aim is to show that an o-minimal robust class can support some algebraic structure, despite the discreteness of finite total orders, and thus approximate an infinite o-minimal structure on which at least one algebraic operation is defined. The goal here is to construct an o-minimal robust class whose direct limit is $(\mathbb{Q}, <, +)$. The dense ordering can easily be constructed by dovetailing embeddings as in Example 4.8.6; the difficulty lies in incorporating the group operation. Since ordered groups are torsion-free, hence not locally finite, the group operation must be given by a ternary relation symbol.

Theorem 4.9.1 *Let $L = \{<, 0, -, R\}$, where $-$ is a unary function, and R is a ternary relation. Then there is an o-minimal robust class $\mathcal{C}$ such that $T_{\mathcal{C}}^{as}$ is near model complete, and $T_{\mathcal{C}}^{\lim}$ is the theory of divisible ordered abelian groups, with the addition function parsed as a ternary relation R.*

Proof. For each $n > 0$, let H_n be the L-structure on $\{-n, -n+1, \ldots, 0, \ldots, n\}$, with 0, $<$, $-$ interpreted naturally, and such that $R(x, y, z)$ holds for $x, y, z \in H_n$ if and only if $x + y = z$. Let $\mathcal{H} = \{H_n : n > 0\}$.

A sequence $(a_k)_{k>0}$ of natural numbers is called *legal* if

a. $(\forall k > 0)\, 0 \le a_k < k$
b. $(\forall k > 0)(\exists \ell > k)(\forall r \le k)\ \ell \equiv a_r \pmod{r}$.

It is not difficult to show that legal sequences exist. For example, one may argue by induction employing the fact (see, e.g., [50, Section 2.3, Exercise 23]) that for arbitrary integers $m_1, \ldots, m_r$ the system

$$x \equiv a_1 \pmod{m_1}, \ldots, x \equiv a_r \pmod{m_r}$$

has a solution if and only if $a_i \equiv a_j \pmod{(m_i, m_j)}$ for all $1 \le i < j \le r$. Note also that the constant sequence $(0)_{k>0}$ is legal. Let $\mathcal{S}$ be the set of all legal sequences.

Let $S = (a_k)_{k>0} \in \mathcal{S}$. From S, we can obtain a sequence $(n_k)_{k>0}$ such that $n_{k+1} \gg n_k{}^2$ that further satisfies the condition that

$$(\forall r)\,(\exists \ell)\,(\forall k \ge \ell)\,(\forall s \le r)\, n_k \equiv a_s \pmod{s}.$$

Let $\mathcal{C}_S := \{H_{n_k} : k > 0\}$. To make $\mathcal{C}_S$ into a chain, we must define how H_{n_k} is embedded into $H_{n_{k+1}}$ for all k. To this end, let $(d_k)_{k>0}$ be a sequence of integers so that $2 \le d_k \le n_k$ for all k and in which every $m \ge 2$ appears infinitely often. Then we embed H_{n_k} into $H_{n_{k+1}}$ via the mapping $i \mapsto d_k i$. The intent of this is to ensure that the direct limit structure of $\mathcal{C}_S$ is isomorphic to $(\mathbb{Q}, <, +)$, with $x + y = z$ parsed as the relation $R(x, y, z)$. Finally, let T_S^{as} be the asymptotic theory of $\mathcal{C}_S$, and $T_S^{\lim}$ be the limit theory. Observe that if S, S' are distinct legal sequences, then $T_S^{\mathrm{as}} \ne T_{S'}^{\mathrm{as}}$.

We shall prove that each theory T_S^{as} is near model complete. It follows by Proposition 4.7.4 that $\mathcal{C}_S$ is robust and, as $\mathcal{C}_S$ is totally ordered, that T_S^{as} is complete. Lastly, we prove that $\mathcal{C}_S$ is o-minimal.

For each $M \in \mathcal{C}_S$, we define the *span* of M, span(M), and the *hull* of M, hull(M). In fact, span(M) is isomorphic to $(\mathbb{Z}, <, R, 0)$, and hull(M) is isomorphic to $(\mathbb{Q}, <, R, 0)$ – that is, the divisible hull of $(\mathbb{Z}, <, R, 0)$ – with R interpreted as the graph of the group operation. In this proof, however, we shall view span(M) and hull(M) as many sorted structures, with infinitely many sorts, to ensure that the span and hull of every model of T_S^{as} (and its substructures) is well-defined.

First, $0 \in$ span(M) and each positive element of span(M) has, for some $r > 0$, a representative of the form $(x_1, \ldots, x_r)$, where $x_1, \ldots, x_r \in M^{>0}$. For intuition, the reader may interpret $(x_1, \ldots, x_r)$ as the sum $x_1 + \cdots + x_r$. Recursively on $r + s$, we define $(x_1, \ldots, x_r)E(y_1, \ldots, y_s)$ as the symmetric closure of the relation defined to hold if $x_r \geq y_s$ and $(x_1, \ldots, x_{r-1}, x_r - y_s)E(y_1, \ldots, y_{s-1})$. For each $x_1, \ldots, x_r \in M^{>0}$ as above, span(M) also contains an element $-(x_1, \ldots, x_r)$ with representative $(-x_1, \ldots, -x_r)$. Addition is defined on E-classes by putting

$$(x_1, \ldots, x_r)/E + (y_1, \ldots, y_s)/E = (x_1, \ldots, x_r, y_1, \ldots, y_s)/E$$

and

$$-(x_1, \ldots, x_r)/E + -(y_1, \ldots, y_s)/E = -(x_1, \ldots, x_r, y_1, \ldots, y_s)/E.$$

Lastly, if $x_1, \ldots, x_r > 0$ and $y_1, \ldots, y_s < 0$, then $(x_1, \ldots, x_r)/E + (y_1, \ldots, y_s)/E = (z_1, \ldots, z_t)/E$, where $z_1, \ldots, z_t > 0$, if and only if $(x_1, \ldots, x_r)/E = (-y_1, \ldots, -y_s)/E + (z_1, \ldots, z_t)/E$.

It is convenient to extend the definitions to sequences $(x_1, \ldots, x_r)$ where some x_i are positive and some negative. First, for all permutations $\pi \in$ Sym(r) we define $(x_1, \ldots, x_r)E(x_{\pi(1)}, \ldots, x_{\pi(r)})$. Then if $x_1, \ldots, x_r > 0$ and $x_{r+1}, \ldots, x_s < 0$, define $(x_1, \ldots, x_s)E(y_1, \ldots, y_t)$, where $y_i > 0$ for all i, to hold if $(x_1, \ldots, x_r)/E = -(x_{r+1}, \ldots, x_s)/E + (y_1, \ldots, y_t)/E$.

The group span(M) is an ordered group, with the ordering defined sortwise by $(x_1, \ldots, x_r)/E > 0$ if and only if there are $y_1, \ldots, y_s > 0$ with $(x_1, \ldots, x_r)E(y_1, \ldots, y_s)$. It is now easily checked that span(M) is a $\mathbb{Z}$-group. The key point is that for every $n > 1$ and $x_1, \ldots, x_r > 0$, there are $i \in \{0, \ldots, n - 1\}$ and $y_1, \ldots, y_{r'} > 0$ such that $n(y_1, \ldots, y_{r'}) = (x_1, \ldots, x_{r-1}, x_r - i)$. Also M clearly is convex in span(M).

Now we define hull(M) to be the divisible hull of span(M). Since the above definitions are uniform across $\mathcal{C}_S$, they carry across to models M of T_S^{as}: that is, we may talk of span(M) and hull(M), defined as above. As the definitions are quantifier-free, they apply also to substructures.

Model-theoretically, we remark that this construction may be understood as analogous to that of the algebraic closure of a field F: each finite extension of F is interpretable in F, but the full algebraic closure lives on the union of infinitely many sorts of F^{eq}.

The key element in the proof of the theorem is the following assertion.

Claim 4.9.2 *Let $M, N \models T_S^{as}$, and suppose that M, N have a common $\mathcal{L}$-substructure A. Then $M \equiv_A N$ if and only if $M \cap \text{hull}(A) \cong_A N \cap \text{hull}(A)$.*

Proof. The left-to-right direction is immediate, since elements of hull(A) are quantifier-free definable over A.

For the right-to-left direction, we assume that $M \cap \text{hull}(A) = N \cap \text{hull}(A)$ and that M and N are saturated of the same cardinality $> |A|$. We build an isomorphism $\varphi : M \to N$ that extends $\varphi_0 := \text{id}_A$. We further suppose that φ_0 is extended to $\text{id}_{\text{hull}(A)}$.

The construction of the isomorphism proceeds through a series of steps.

Step 1 Let $1_M := \min\{x \in M : x > 0\}$ and $1_N := \min\{x \in N : x > 0\}$. Define $\varphi(1_M) = 1_N$. Then extend φ to hull$(A \cup \{1_M\})$.

Step 2 Define $\varphi(\max(M)) = \max(N)$, and extend φ to hull$(A \cup \{\max(M)\})$.

In Step 1, we must check for all $x \in$ hull$(A \cup 1_M)$ that $x \in M$ if and only if $\varphi(x) \in N$; likewise, in Step 2, with $\max(M)$ in place of 1_M. This is done carefully in the last paragraph of the argument in Step 3 below, so we omit the details in this case. Note that it is needed here that $\max(M)$ and $\max(N)$ satisfy the same congruence conditions, which holds as $M, N \models T_S^{\mathrm{as}}$.

Once Steps 1 and 2 have been completed, we may suppose that initially $1_M = 1_N \in$ hull(A), which we denote by 1 in what follows, and that $\max(M) = \max(N) \in$ hull(A). The remaining task is the following:

Step 3 We must extend φ to $x \in M \setminus \mathrm{dom}(\varphi)$.

Put $C := \{y \in \mathrm{hull}(A) : y < x\}$ and $D := \{y \in \mathrm{hull}(A) : x < y\}$. We first assert that for each $z \in C$ there is some $a_z \in \mathrm{span}(A)$ with $z \leq a_z < x$. Indeed, using the fact that for every n and every $w \in \mathrm{span}(M)$, among the elements $w, w + 1, \ldots, w + (n - 1)$ there is an element that is divisible by n, it is easy to see that if the assertion were false then we must have $z < x < z + 1$. With $z = z'/q$, where $z' \in \mathrm{span}(A), q \in \mathbb{Z} \setminus \{0\}$, it follows that $z' < qx < z' + q$. Then $z', z' + q \in \mathrm{span}(A) \subseteq \mathrm{dom}(\varphi)$, whence, as $\mathrm{span}(M)$ is a $\mathbb{Z}$-group, $qx \in \mathrm{dom}(\varphi)$. It follows that $x \in \mathrm{dom}(\varphi)$, a contradiction. A similar argument shows that for each $z \in D$ there is some $b_z \in \mathrm{span}(A)$ with $x < b_z \leq z$.

Now let $C' := \{a_z : z \in C\}$ and $D' := \{b_z : z \in D\}$. So $C' < x < D'$, and $C', D' \subset \mathrm{span}(A) \cap M$. The preceding paragraph shows that C' has no greatest element and D' no least element. By saturation of N we see that there is some $y \in N$ such that $C' < y < D'$ and $x \equiv y \pmod{k}$ for each positive integer k. We put $\varphi(x) = y$, and extend φ to hull$(A \cup \{x\})$.

A typical element of hull$(A \cup \{x\})$ has the form $\frac{1}{q}(e_1 + \cdots + e_r + nx)$, where $q \in \mathbb{Z}$ and $q \neq 0$, n is a positive integer, and $e_1, \ldots, e_r \in A$ (not necessarily distinct). It remains to check that $\frac{1}{q}(e_1 + \cdots + e_r + nx) \in M$ if and only if $\frac{1}{q}(e_1 + \cdots + e_r + ny) \in N$. Observe first that both $e_1 + \cdots + e_r + nx \in \mathrm{span}(M)$ and $e_1 + \cdots + e_r + ny \in \mathrm{span}(N)$, and that $e_1 + \cdots + e_r + nx > \max(M)$ if and only if $e_1 + \cdots + e_r + ny > \max(N)$. Thus, it suffices to see that $e_1 + \cdots + e_r + nx \equiv 0 \pmod{q}$ in M if and only if $e_1 + \cdots + e_r + ny \equiv 0 \pmod{q}$ in N. This holds since x and y satisfy the same congruences. With this, the proof of the claim is complete. $\square$

We now finish the proof of the theorem. First, we assert that T_S^{as} is near model complete. By Claim 4.9.2, it follows for all $\bar{a} \in M \models T_S^{as}$ and substructures A of M that $\text{tp}(\bar{a}/A)$ is determined by the collection of formulas which describe the quantifier-free type of $\bar{a}$ over $\text{hull}(A)$. Such formulas are boolean combinations of existential formulas over A. From this and Proposition 4.7.4, it follows that C_S is robust. As C_S is totally ordered, we also see that T_S^{as} is complete.

It remains only to check that C_S is o-minimal. Let $G := \lim C_S$, with $x + y = z$ parsed as $R(x, y, z)$. It is evident that G is isomorphic to $(\mathbb{Q}, <, +)$, and hence is o-minimal. The o-minimality of C_S then follows immediately from the near model completeness of T_S^{as} and Proposition 4.7.7 . $\qquad\square$

Remark 4.9.3 By varying the embeddings, we may find, for any $S \in \mathcal{S}$ other robust classes C' with the same asymptotic theory T_S^{as} but different limits theories. For example, we may realize the infinite cyclic group $\mathbb{Z}$ as $\lim C'$, by putting $C' = C$, equipped with the identity embeddings.

We expect that a variation of this argument should allow us to construct $(\mathbb{Q}, +)$ as the limit of a strongly minimal robust class. It is an open problem to construct an o-minimal robust chain whose limit is a real closed field.

References

[1] G. Ahlbrandt, M. Ziegler, 'Quasifinitely axiomatizable totally categorical theories', Ann. Pure Appl. Logic 30 (1986), 63–82.

[2] J. Allsup, R. W. Kaye, 'Normal subgroups of non-standard symmetric and alternating groups', Arch. Math. Logic 46 (2007), 107–121.

[3] M. Aschbacher, *Finite Group Theory*, 2nd Ed., Cambridge University Press, Cambridge, 2000.

[4] J. Ax, 'The elementary theory of finite fields', Ann. Math. 88 (1968), 239–271.

[5] J.T. Baldwin, S. Shelah, 'Randomness and semigenericity', Trans. Amer. Math. Soc. 349 (1997), 1359–1376.

[6] G. Bergman, H.W. Lenstra, 'Subgroups close to normal subgroups', J. Algebra 127 (1989), 80–97.

[7] O. Beyarslan, 'Random Hypergraphs in Pseudofinite Fields', J. Inst. Math. Jussieu 9 (2010), 29–47.

[8] T. Blossier, 'Ensembles minimaux localement modulaires', Ph.D. thesis, Université Paris VII, 2001.

[9] B. Bollobás, A. Thomason, 'Graphs which contain all small graphs', Europ. J. Comb. 2 (1981), 13–15.

[10] B. Bollobás, *Random Graphs*, Academic Press, New York, 1985.

[11] S. Koppelberg, *Handbook of Boolean Algebras* Vol. 1 (Eds. J.D. Monk, R. Bonnet), North-Holland, 1989.

[12] E.I. Bunina, A.V. Mikhalev, 'Elementary properties of linear groups and related problems', J. Math. Sci. 123 (2004), 3921–3985.

[13] Z. Chatzidakis, L. van den Dries, A.J. Macintyre, 'Definable sets over finite fields', J. Reine Angew. Math. 427 (1992), 107–135.

[14] Z. Chatzidakis, 'Model theory of finite and pseudo-finite fields', Ann. Pure Apl. Logic 88 (1997), 95–108.

[15] G. Cherlin, L. Harrington, and A.H. Lachlan, '$\aleph_0$-categorical, $\aleph_0$-stable structures', Ann. Pure Appl. Logic 28 (1985), 103–135

[16] G. Cherlin, A.H. Lachlan, 'Stable finitely homogeneous structures', Trans. Amer. Math. Soc. 296 (1986), 815–850.

[17] G. Cherlin, 'Large finite structures with few types', in *Algebraic model theory* (Eds. B. Hart, A.H. Lachlan, M. Valeriote), NATO ASI Series C, vol. 496, Kluwer, Dordrecht, 1997, pp. 53–105.

174

[18] G. Cherlin, E. Hrushovski, *Finite structures with few types*, Annals of Mathematics Studies No. 152, Princeton University Press, Princeton, 2003.

[19] P. Dello Stritto, 'Asymptotic classes of finite Moufang polygons', preprint.

[20] L. van den Dries, *Tame topology and o-minimal structures*, London Math. Soc Lecture Notes 248, Cambridge University Press, 1998.

[21] H-D. Ebbinghaus, J. Flum, *Finite model theory*, Springer, 1999.

[22] R. Elwes, 'Asymptotic classes of finite structures', J. Symb. Logic 72 (2007), 418–438.

[23] R. Elwes, *Dimension and measure in finite first order structures*, Ph.D. thesis, University of Leeds, 2005.

[24] R. Elwes, E. Jaligot, H.D. Macpherson, M. Ryten, 'Groups in supersimple and pseudofinite theories', preprint.

[25] R. Elwes, H.D. Macpherson, 'A survey of asymptotic classes and measurable structures', in *Model theory with Aplications to Algebr and Analysis*, vol. 2 (Eds. Z. Chatzidakis, H.D. Macpherson, A. Pillay, A.J. Wilkie), London Math. Soc. Lecture Notes no. 350, 2008, pp. 125–159.

[26] R. Elwes, M. Ryten, 'Measurable groups of low dimension', Math. Logic Quarterly 54 (2008), 374–386.

[27] Y.L. Ershov, 'Undecidability of theories of symmetric and finite simple groups', Dokl. Akad. Nauk SSSR 158 (1964), 777–779.

[28] H. Gaifman, 'On local and non-local properties', *Proceedings of the Herbrand Symposium, Logic Colloquium '81* (Ed. J. Stern), Studies in Logic and the Foundation of Mathematics, vol. 107. Amsterdam: North Holland, 1982, 105–135.

[29] R. L. Graham, J. H. Spencer, 'A constructive solution to a tournament problem', Canad. Math. Bull. 14 (1971), 45–48.

[30] W. Hodges, 'Encoding orders and trees in binary relations', Mathematika 28 (1981), 67–71.

[31] W. Hodges, *Model Theory*, Cambridge University Press, Cambridge, 1993.

[32] E. Hrushovski, 'Totally categorical structures', Trans. Amer. Math. Soc. 313 (1989), 131–159.

[33] E. Hrushovski, 'Unimodular minimal structures', J. London Math. Soc. (2) 46 (1992), 385–396.

[34] E. Hrushovski, 'A new strongly minimal set', Ann. Pure Appl. Logic 62 (1993), 147–166.

[35] E. Hrushovski, 'Finite structures with few types', in *Finite and infinite combinatorics in sets and logic* (Eds. N.W. Sauer, R.E. Woodrow, B. Sands), NATO ASI Series C vol. 411, Kluwer, Dordrecht, 1993, pp. 175–187.

[36] E. Hrushovski, 'Pseudofinite fields and related structures', in *Model theory and applications* (Eds. L. Bélair, Z. Chatzidakis, P. D'Aquino, D. Marker, M. Otero, F. Point, A. Wilkie), Quaderni di Matematica, vol. 11, Caserta, 2005, 151–212.

[37] E. Hrushovski, 'The first order theory of the Frobenius', preprint, available at http://front.math.ucdavis.edu/math.LO/0406514.

[38] E. Hrushovski, A. Pillay, 'Definable subgroups of algebraic groups over finite fields', J. Reine Angew. Math. 462 (1995), 69–91.

[39] W.M. Kantor, M.W. Liebeck, H.D. Macpherson, '$\aleph_0$-categorical structures smoothly approximated by finite substructures', Proc. London Math. Soc. (3) 59 (1989), 439–463.

[40] C. Kiefe, 'Sets definable over finite fields: their zeta functions', Trans. Amer. Math. Soc. 223 (1976), 45–59.

[41] B. Kim, A. Pillay, 'Simple theories', Ann. Pure Appl. Logic 88 (1997), 149–164.

[42] L. Kramer, G. Röhrle, K. Tent, 'Defining k in $G(k)$', J. Alg. 216 (1999), 77–85.

[43] A.H. Lachlan, 'Two conjectures regarding the stability of ω-categorical theories', Fund. Math. 81 (1974), 133–145.

[44] A.H. Lachlan, 'On countable stable structures which are homogeneous for a finite relational language', Israel J. Math. 49 (1984), 69–153.

[45] S. Lang, A. Weil, 'Number of points of varieties in finite fields', Am. J. Math. 76 (1954), 819–827.

[46] L. Libkin, *Elements of Finite Model Theory*, Springer, Heidelberg, 2004.

[47] M.W. Liebeck, H.D. Macpherson, K. Tent, 'Primitive permutation groups of bounded orbital diameter', Proc. London Math. Soc. (3) 100 (2010), 216–248.

[48] H.D. Macpherson, C. Steinhorn, 'One-dimensional asymptotic classes of finite structures', Trans. Amer. Math.Soc. 360 (2008), 411–448.

[49] R. Marshall, *Robust classes of finite structures*, Ph.D thesis, University of Leeds, 2008.

[50] I. Niven, H. Zuckerman, H. Montgomery, *An Introduction to the Theory of Numbers* (Fifth Edition), John Wiley & Sons, 1991.

[51] A. Pillay and C. Steinhorn, 'On discrete o-minimal structures', J. Pure and Applied Logic 34 (1987), 275–289.

[52] M. Ryten, *Results around asymptotic and measurable groups*, Ph.D. thesis, University of Leeds, 2007.

[53] M. Ryten, I. Tomašić, 'ACFA and measurability', Selecta Math. (New series) 11 (2005), 523–537.

[54] T. Scanlon, 'Fields admitting nontrivial strong ordered Euler characteristics are quasifinite', unpublished manuscript, available at http://math.berkeley.edu/ scanlon/papers/papers.html.

[55] S. Shelah, 'Simple unstable theories', Ann. Pure Appl. Logic 19 (1980), 177–203.

[56] W. Szmielew, 'Elementary properties of abelian groups', Fund. Math. 41 (1955), 203–271.

[57] T. Szönyi, 'Some applications of algebraic curves in finite geometry and combinatorics', in *Surveys in Combinatorics* (Ed. R. Bailey), London Math. Society Lecture Notes 241, Cambridge University Press, Cambridge, 1997, pps. 197–236.

[58] A. Tarski, 'Arithmetical classes and types of boolean algebras', Bull. of the A.M.S. 55 (1949), 63.

[59] S. Thomas, *Classification theory of simple locally finite groups*, Ph.D. thesis, University of London, 1983.

[60] S. Thomas, 'Reducts of random hypergraphs', Ann. Pure Appl. Logic 80 (1996), 165–193.

[61] J. Tits, R. M. Weiss, *Moufang polygons*, Springer, 2002.

[62] F.O. Wagner, *Simple theories*, Kluwer, Dordrecht, 2000.

[63] J. Wiegold, 'Groups with boundedly many finite classes of conjugate elements', Proc. Royal Soc. A 238 (1956), 389–401.

5

Algorithmic meta-theorems

STEPHAN KREUTZER[a]

Abstract

Algorithmic meta-theorems are general algorithmic results applying to a whole range of problems, rather than just to a single problem alone. They often have a *logical* and a *structural* component, that is they are results of the form: *every computational problem that can be formalised in a given logic $\mathcal{L}$ can be solved efficiently on every class $\mathcal{C}$ of structures satisfying certain conditions.*

This paper gives a survey of algorithmic meta-theorems obtained in recent years and the methods used to prove them. As many meta-theorems use results from graph minor theory, we give a brief introduction to the theory developed by Robertson and Seymour for their proof of the graph minor theorem and state the main algorithmic consequences of this theory as far as they are needed in the theory of algorithmic meta-theorems.

5.1 Introduction

Algorithmic meta-theorems are general algorithmic results applying to a whole range of problems, rather than just to a single problem alone. In this paper we will concentrate on meta-theorems that have a *logical* and a *structural* component, that is on results of the form: *every computational problem that can be formalised in a given logic $\mathcal{L}$ can be solved efficiently on every class $\mathcal{C}$ of structures satisfying certain conditions.*

The first such theorem is Courcelle's well-known result [13] stating that every problem definable in monadic second-order logic can be solved efficiently

[a] Oxford University Computing Laboratory, University of Oxford, Oxford, OX1 3QD, UK
stephan.kreutzer@comlab.ox.ac.uk

on any class of graphs of bounded tree-width[1]. Another example is a much more recent result stating that every first-order definable optimisation problem admits a polynomial-time approximation scheme on any class C of graphs excluding at least one minor (see [22]).

Algorithmic meta-theorems lie somewhere between computational logic and algorithm or complexity theory and in some sense form a bridge between the two areas. In algorithm theory, an active research area is to find efficient solutions to otherwise intractable problems by restricting the class of admissible inputs. For instance, while the dominating set problem is NP-complete in general, it can be solved in polynomial time on any class of graphs of bounded tree-width.

In this line of research, algorithmic meta-theorems provide a simple and easy way to show that a certain problem is tractable on a given class of structures. Formalising a problem in MSO yields a formal proof for its tractability on classes of structures of bounded tree-width, avoiding the task of working out the details of a solution using dynamic programming – something that is not always trivial to do but often enough solved by hand-wavy arguments such as "using standard techniques from dynamic programming . . .".

Another distinguishing feature of logic based algorithmic meta-theorems is the observation that for a wide range of problems, such as covering or colouring problems, their precise mathematical formulation can often directly be translated into monadic second-order logic. Hence, ideally, instead of having to design an explicit algorithm for solving a problem on bounded tree-width graphs, one can read off tractability results directly from the problem description.

Finally, algorithmic meta-theorems yield tractability results for a whole class of problems providing valuable insight into how far certain algorithmic techniques range. On the other hand, in their negative form of intractability results, they also exhibit some limits to applications of certain algorithmic techniques.

In logic, one of the core tasks is the evaluation of logical formulas in structures – a task underlying problems in a wide variety of areas in computer science from database theory, artificial intelligence to verification and finite model theory.

Among the important logics studied in this context is first-order logic and its various fragments, such as its existential conjunctive fragment known as conjunctive queries in database theory. Whereas first-order model-checking is PSPACE-complete in general, even on input structures with only two elements, it becomes polynomial time for every fixed formula. So what can we possibly

---

[1] The definition of tree-width and the other graph parameters and logics mentioned in the introduction will be presented formally in the following sections.

gain from restricting the class of admissible structures, if the problem is hard as soon as we have two elements and becomes easy if we fix the formula? Not much, if the distinction is only between taking the formula as full part of the input or keeping it fixed.

A finer analysis of first-order model-checking can be obtained by studying the problem in the framework of parameterized complexity (see [36, 46, 69]). The idea is to isolate the dependence of the running time on a certain part of the input, called the *parameter*, from the dependence on the rest. We will treat parameterized complexity formally in Section 5.2.4. The parameterized first-order evaluation problem is the problem, given a structure A and a sentence $\varphi \in$ FO, to decide whether $A \models \varphi$. The parameter is $|\varphi|$, the length of the formula. It is called *fixed parameter tractable* (FPT) if it can be solved in time $f(|\varphi|) \cdot |A|^c$, for some fixed constant c and a computable function $f : \mathbb{N} \to \mathbb{N}$. While first-order model-checking is unlikely to be fixed-parameter tractable in general (unless unexpected results in parameterized complexity happen), Courcelle's theorem shows that even the much more expressive monadic second-order logic becomes FPT on graph classes of bounded tree-width. Hence, algorithmic meta-theorems give us a much better insight into the structure of model-checking problems taking structural information into account.

In this paper we will give an overview of algorithmic meta-theorems obtained so far and present the main methods used in their proofs. As mentioned before, these theorems usually have a logical and a structural component. As for the logic, we will primarily consider first-order and monadic second-order logic (see Section 5.2). As for the structural component, most meta-theorems have been proved relative to some structure classes based on graph theory, in particular on graph minor theory, such as classes of graphs of bounded tree-width, planar graphs, or H-minor free graphs. We will therefore present the relevant parts of graph structure theory needed for the proofs of the theorems presented here.

The paper is organised as follows. In Section 5.2, we present basic notation used throughout the paper. In Section 5.2.3 we present the relevant logics and give a brief overview of their model-checking problem. Section 5.2.4 contains an introduction to parameterized complexity. In Section 5.3, we introduce the notion of the tree-width of a graph and establish some fundamental properties. We then state and prove theorems by Seese and Courcelle establishing tractability results for monadic second-order logic on graph classes of bounded tree-width. In Section 5.4 we present an extension of tree-width called clique-width and a more recent, broadly equivalent notion called rank-width. Again we will see that monadic second-order model checking and satisfiability is tractable on graph classes of bounded clique-width. Section 5.5 contains a brief

introduction to the theory of graph minors to the extent needed in later sections
of the paper. The results presented in this section are then used in Section 5.7
to obtain tractability results on graph classes excluding a minor. In Section 5.7,
we also consider the concept of localisation of graph invariants and use it to
obtain further tractability results for first-order model checking. But before,
in Section 5.6, we use the results obtained in Section 5.5 to show limits to
MSO-tractability. Finally, we conclude the paper in Section 5.9.

Remark An excellent survey covering similar topics to this paper has recently
been written by Martin Grohe as a contribution to a book celebrating Wolfgang
Thomas' 60th birthday [53]. While the two papers share a common core of
results, they present the material in different ways and with a different focus.

5.2 Preliminaries

In this section we introduce basic concepts from logic and graph theory and fix
the notation used throughout the paper. The reader may safely skip this section
and come back to it whenever notation is unclear.

5.2.1 Sets

By $\mathbb{N} := \{0, 1, 2, \ldots\}$ we denote the set of non-negative integers and by $\mathbb{Z}$ the
set of integers. For $k \in \mathbb{N}$ we write $[k]$ for the set $[k] := \{0, \ldots, k - 1\}$. For a
set M and $k \in \mathbb{N}$ we denote by $[M]^k$ and $[M]^{\leq k}$ the set of all subsets of M of
size k and size $\leq k$, respectively, and similarly for $[M]^{<k}$.

margin notes: $[k]$, $[M]^k$, $[M]^{\leq k}$

5.2.2 Graphs

A *graph* G is a pair consisting of a set $V(G)$ of *vertices* and a set $E(G) \subseteq
[V(G)]^2$ of *edges*. All graphs in this paper are finite, simple, i.e. no multiple
edges, undirected and loop-free. We will sometimes write $G := (V, E)$ for a
graph G with vertex set V and edge set E. We denote the class of all (finite)
graphs by GRAPH.

An edge $e := \{u, v\}$ is *incident* to its end vertices u and v and u, v
are *adjacent*. If G is a graph then $|G| := |V(G)|$ is its *order* and $\|G\| :=
\max\{|V(G)|, |E(G)|\}$ its *size*.

For graphs H, G we define the *disjoint union* $G \dot\cup H$ as the graph obtained as
the union of H and an isomorphic copy G' of G such that $V(G') \cap V(H) = \varnothing$.

margin notes: $V(G)$, $E(G)$, GRAPH, *incident,* *adjacent* $|G|$, $\|G\|$

Subgraphs. A graph H is a *subgraph* of G, written as $H \subseteq G$, if $V(H) \subseteq$ $H \subseteq G$
$V(G)$ and $E(H) \subseteq E(G) \cap [V(H)]^2$. If $E(H) = E(G) \cap [V(H)]^2$ we call H
an *induced* subgraph.

Let G be a graph and $U \subseteq V(G)$. The subgraph $G[U]$ *induced by* U *in* G is $G[U]$
the graph with vertex set U and edge set $E(G) \cap [U]^2$.

For a set $U \subseteq V(G)$, we write $G - U$ for the graph induced by $V(G) \setminus U$. $G - U$
Similarly, if $X \subseteq E(G)$ we write $G - X$ for the graph $(V(G), E(G) \setminus X)$. $G - X$
Finally, if $U := \{v\} \subseteq V(G)$ or $X := \{e\} \subseteq E(G)$, we simplify notation and $G -$
write $G - v$ and $G - e$. $v, \quad G - e$

Degree and neighbourhood. Let G be a graph and $v \in V(G)$. The *neigh-*
bourhood $N_G(v)$ of v in G is defined as $N_G(v) := \{u \in V(G) : \{u, v\} \in E(G)\}$.
The *distance* $d^G(u, v)$ between two vertices $u, v \in V(G)$ is the length of the $N_G(v)$
shortest path from u to v or ∞ if there is no such path. For every $v \in V(G)$ and
$r \in \mathbb{N}$ we define the *r-neighbourhood* of v in G as the set

$$N_r^G(v) := \{w \in V(G) : d^G(v, w) \le r\}.$$

of vertices of distance at most r from v. For a set $W \subseteq V(G)$ we define
$N_r^G(W) := \bigcup_{v \in W} N_r^G(v)$. We omit the index $\cdot^G$ whenever G is clear from the
context.

The *degree* of v is defined as $d_G(v) := |N_G(v)|$. We will drop the index $d_G(v)$
G whenever G is clear from the context. Finally, $\Delta(G) := \max\{d(v) : v \in V\}$
denotes the *maximal degree*, or just *degree*, of G and $\delta(G) := \min\{d(v) : v \in$ $\Delta(G)$
$V\}$ the *minimal degree*. $\delta(G)$

Paths and walks. A walk P in G is a sequence $x_1, e_1, \ldots, x_n, e_n, x_{n+1}$
such that $e_i := \{x_i, x_{i+1}\} \in E(G)$ and $x_i \in V(G)$. The *length* of P is n, i.e. the
number of edges. A *path* is a walk without duplicate vertices, i.e. $v_i \ne v_j$
whenever $i \ne j$. We find it convenient to consider paths as subgraphs and
hence use $V(P)$ and $E(P)$ to refer to its set of vertices and edges, resp. An
$X - Y$-path, for $X, Y \subseteq V(G)$, is a path with first vertex in X and last vertex
in Y. If $X := \{s\}$ and $Y := \{t\}$ are singletons, we simply write $s - t$-path.

A graph is *connected* if it is non-empty and between any two vertices s and
t there is an $s - t$-path. A *connected component* of a graph G is a maximal
connected subgraph of G.

Special graphs. For $n, m \ge 1$ we write K_n for the complete graph on n K_n
vertices and $K_{n,m}$ for the complete bipartite graph with one partition of order n $K_{n,m}$
and one of order m. Furthermore, if X is a set then $K[X]$ denotes the complete $K[X]$
graph with vertex set X.

For $n, m \ge 1$, the $n \times m$-grid $G_{n \times m}$ is the graph with vertex set $\{(i, j) : 1 \le$ $G_{n \times m}$
$i \le n, 1 \le j \le m\}$ and edge set $\{((i, j), (i', j')) : |i - i'| + |j - j'| = 1\}$. For

Figure 5.1 A 3×4-grid

$i \geq 1$, the subgraph induced by $\{(i, j) : 1 \leq j \leq m\}$ is called the i*th row* of $G_{n \times m}$ and for $j \geq 1$ the subgraph induced by $\{(i, j) : 1 \leq i \leq n\}$ is called the j*th column*. See Figure 5.1 for a 3×4-grid.

Trees. A *tree* T is a connected acyclic graph. Often we will work with *rooted trees* T with a distinguished vertex r, the *root* of T. A *leaf* in T is a vertex of degree 1, all other vertices are called *inner vertices*. A tree is *subcubic*, if all vertices have degree at most 3. It is *cubic* if every vertex has degree 3 or 1.

A *directed tree* is a rooted tree where all edges are directed away from the root. A *binary tree* is a directed tree where every vertex has at most two outgoing edges. In directed graphs, we view edges as tuples (u, v), where u is the tail and v is the head of the edge, rather than sets $\{u, v\}$.

Coloured graphs. Let Σ be an alphabet. A Σ-*labelled tree* is a pair (T, λ), where T is a tree and $\lambda : V(T) \to \Sigma$ is a labelling function. Often, Σ will be a set C of colours and then we call C-labelled trees C-*coloured*, or just *coloured*. A Σ-*tree* is a Σ-labelled tree.

5.2.3 Logic

I assume familiarity with basic notions from mathematical logic. See e.g. [38, 57] for an introduction to mathematical logic.

signature A *signature* $\sigma := \{R_1, \ldots, R_k, c_1, \ldots, c_q\}$ is a finite set of relation symbols R_i and constant symbols c_i. To each relation symbol $R \in \sigma$ we assign
ar(R) an *arity* ar(R). A σ-*structure* A is a tuple $A := \big(V(A), R_1(A), \ldots, R_k(A),$ $c_1(A), \ldots, c_q(A)\big)$ consisting of a set $V(A)$, the *universe*, for each $R_i \in \sigma$ of arity ar(R_i) $:= r$ a set $R_i(A) \subseteq V(A)^r$ and for each $c_i \in \sigma$ a constant $c_i(A) \in V(A)$. We will usually use letters $A, B, \ldots$ for structures. Their universe is denoted as $V(A)$ and for each $R \in \sigma$ we write $R(A)$ for the relation R in the structure A and similarly for constant symbols $c \in \sigma$.

$\bar{a}$ Tuples of elements are denoted by $\bar{a} := a_1, \ldots a_k$. We will frequently write $\bar{a}$ without stating its length explicitly, which will then be understood or not relevant. Abusing notation, we will treat tuples sometimes as sets and write $a \in \bar{a}$, with the obvious meaning, and also $\bar{a} \subseteq \bar{b}$ to denote that every element in $\bar{a}$ also occurs in $\bar{b}$.

Two σ-structures A, B are *isomorphic*, denoted $A \cong B$, if there is a bijection $\quad A \cong B$
$\pi : V(A) \to V(B)$ such that

- for all relation symbols $R \in \sigma$ of arity $r := \mathrm{ar}(R)$ and all $\bar{a} \in V(A)^r$, $\bar{a} \in$
  $R(A)$ if, and only if, $(\pi(a_1), \ldots, \pi(a_r)) \in R(B)$ and
- for all constant symbols $c \in \sigma$, $c(B) = \pi(c(A))$.

Let σ be a signature. We assume a countably infinite set of first-order variables $x, y, \ldots$ and second-order variables $X, Y, \ldots$. A σ-*term* is a first-order variable or a constant symbol $c \in \sigma$. The class of formulas of *first-order logic* over σ, denoted FO$[\sigma]$, is inductively defined as follows. If $R \in \sigma$ and $\bar{x}$ is a tuple of σ-terms of length $\mathrm{ar}(R)$, then $R\bar{x} \in$ FO$[\sigma]$ and if t and s are terms then $t = s \in$ FO$[\sigma]$. Further, if $\varphi, \psi \in$ FO$[\sigma]$, then so are $(\varphi \wedge \psi)$, $(\varphi \vee \psi)$ and $\neg\varphi$. Finally, if $\varphi \in$ FO$[\sigma]$ and x is a first-order variable, then $\exists x \varphi \in$ FO$[\sigma]$ and $\forall x \varphi \in$ FO$[\sigma]$.

The class of formulas of *monadic second-order logic* over σ, denoted MSO$[\sigma]$, is defined by the rules for first-order logic with the following additional rules: if X is a second-order variable and $\varphi \in$ MSO$[\sigma \dot{\cup} \{X\}]$, then $\exists X \varphi \in$ MSO$[\sigma]$ and $\forall X \varphi \in$ MSO$[\sigma]$. Finally, we define FO $:= \bigcup_{\sigma \text{ signature}}$ FO$[\sigma]$ and likewise for MSO.

First-order variables range over elements of σ-structures and monadic second-order variables X range over sets of elements. Formulas $\varphi \in$ FO$[\sigma]$ are interpreted in σ-structures A in the obvious way, where atoms $R\bar{x}$ denote containment in the relation $R(A)$, $=$ denotes equality of elements, $\vee, \wedge, \neg$ denote disjunction, conjunction and negation and $\exists x \varphi$ is true in A if there is an element $a \in V(A)$ such that φ is true in A if x is interpreted by a. Analogously, $\forall x \varphi$ is true in A if φ is true in A for all interpretations of x by elements $a \in V(A)$.

For MSO$[\sigma]$-formulas, $\exists X \varphi$ is true in A if there is a set $U \subseteq V(A)$ such that φ is true if X is interpreted by U and analogously for $\forall X \varphi$.

The set of *free* variables of a formula is defined in the usual way. We will write $\varphi(\bar{x})$ to indicate that the variables in $\bar{x}$ occur free in φ. Formulas without free variables are called *sentences*. If φ is a sentence we write $A \models \varphi$ if φ is $\quad A \models \varphi$
true in A. If $\varphi(\bar{x})$ has free variables $\bar{x}$ and $\bar{a}$ is a tuple of the same length as $\bar{x}$, we write $A \models \varphi(\bar{a})$ or $(A, \bar{a}) \models \varphi$ if φ is true in A where the free variables $\quad A \models \varphi(\bar{a})$
$\bar{x}$ are interpreted by the elements in $\bar{a}$ in the obvious way. We will sometimes $\quad (A, \bar{a}) \models \varphi$
consider formulas $\varphi(X)$ with a free second-order variable X. The notation extends naturally to free second-order variables.

We will use obvious abbreviations in formulas, such as $\rightarrow$ (implication), $x \neq y$ instead of $\neg x = y$ and $\bigvee_{i=1}^{k} \varphi_i$ and $\bigwedge_{i=1}^{k} \varphi_i$ for disjunctions and conjunctions over a range of formulas.

independent   **Example 5.2.1** *1. An* independent set, *or* stable set, *in a graph G is a*
set   *set $X \subseteq V(G)$ such that $\{u, v\} \notin E$ for all $u, v \in X$. The first-order*
                 sentence

$$\varphi_k := \exists x_1 \ldots \exists x_k \bigwedge_{1 \leq i < j \leq k} \left(x_i \neq x_j \wedge \neg E x_i x_j \right)$$

is true in a graph G (considered as an $\{E\}$-structure in the obvious way) if, and only if, G contains an independent set of size k.

dominating   *2. A* dominating set *in a graph G is a set $X \subseteq V(G)$ such that for all*
set   *$v \in V(G)$, either $v \in X$ or there is a $u \in X$ such that $\{v, u\} \in E(G)$. The*
                 formula

$$\varphi(X) := \forall x \left(X x \vee \exists z (E x z \wedge X z) \right)$$

states that X is a dominating set. Precisely, a set $U \subseteq V(G)$ is a dominating set in G if, and only if, $(G, U) \models \varphi$.

To say that a graph contains a dominating set of size k we can use the formula $\exists x_1 \ldots \exists x_k \forall y \bigvee_{i=1}^{k} \left(y = x_i \vee E x_i y \right)$.   ⊣

Note the difference between the formulas defining an independent set and a dominating set: whereas an independent set of size k can be defined by a formula using existential quantifiers only, i.e. without alternation between existential and universal quantifiers, the formula defining a dominating set of size k contains one alternation of quantifiers. This indicates that the independent set problem might be simpler than the dominating set problem, a realisation that is reflected in the parameterized complexity of the problem as discussed later (see Proposition 5.2.10).

Example 5.2.2 *1. Consider the following* MSO-*formula*

$$\varphi := \forall X \left(\left(\exists x X x \wedge \forall x \forall y (X x \wedge E x y \rightarrow X y) \right) \rightarrow \forall x X x \right).$$

The formula says of a graph G that all sets $X \subseteq V(G)$ which are non-empty ($\exists x X x$) and have the property that whenever $v \in X$ and $\{v, u\} \in E(G)$ then also $u \in X$, already contain the entire vertex set of G.

Clearly, $G \models \varphi$ if, and only if, G is connected, as the vertex set of any connected component satisfies $\left(\exists x X x \wedge \forall x \forall y (X x \wedge E x y \rightarrow X y) \right)$.

2. *A 3-colouring of a graph G is a function $f : V(G) \to \{1, 2, 3\}$ such that $f(u) \neq f(v)$ for all $\{u, v\} \in E(G)$. The formula*

$$\varphi := \exists C_1 \exists C_2 \exists C_3 \Big(\forall x \bigvee_{i=1}^{3} C_i x \Big) \wedge \forall x \forall y \Big(Exy \to \bigwedge_{i=1}^{3} \neg (C_i x \wedge C_i y) \Big)$$

is true in a graph G if, and only if, G is 3-colourable. ⊣

With any logic $\mathcal{L}$, we can naturally associate the following decision problem, called the *model-checking problem* of $\mathcal{L}$.

MC($\mathcal{L}$)
    Input: Structure A and sentence $\varphi \in \mathcal{L}$.
    Problem: Decide $A \models \varphi$.

Much of this paper will be devoted to studying the complexity of model-checking problems on various classes of graphs, primarily in the parameterized setting introduced in the next section.

Another natural problem associated with any logic is its *satisfiability problem*   *satisfiability* defined as the problem to decide for a given sentence $\varphi \in \mathcal{L}$ whether it has a model. We will study this problem relative to a given class $\mathcal{C}$ of structures. This is equivalent to asking whether the $\mathcal{L}$-*theory* of $\mathcal{C}$, i.e. the class of all formulas $\varphi \in \mathcal{L}$ which are true in every structure $A \in \mathcal{C}$, is decidable.

The *quantifier rank* of a formula φ, denoted $qr(\varphi)$, is the maximal number   *quantifier rank* of quantifiers in φ nested inside each other. If $\varphi \in$ MSO, we count first- and $qr(\varphi)$ second-order quantifiers. For instance, the formula in Example 5.2.2 (1) has quantifier rank 3.

Let A be a structure and $v_1, \dots, v_k$ be elements in $V(A)$. For $q \geq 0$, the *first-*   *first-order* *order q-type* $\mathrm{tp}_q^{\mathrm{FO}}(A, \bar{v})$ of $\bar{v}$ is the class of all FO-formulas $\varphi(\bar{x})$ of quantifier-   *type* $\mathrm{tp}_q^{\mathrm{FO}}(A, \bar{v})$ rank $\leq q$ such that $A \models \varphi(\bar{v})$. *Monadic second-order types* $\mathrm{tp}_q^{\mathrm{MSO}}(A, \bar{v})$ are $\mathrm{tp}_q^{\mathrm{MSO}}(A, \bar{v})$ defined analogously.

By definition, types are infinite. However, it is well known that there are only finitely many FO or MSO-formulas of quantifier rank $\leq q$ which are pairwise not equivalent. Furthermore, we can effectively *normalise* formulas in such a way that equivalent formulas are normalised syntactically to the same formula. Hence, we can represent types by their finite set of normalised formulas.

This has a number of algorithmic applications. For instance, it is decidable if two types are the same and whether a formula φ is contained in a type Θ: we simply normalise φ to a formula ψ and check whether $\psi \in \Theta$. Note, however, that it is undecidable whether a set of normalised formulas is a type:

by definition, types are satisfiable and satisfiability of first-order formulas is undecidable.

The following lemma, which essentially goes back to Feferman and Vaught will be used frequently later on. We refer the reader to [53] or [64] for a proof.

Lemma 5.2.3 *Let* tp *be either* $\mathrm{tp}^{\mathrm{MSO}}$ *or* $\mathrm{tp}^{\mathrm{FO}}$ *and let* H, G *be graphs such that* $V(H) \cap V(G) = \{\overline{v}\}$. *Let* $\overline{u} \in V(H)$ *and* $\overline{w} \in V(G)$.

For all $q \geq 0$, $\mathrm{tp}_q(G \cup H, \overline{vuw})$ *is uniquely determined by* $\mathrm{tp}_q(G, \overline{vw})$ *and* $\mathrm{tp}_q(H, \overline{uv})$ *and this is effective, i.e. there is an algorithm that computes* $\mathrm{tp}_q(G \cup H, \overline{vuw})$ *given* $\mathrm{tp}_q(G, \overline{vw})$ *and* $\mathrm{tp}_q(H, \overline{uv})$.

Suppose $G = H_1 \cup H_2$ can be decomposed into subgraphs H_1, H_2 such that $V(H_1 \cap H_2) = \overline{v}$. The importance of the lemma is that it allows us to infer the truth of a formula in G from the q-type of $\overline{v}$ in H_1 and H_2, where $q := \mathrm{qr}(\varphi)$. Hence, if G is decomposable in this way, we can reduce the question $G \models \varphi$ to the question on smaller graphs H_1, H_2. This will be of importance when we study graph-decompositions such as tree-decompositions and similar concepts in Section 5.3 and 5.4.

MSO-Interpretations

Let $\mathcal{C}$ be a class of σ-structures and $\mathcal{D}$ be a class of τ-structures. Suppose we know already that MSO-model-checking is tractable on $\mathcal{C}$ and we want to show that it is tractable on $\mathcal{D}$ also. Here is one way of doing this: find a way to "encode" a given graph $G \in \mathcal{D}$ in a graph $G' \in \mathcal{C}$ and also to "rewrite" the formula $\varphi \in \mathrm{MSO}[\tau]$ into a new formula $\varphi' \in \mathrm{MSO}[\sigma]$ so that $G \models \varphi$ if, and only if, $G' \models \varphi'$. Then tractability of MSO-model checking on $\mathcal{D}$ follows immediately from tractability on $\mathcal{C}$ – provided the encoding is efficient.

MSO-interpretations help us in doing just this: they provide a way to rewrite the formula φ speaking about $\mathcal{D}$ to a formula φ' speaking about $\mathcal{C}$ and also give us a translation of graphs "in the other direction", namely a way to translate a graph $G' \in \mathcal{C}$ to a graph $G := \Gamma(G') \in \mathcal{D}$ so that $G' \models \varphi'$ if, and only if, $G \models \varphi$. Hence, to reduce the model checking problem for MSO on $\mathcal{D}$ to the problem on $\mathcal{C}$, we have to find an interpretation Γ to translate the formulas from $\mathcal{D}$ to $\mathcal{C}$ and an encoding of graphs $G \in \mathcal{D}$ to graphs $G' \in \mathcal{C}$ so that $\Gamma(G') \cong G$. Figure 5.2 demonstrates the way interpretations are used as reductions.

We will first define the notion of interpretations formally and then demonstrate the concept by giving an example.

Definition 5.2.4 *Let* $\sigma := \{E, P_1, \ldots, P_k\}$ *and* $\tau := \{E\}$ *be signatures, where* E *is a binary relation symbol and the* P_i *are unary. A (one-dimensional)*

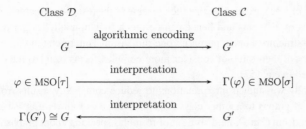

Class $\mathcal{D}$ Class $\mathcal{C}$

algorithmic encoding
$G \xrightarrow{\hspace{4cm}} G'$

interpretation
$\varphi \in \text{MSO}[\tau] \xrightarrow{\hspace{4cm}} \Gamma(\varphi) \in \text{MSO}[\sigma]$

interpretation
$\Gamma(G') \cong G \xleftarrow{\hspace{4cm}} G'$

Figure 5.2 Using interpretations as reductions between problems

MSO interpretation *from σ-structures to τ-structures is a triple* $\Gamma :=$ MSO-
$(\varphi_{univ}, \varphi_{valid}, \varphi_E)$ *of* MSO$[\sigma]$*-formulas.*   *interpretation*
*For every σ-structure T with $T \models \varphi_{valid}$ we define a graph (i.e. τ-structure)
$G := \Gamma(T)$ as the graph with vertex set $V(G) := \{u \in V(T) : T \models \varphi_{univ}(v)\}$
and edge set*

$$E(G) := \{\{u, v\} \in V(G) : T \models \varphi_E(u, v)\}.$$

If $\mathcal{C}$ is a class of σ-structures we define $\Gamma(\mathcal{C}) := \{\Gamma(T) : T \in \mathcal{C}, T \models \varphi_{valid}\}$.

Every interpretation naturally defines a mapping from MSO$[\tau]$-formulas φ
to MSO$[\sigma]$-formulas $\varphi^* := \Gamma(\varphi)$. Here, φ^* is obtained from φ by recursively
replacing

- first-order quantifiers $\exists x\varphi$ and $\forall x\varphi$ by $\exists x(\varphi_{univ}(x) \wedge \varphi^*)$ and $\forall x(\varphi_{univ}(x)$
  $\to \varphi^*)$ respectively,
- second-order quantifiers $\exists X\varphi$ and $\forall X\varphi$ by $\exists X(\forall y(Xy \to \varphi_{univ}(y)) \wedge \varphi^*)$
  and $\forall X(\forall y(Xy \to \varphi_{univ}(y)) \to \varphi^*)$ respectively and
- atoms $E(x, y)$ by $\varphi_E(x, y)$.

The following lemma is easily proved (see [57]).

Lemma 5.2.5 (interpretation lemma) *Let Γ be an* MSO*-interpretation from
σ-structures to τ-structures. Then for all* MSO$[\tau]$*-formulas and all σ-structures
$G \models \varphi_{valid}$*

$$G \models \Gamma(\varphi) \iff \Gamma(G) \models \varphi.$$

Note that here we are using a restricted form of interpretations. In particular, we
only allow one free variable in the formula $\varphi_{univ}(x)$ defining the universe of the
resulting graph. A consequence of this is that in any such an interpretation Γ,
we always have $|\Gamma(G)| \leq |G|$. In general interpretations, $\varphi_{univ}(\overline{x})$ can have any

number of free variables, so that the universe of the resulting structure consists of tuples of elements and hence can be much (polynomially) larger than the original structure. For our purposes, one-dimensional interpretations are enough and we will therefore not consider more complex forms of interpretations as discussed in e.g. [57].

Initially we studied interpretations to reduce complexity results from one class C of graphs to another class D. This is done as follows. Let Γ be interpretation from C in D, i.e. Γ is a set of formulas speaking about graphs in C so that for all $G \in C$, $\Gamma(G) \in D$.

We first design an algorithm that encodes a given graph $G \in D$ in a graph $G' \in C$ so that $\Gamma(G') \cong G$. Now, given $G \in D$ and $\varphi \in$ MSO as input, we translate G to a graph $G' \in C$ and use the interpretation Γ to obtain $\varphi' \in$ MSO$[\sigma]$ such that $G' \models \varphi'$ if, and only if, $G \models \varphi$. Then we can check – using the model-checking algorithm for C – whether $G' \models \varphi'$.

Example 5.2.6 *Let C be the class of finite paths and D be the class of finite cycles. Then $\Gamma(C) = D$ for the following interpretation $\Gamma := (\varphi_{univ}, \varphi_{valid}, \varphi_E)$: $\varphi_{univ}(x) = \varphi_{valid} := true$ and*

$$\varphi_E(x, y) := Exy \lor \neg \exists z_1 \exists z_2 (z_1 \neq z_2 \land ((Exz_1 \land Exz_2) \lor (Eyz_1 \land Eyz_2)))$$

The formula is true for a pair x, y if there is an edge between x and y or if neither x nor y have two different neighbours. Hence, if $P \in C$ is a path then $G := \Gamma(P)$ is the cycle obtained from P by connecting the two endpoints.

Now, if we know that MSO*-model-checking is tractable on C then we can infer tractability on D is follows. Given $C \in D$ and $\varphi \in$* MSO*, delete an arbitrary edge from C to obtain a path $P \in C$ and construct $\varphi' := \Gamma(\varphi)$. Obviously, $\Gamma(P) \cong C$ and hence $P \models \varphi'$, if and only if, $C \models \varphi$.*   ⊣

5.2.4 Complexity

We assume familiarity with basic principles of algorithm design and analysis, in particular Big-O notation, as can be found in any standard textbook on algorithms, e.g. [11]. Also, we assume familiarity with basic complexity classes such as PTIME, NP and PSPACE and standard concepts from complexity theory such as polynomial-time reductions as can be found in any text book on complexity theory, e.g. [72]. By reductions we will generally mean polynomial-time many-one reductions, unless explicitly stated otherwise.

The following examples introduce some of the problems we will be considering throughout the paper.

Example 5.2.7 *1. Recall from Example 5.2.1 that an* independent set *in an* *independent*
graph G is a set $X \subseteq V(G)$ *such that* $\{u, v\} \notin E$ *for all* $u, v \in X$. *The* *set*
independent set *problem is defined as*

INDEPENDENT SET
   Input: A graph G and $k \in \mathbb{N}$.
   Problem: Decide if G contains an independent set of size k.
---

2. Recall from Example 5.2.1 that dominating set *in a graph G is a set* $X \subseteq$ *dominating*
$V(G)$ *such that for all* $v \in V(G)$, *either* $v \in X$ *or there is a* $u \in X$ *such that* *set*
$\{v, u\} \in E(G)$. *The* dominating set *problem is defined as*

DOMINATING SET
   Input: A graph G and $k \in \mathbb{N}$.
   Problem: Decide if G contains a dominating set of size k.
---

3. A k-colouring of a graph G is a function $f : V(G) \to \{1, \dots, k\}$ *such that*
$f(u) \neq f(v)$ *for all* $\{u, v\} \in E(G)$. *Of particular interest for this paper is*
the problem to decide if a graph can be coloured by three colours.

3-COLOURING
   Input: A graph G.
   Problem: Decide if G has a 3-colouring.

⊣

It is well known that all three problems in the previous example are NP-
complete. Furthermore, we have already seen that the dominating set problem
can be reduced to first-order model-checking MC(FO). Hence, the latter is NP-
hard as well. However, as the following lemma shows, MC(FO) is (presumably)
even much harder than DOMINATING SET.

Lemma 5.2.8 *(Vardi [88])* MC(FO) *and* MC(MSO) *are* PSPACE-*complete.*

Proof (sketch). It is easily seen that MC(MSO), and hence MC(FO) is in
PSPACE: given A and $\varphi \in$ MSO, simply try all possible interpretations for the
variables quantified in φ. This requires only polynomial space.

Hardness of MC(FO) follows easily from the fact that QBF, the problem to
decide whether a quantified Boolean formula is satisfiable, is PSPACE-complete.
Given a QBF-formula $\varphi := Q_1 X_1 \dots Q_k X_k \psi$, where ψ is a formula in propositional logic over the variables $X_1 \dots X_k$ and $Q_i \in \{\exists, \forall\}$, we compute the first-
order formula $\varphi' := \exists t \exists f (t \neq f \wedge Q_1 x_1 \dots Q_k x_k \psi')$, where ψ' is obtained
from ψ by replacing each positive literal X_i by $x_i = t$ and each negative literal

$\neg X_i$ by $x_i = f$. Here, the variables t, f represent the truth values *true* and *false*. Clearly, for every structure A with at least two elements, $A \models \varphi'$ if, and only if, φ is satisfiable. $\qquad\square$

An immediate consequence of the proof is that MC(FO) is hard even for very simple structures: they only need to contain at least two elements. An area of computer science where evaluation problems for logical systems have intensively been studied is database theory, where first-order logic is the logical foundation of the query language SQL. A common assumption in database theory is that the size of the query is relatively small compared to the size of the database. Hence, giving the same weight to the database and the query may not truthfully reflect the complexity of query evaluation. It has therefore become standard to distinguish between three ways of measuring the complexity of logical systems:

- *combined complexity*: given a structure A and a formula φ as input, what is the complexity of deciding $A \models \varphi$ measured in the size of the structure and the size of the formula?
- *data complexity*: fix a formula φ. Given a structure A as input, what is the complexity of deciding $A \models \varphi$ measured in the size of the structure only?
- *expression complexity*: fix a structure A. Given a formula φ as input, what is the complexity of deciding $A \models \varphi$ measured in the size of the formula only?

As seen in Lemma 5.2.8, the combined complexity of first-order logic is PSPACE-complete. Furthermore, the proof shows that even the expression complexity is PSPACE-complete, as long as we fix a structure with at least two elements. On the other hand, it is easily seen that for a fixed formula φ, checking whether $A \models \varphi$ can be done in time $|A|^{O(|\varphi|)}$. Hence, the data complexity of first-order logic is in PTIME.

Besides full first-order logic, various fragments of FO have been studied in database theory and finite model theory. For instance, the combined complexity of the *existential conjunctive fragment* of first-order logic – known as *conjunctive queries* in database theory – is NP-complete. And if we consider the *bounded variable fragment* of first-order logic, the combined complexity is PTIME [89].

Much of this paper is devoted to studying model-checking problems for a logic $\mathcal{L}$ on restricted classes $\mathcal{C}$ of structures or graphs, i.e. to study the problem

---

$\mathrm{MC}(\mathcal{L}, \mathcal{C})$

    Input: $A \in \mathcal{C}$ and $\varphi \in \mathcal{L}$.

    Problem: Decide $A \models \varphi$.

In Example 5.2.2, we have already seen that 3-colourability is definable by a fixed sentence $\varphi \in$ MSO. As the problem is NP-complete, this shows that the data-complexity of MSO is NP-hard. In fact, it is complete for the polynomial time hierarchy. There are, however, interesting classes of graphs on which the data-complexity of MSO is PTIME. One example is the class of trees, another are classes of graphs of bounded tree-width.

For first-order logic there is not much to classify in terms of input classes $\mathcal{C}$, as the combined complexity is PSPACE-complete as soon as we have at least one structure of size ≥ 2 in $\mathcal{C}$ and the data complexity is always PTIME. Hence, the classification into expression and data complexity is too coarse for an interesting theory. However, polynomial time data complexity is somewhat unsatisfactory, as it does not tell us much about the degree of the polynomials. All it says is that for every fixed formula φ, deciding $A \models \varphi$ is in polynomial time. But the running time of the algorithms depends exponentially on $|\varphi|$ – and this is unacceptably high even for moderate formulas. Hence, the distinction between data and expression complexity is only of limited value for classifying tractable and intractable instances of the model checking problem.

A framework that allows for a much finer classification of model-checking problems is *parameterized complexity*, see [36, 46, 69]. A *parameterized problem* is a pair (P, χ), where P is a decision problem and χ is a polynomial time computable function that associates with every instance w of P a positive integer, called the *parameter*. Throughout this paper, we are mainly interested in parameterized model-checking problems. For a given logic $\mathcal{L}$ and a class $\mathcal{C}$ of structures we define[2]

MC($\mathcal{L}, \mathcal{C}$)	
Input:	Given $A \in \mathcal{C}$ and $\varphi \in \mathcal{L}$.
Parameter:	$|\varphi|$.
Problem:	Decide $A \models \varphi$.

A parameterized problem is *fixed-parameter tractable*, or in the complexity class FPT, if there is an algorithm that correctly decides whether an instance w is in P in time

$$f(\chi(w)) \cdot |w|^{\mathcal{O}(1)},$$

for some computable function $f : \mathbb{N} \to \mathbb{N}$. An algorithm with such a running time is called an *fpt algorithm*. Sometimes we want to make the exponent

FPT

*fpt
algorithm*

---

[2] We abuse notation here and also refer to the parameterized problem as MC($\mathcal{L}, \mathcal{C}$). As we will not consider the classical problem anymore, there is no danger of confusion.

of the polynomial explicit and speak of *linear fpt algorithm*, if the algorithm achieves a running time of $f(\chi(w)) \cdot |w|$, and similarly for quadratic and cubic fpt algorithms. We will sometimes relax the definition of parameterized problems slightly by considering problems (P, χ) where the function χ is no longer polynomial time computable, but is itself fixed-parameter tractable. For instance, this will be the case for problems where the parameter is the treewidth of a graph (see Section 5.3.1), a graph parameter that is computable by a linear fpt-algorithm but not in polynomial time (unless PTIME = NP). Everything we need from parameterized complexity theory in this paper generalises to this parametrization also. See [46, Chapter 11.4] for a discussion of this issue.

In the parameterized world, FPT plays a similar role to PTIME in classical complexity – a measure of tractability. Hence, much work has gone into classifying problems into those which are fixed-parameter tractable and those which are not, i.e. those that can be solved by algorithms with a running time such as $\mathcal{O}(2^{k^2} n^2)$ and those which require something like $\mathcal{O}(n^k)$, where k is the parameter. Running times of the form $\mathcal{O}(n^k)$ yield the parameterized complexity class XP, defined as the class of parameterized problems that can be solved in time $\mathcal{O}(n^{f(k)})$, for some computable function $f : \mathbb{N} \to \mathbb{N}$.

In terms of model-checking problems, a model-checking problem MC($\mathcal{L}$, $\mathcal{C}$) is in XP if, and only if, the data complexity of $\mathcal{L}$ on $\mathcal{C}$ is PTIME. Obviously, FPT $\subseteq$ XP and this inclusion is strict, as can be proved using the time hierarchy theorem. If FPT is the parameterized analogue of PTIME then XP can be seen as the analogue of EXPTIME. And again, similar to classical complexity, there are hierarchies of complexity classes in between FPT and XP. For our purpose, the most important class is called W[1], which is the first level of the *W-hierarchy* formed by classes W[i], for all $i \geq 1$. We refrain from giving the precise definition of W[1] and the W-hierarchy and refer the reader to the monograph [46]. For our purposes, it suffices to know that FPT, XP and the W[i]-classes form the following hierarchy

$$\text{FPT} \subseteq W[1] \subseteq W[2] \subseteq \cdots \subseteq \text{XP}.$$

In some sense, W[1] plays a similar role in parameterized complexity as NP in classical complexity, in that it is generally believed that FPT $\neq$ W[1] (as far as these beliefs go) and proving that a problem is W[1]-hard establishes that it is unlikely to be fixed-parameter tractable, i.e. efficiently solvable in the parameterized sense. The notion of reductions used here is *fpt-reduction*. Again, we refer to [46].

We close the section by stating the parameterized complexity of some problems considered in this paper.

Definition 5.2.9 *1. The p-*DOMINATING SET *problem is the problem, given a graph G and k $\in$ $\mathbb{N}$, to decide whether G contains a dominating set of size k. The parameter is k.*

2. *The p-*INDEPENDENT SET *problem is the problem, given a graph G and k $\in$ $\mathbb{N}$, to decide whether G contains an independent set of size k. The parameter is k.*

3. *The p-*CLIQUE *problem is the problem, given a graph G and k $\in$ $\mathbb{N}$, to decide whether G contains a clique of size k. The parameter is k.*

In the sequel, we will usually drop the prefix $p-$ and simply speak about the DOMINATING SET problem. It will always be clear from the context whether we are referring to the parameterized or the classical problem.

Lemma 5.2.10 (*Downey, Fellows* [34, 35]) *1. p-*DOMINATING SET *is W[2]-complete (see [34]).*

2. *p-*INDEPENDENT SET *is W[1]-complete (see [35]).*

3. *p-*CLIQUE *is W[1]-complete (see [35]).*

We have already seen that dominating and independent sets of size k can uniformly be formalised in first-order logic. Hence MC(FO) is W[2]-hard as well. In fact, it is complete for the parameterized complexity class AW[$*$], which contains all levels of the W-hierarchy and is itself contained in XP. Finally, as 3-colourability is expressible in MSO, MSO model-checking is not in XP unless NP = PTIME.

5.3 Monadic Second-Order Logic on Tree-Like Structures

It is a well-known fact, based on the close relation between monadic second-order logic and finite tree- and word-automata (see e.g. [9, 31, 85, 86, 10, 46, 63]), that model-checking and satisfiability for very expressive logics such as MSO becomes tractable on the class of finite trees. At the core of these results is the observation that the validity of an MSO sentence at the root of a tree can be inferred from the label of the root and the MSO-types realised by its successors. There are various ways in which this idea can be turned into a proof or algorithm: we can use effective versions of Feferman-Vaught style theorems (see e.g. [64]) or we can convert formulas into suitable tree-automata and let them run on the trees. The aim of the following sections is to extend the results for MSO and FO from trees to more general classes of graphs. The aforementioned composition methods will in most cases provide the key to obtaining these stronger results.

In this section we generalise the results for MSO model-checking and satisfiability from trees to graphs that are no longer trees but still tree-like enough so that model-checking and satisfiability testing for such graphs can be reduced to the case of trees.

5.3.1 Tree-Width

The precise notion for "tree-likeness" we use is the concept of tree-width. We first introduce tree-decompositions, establish some closure properties and then comment on algorithmic problems in relation to tree-width.

Tree-Decompositions

tree- **Definition 5.3.1** *A* tree-decomposition *of a graph* G *is a pair* $\mathcal{T} :=$
decomposition $(T, (B_t)_{t \in V(T)})$ *consisting of a tree* T *and a family* $(B_t)_{t \in V(T)}$ *of sets* $B_t \subseteq V(G)$
such that

$B^{-1}(v)$ *1. for all* $v \in V(G)$ *the set*

$$B^{-1}(v) := \{t \in V(T) : v \in B_t\}$$

is non-empty and connected in T *and*

2. for every edge $e \in E(G)$ *there is a* $t \in V(T)$ *with* $e \subseteq B_t$.

tree-width, *The* width $w(\mathcal{T})$ *of* $\mathcal{T}$ *is* $w(\mathcal{T}) := \{|B_t| - 1 : t \in V(T)\}$ *and the* tree-width *of*
$w(\mathcal{T})$ G *is defined as the minimal width of any of its tree-decompositions.*

bags   We refer to the sets B_t of a tree-decomposition as *bags*. For any edge
cut $e := \{s, t\} \in E(T)$ we call $B_s \cap B_t$ the *cut* at or *along* the edge e. (The reason for this will become clear later. See Lemma 5.3.13.)

Example 5.3.2 *Consider the graph in Figure 5.3 a). A tree-decomposition of this graph is shown in Figure 5.3 b).*   ⊣

Example 5.3.3 *Trees have tree-width* 1. *Given a tree* T, *the tree-decomposition has a node* t *for each edge* $e \in E(T)$ *labelled by* $B_t := e$ *and suitable edges connecting the nodes.*   ⊣

series- **Example 5.3.4** *The class of* series-parallel *graphs* (G, s, t) *with source s and*
parallel *sink t is inductively defined as follows.*

1. Every edge $\{s, t\}$ *is series-parallel.*
2. If (G_1, s_1, t_1) *and* (G_2, s_2, t_2) *are series parallel with* $V(G_1) \cap V(G_2) = \varnothing$, *then so are the following graphs:*

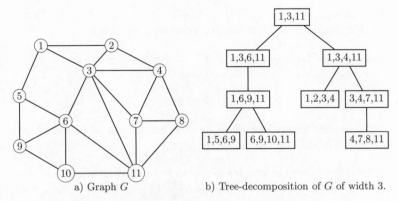

a) Graph G b) Tree-decomposition of G of width 3.

Figure 5.3 Graph and tree-decomposition from Example 5.3.2

a) the graph (G, s, t) obtained from $G_1 \cup G_2$ by identifying t_1 and s_2 and setting $s = s_1$ and $t = t_2$ (serial composition).

b) the graph (G, s, t) obtained from $G_1 \cup G_2$ by identifying s_1 and s_2 and also t_1 and t_2 and setting $s = s_1$ and $t = t_2$ (parallel composition).

The class of series-parallel graphs has tree-width 2. Following the inductive definition of series-parallel graphs one can easily show that every such graph (G, s, t) has a tree-decomposition of width 2 containing a node labelled by $\{s, t\}$. This is trivial for edges. For parallel and serial composition the tree-decompositions of the individual parts can be glued together at the node labelled by the respective source and sink nodes.   ⊣

The final example shows that grids have very high tree-width. Grids play a special role in relation to tree-width. As we will see later, every graph of sufficiently high tree-width contains a large grid minor. Hence, in this sense, grids are the least complex graphs of high tree-width.

Lemma 5.3.5 *For all $n > 1$, the $n \times n$-grid $G_{n,n}$ has tree-width n.*

In the remainder of this section we will present some basic properties of tree-decompositions and tree-width.

Closure Properties and Connectivity. It is easily seen that tree-width is preserved under taking subgraphs. For, if $(T, (B_t)_{t \in V(T)})$ is a tree-decomposition of width w of a graph G, then $(T, (B_t \cap V(H))_{t \in V(T)})$ is a tree-decomposition of H of width at most w. Further, if G and H are disjoint graphs, we can combine tree-decompositions for G and H to a tree-decomposition of the disjoint union $G \dot\cup H$ by adding one edge connecting the two decompositions.

Lemma 5.3.6 *Let G be a graph. If $H \subseteq G$, then $\mathrm{tw}(H) \leq \mathrm{tw}(G)$.*
Further, if $C_1, \ldots, C_k$ are the components of G, then

$$\mathrm{tw}(G) = \max\{\mathrm{tw}(C_i) : 1 \leq i \leq k\}.$$

To state the next results, we need further notation. Let G be a graph and $(T, (B_t)_{t \in V(T)})$ be a tree-decomposition of G.

$B^{-1}(H)$ 1. If $H \subseteq G$ we define $B^{-1}(H) := \{t \in V(T) : B_t \cap V(H) \neq \varnothing\}$.
$B(U)$ 2. Conversely, for $U \subseteq T$ we define $B(U) := \bigcup_{t \in V(U)} B_t$.

Occasionally, we will abuse notation and use B, B^{-1} for sets instead of subgraphs. The next lemma is easily proved by induction on $|H|$ using the fact that for each vertex $v \in V(G)$ the set $B^{-1}(v)$ is connected in any tree-decomposition $\mathcal{T}$ of G and that edges $\{u, v\} \in E(G)$ are covered by some bag B_t for $t \in V(T)$. Hence, $B^{-1}(u) \cup B^{-1}(v)$ is connected in $\mathcal{T}$ for all $\{u, v\} \in E(H)$.

Lemma 5.3.7 *Let G be a graph and $\mathcal{T} := (T, (B_t)_{t \in V(T)})$ be a tree-decomposition of G. If $H \subseteq G$ is connected, then so is $B^{-1}(H)$ in $\mathcal{T}$.*

Small tree-decompositions.   A priori, by duplicating nodes, tree-decompositions of a graph can be arbitrarily large (in terms of the number of nodes in the underlying tree). However, this is not very useful and we can always avoid this from happening. We will now consider tree-decompositions which are *small* and derive various useful properties from this.

small tree- **Definition 5.3.8** *A tree-decomposition $(T, (B_t)_{t \in V(T)})$ is small, if $B_t \not\subseteq B_u$ for*
decompositions all $u, t \in V(T)$ with $t \neq u$.

The next lemma shows that we can easily convert every tree-decomposition to a small one in linear time.

Lemma 5.3.9 *Let G be a graph and $\mathcal{T} := (T, (B_t)_{t \in V(T)})$ a tree-decomposition of G.*
Then there is a small tree-decomposition $\mathcal{T}' := \left(T', (B'_t)_{t \in V(T')}\right)$ of G of the same width and with $V(T') \subseteq V(T)$ and $B'_t = B_t$ for all $t \in V(T')$.

Proof. Suppose $B_s \subseteq B_t$ for some $s \neq t$. Let $s = t_1 \ldots t_n = t$ be the nodes of the path from s to t in T. Then $B_s \subseteq B_{t_2}$, by definition of tree-decompositions.

But then, $(T', (B_t)_{t \in V(T')})$ with $V(T') := V(T) \setminus \{s\}$ and

$$E(T') := \begin{array}{l} \big(E(T) \setminus \{\{v, s\} : \{v, s\} \in E(T)\}\big) \cup \\ \{\{v, t_2\} : \{v, s\} \in E(T) \text{ and } v \neq t_2\}. \end{array}$$

is a tree-decomposition of G with $V(T)' \subset V(T)$. We repeat this until T is small. $\qquad\square$

A consequence of this is the following result, which implies that in measuring the running time of algorithms on graphs whose tree-width is bounded by a constant k, it is sufficient to consider the order of the graphs rather than their size.

Lemma 5.3.10 *Every (non-empty) graph of tree-width at most k contains a vertex of degree at most k.*

Proof. Let G be a graph and let $\mathcal{T} := (T, (B_t)_{t \in V(T)})$ be a small tree-decomposition of G of width $k := \mathrm{tw}(G)$. If $|T| = 1$, then $|G| \leq k + 1$ and there is nothing to show. Otherwise let t be a leaf of T and s be its neighbour in T. As $\mathcal{T}$ is small, $B_t \not\subseteq B_s$ and hence there is a vertex $v \in B_t \setminus B_s$. By definition of tree-decompositions, v must have all its neighbours in B_t and hence has degree at most k. $\qquad\square$

Corollary 5.3.11 *Every graph G of tree-width $\mathrm{tw}(G) \leq k$ has at most $k \cdot |V(G)|$ edges, i.e., for $k > 0$, $||G|| \leq k \cdot |G|$.*

Separators. We close this section with a characterisation of graphs of small tree-width in terms of separators. This separation property allows for the aforementioned applications of automata theory or Feferman-Vaught style theorems.

Definition 5.3.12 *Let G be a graph.*

(i) *Let $X, Y \subseteq V(G)$. A set $S \subseteq V(G)$ separates X and Y, or is a* separator *for X and Y, if every path containing a vertex of Y and a vertex of Z* separator *also contains a vertex of S. In other words, X and Y are disconnected in $G - S$.*

(ii) *A separator of G is a set $S \subseteq V(G)$, so that $G - S$ has more than one component, i.e. there are sets $X, Y \subseteq V(G)$ such that S separates X and Y and $X \setminus S \neq \varnothing$ and $Y \setminus S \neq \varnothing$.*

Lemma 5.3.13 *Let $(T, (B_t)_{t \in V(T)})$ be a small tree-decomposition of a graph G.*

(i) *If* $e := \{s, t\} \in E(T)$ *and* T_1, T_2 *are the components of* $T - e$, *then* $B_t \cap B_s$ *separates* $B(T_1)$ *and* $B(T_2)$.

(ii) *If* $t \in V(T)$ *is an inner vertex and* $T_1, \ldots, T_k$ *are the components of* $T - t$ *then* B_t *separates* $B(T_i)$ *and* $B(T_j)$, *for all* $i \neq j$.

Proof. Let $e := \{s, t\} \in E(T)$ and let T_1, T_2 be the components of $T - e$. As T is small, $X := B(T_1) \setminus B(T_2) \neq \varnothing$ and $Y := B(T_2) \setminus B(T_1) \neq \varnothing$. Suppose there was an $X - Y$-path P in G not using any vertex from $B_t \cap B_s$. By Lemma 5.3.7, $B^{-1}(P)$ is connected and hence there is a path in T from T_1 to T_2 not using the edge e (as $V(P) \cap B_t \cap B_s = \varnothing$), in contradiction to T being a tree.

Part (ii) can be proved analogously.   □

Recall from the preliminaries that for an edge $e := \{s, t\} \in E(T)$ we refer to the set $B_s \cap B_t$ as the *cut* at the edge e. The previous lemma gives justification to this terminology, as the cut at an edge separates the graph. A simple consequence of this lemma is the following observation, that will be useful later on.

Corollary 5.3.14 *Let* G *be a graph and* $\mathcal{T} := (T, (B_t)_{t \in V(T)})$ *be a tree-decomposition of* G. *If* $X \subseteq V(G)$ *is the vertex set of a complete subgraph of* G, *then there is a* $t \in V(T)$ *such that* $X \subseteq B_t$.

Proof. By Lemma 5.3.9, there is a small tree-decomposition $\mathcal{T}' := (T', (B'_t)_{t \in V(T')})$ such that $V(T') \subseteq V(T)$ and $B'_t = B_t$ for all $t \in V(T')$. Hence, w.l.o.g. we may assume that $\mathcal{T}$ is small.

By Lemma 5.3.13, every cut at an edge $e \in E(T)$ is a separator of the graph G. Hence, as $G[X]$ is complete, if $e \in E(T)$ and T_1, T_2 are the two components of $T - e$, then either $X \subseteq B(T_1)$ or $X \subseteq B(T_2)$ but not both. We orient every edge $e \in E(T)$ so that it points towards the component of $T - e$ containing all of X. As T is acyclic, there is a node $t \in V(T)$ with no outgoing edge. By construction, $X \subseteq B_t$.   □

Corollary 5.3.15 $\mathrm{tw}(K_k) = k - 1$ *for all* $k \geq 1$.

Algorithms and Complexity

The notion of tree-width has been introduced by Robertson and Seymour as part of their proof of the graph minor theorem. Even before that, the notion of *partial k-trees*, broadly equivalent to tree-width, had been studied in the algorithms community. The relevance of tree-width for algorithm design stems from the fact that the tree-structure inherent in tree-decompositions can be used to design bottom-up algorithms on graphs of small tree-width to solve problems efficiently which in general are NP-hard. A key step in designing these algorithms is to compute a tree-decomposition of the input graph. Unfortunately, Arnborg, Corneil, and Proskurowski showed that deciding the tree-width of a graph is NP-complete itself.

Theorem 5.3.16 (*Arnborg, Corneil, Proskurowski* [3]) *The following problem is* NP-*complete.*

> TREE-WIDTH
> *Input:*   Graph $G, k \in \mathbb{N}$.
> *Problem:* $\text{tw}(G) = k$?

However, the problem becomes tractable if the tree-width is not a part of the input, i.e. if we are given a constant upper bound on the tree-width of graphs we are dealing with.

A class $\mathcal{C}$ of graphs has *bounded tree-width*, if there is a $k \in \mathbb{N}$ such that *bounded* $\text{tw}(G) \leq k$ for all $G \in \mathcal{C}$. In [6] Bodlaender proved that for any class of graphs *tree-width* of bounded tree-width tree-decompositions of minimal width can be computed in linear time.

Theorem 5.3.17 (*Bodlaender* [6]) *There is an algorithm which, given a graph G as input, constructs a tree-decomposition of G of width* $k := \text{tw}(G)$ *in time*

$$2^{\mathcal{O}(k^3)} \cdot |G|.$$

The algorithm by Bodlaender is primarily of theoretical interest. We will see later that many NP-complete problems can be solved efficiently on graph classes of bounded tree-width. For these algorithms to work in linear time, it is essential to compute tree-decompositions in linear time as well. From a practical point of view, however, the cubic dependence on the tree-width in the exponent and the complexity of the algorithm itself poses a serious problem. But there are other simpler algorithms with quadratic or cubic running time in the order of the graph but only linear exponential dependence on the tree-width which are practically feasible for small values of k.

5.3.2 Tree-Width and Structures

So far we have only considered graphs and their tree-decompositions. We will do so for most of the remainder, but at least want to comment on tree-decompositions of general structures. We first present the general definition of tree-decompositions of structures and then give an alternative characterisation in terms of the Gaifman- or comparability graph.

Definition 5.3.18 *Let σ be a signature. A* tree-decomposition *of a σ-structure A is a pair* $\mathcal{T} := (T, (B_t)_{t \in V(T)})$, *where T is a tree and* $B_t \subseteq V(A)$ *for all* $t \in V(T)$, *so that*

(i) *for all* $a \in V(A)$ *the set* $B^{-1} := \{t \in V(T) : a \in B_t\}$ *is non-empty and connected in T and*

(ii) *for every* $R \in \sigma$ *and all* $(a_1, \ldots, a_{\mathrm{ar}(R)}) \in R(A)^{\mathrm{ar}(R)}$ *there is a* $t \in V(T)$ *such that* $\{a_1, \ldots, a_{\mathrm{ar}(R)}\} \subseteq B_t$.

The width $\mathrm{w}(\mathcal{T})$ *is defined as* $\max\{|B_t| - 1 : t \in V(T)\}$ *and the* tree-width *of A is the minimal width of any of its tree-decompositions.*

The idea is the same as for graphs. We want the tree-decomposition to contain all elements of the structure and at the same time we want each tuple in a relation to be covered by a bag of the decomposition. It is easily seen that the tree-decompositions of a structure coincide with the tree-decompositions of its Gaifman graph, defined as follows.

Definition 5.3.19 (Gaifman-graph) *Let σ be a signature. The* Gaifman-graph $\mathcal{G}(A)$ *of a σ-structure A is defined as the graph* $\mathcal{G}(A)$ *with vertex set* $V(A)$ *and an edge between* $a, b \in V(A)$ *if, and only if, there is an* $R \in \sigma$ *and* $\overline{a} \in R(A)$ *with* $a, b \in \overline{a}$.

The following observation is easily seen.

Proposition 5.3.20 *A structure has the same tree-decompositions as its Gaifman-graph.*

So far we have treated the notion of graphs informally as mathematical structures. As a preparation to the next section, we consider two different ways of modelling graphs by logical structures. The obvious way is to model a graph σ_{Graph} *G* as a structure *A* over the signature $\sigma_{\mathrm{Graph}} := \{E\}$, where $V(A) := V(G)$ $A(G)$ and $E(A) := \{(a, b) \in V(A) \times V(A) : \{a, b\} \in E(G)\}$. We write $A(G)$ for this encoding of a graph as a structure and refer to it as the *standard encoding*. *incidence* Alternatively, we can model the *incidence graph* of a graph *G* defined as *graph* the graph G_{Inc} with vertex set $V(G) \cup E(G)$ and edges $E(G_{\mathrm{Inc}}) := \{(v, e) : v \in V(G), e \in E(G), v \in e\}$. The incidence graph gives rise to the following encoding of a graph as a structure, which we refer to as the *incidence encoding*.

Definition 5.3.21 *Let* $G := (V, E)$ *be a graph. Let* $\sigma_{\mathrm{inc}} := \{P_V, P_E, I\}$, *where* P_V, P_E *are unary predicates and I is a binary predicate. The* incidence structure $A_I(G)$ *is defined as the* σ_{inc}-*structure* $A := A_I(G)$ *where* $V(A) := V \cup E$,

$P_E(A) := E,\ P_V(A) := V$ *and*

$$I(A) := \{(v, e) : v \in V, e \in E, v \in e\}.$$

The proof of the following lemma is straightforward but may be a good exercise.

Theorem 5.3.22 $\mathrm{tw}(G) = \mathrm{tw}(A_I(G))$ *for all graphs* G.

It may seem to be a mere technicality how we encode a graph as a structure. However, the precise encoding has a significant impact on the expressive power of logics on graphs. For instance, the following $\mathrm{MSO}[\sigma_{\mathrm{inc}}]$-formula defines that a graph contains a Hamilton-cycle using the incidence encoding, a property that is not definable in MSO on the standard encoding (see e.g. [37, Corollary 6.3.5]).

$$\exists U \subseteq P_E \forall v\text{"}v \text{ has degree 2 in } G[U]\text{"} \wedge \varphi_{\mathrm{conn}}(U),$$

where φ_{conn} is a formula saying that the subgraph $G[U]$ induced by U is connected. Clearly, it is MSO-definable that a vertex v is incident to exactly two edges in U, i.e. has degree 2 in $G[U]$. The formula says that there is a set U of edges so that $G[U]$ is connected and that every vertex in $G[U]$ has degree 2. But this means that U is a simple cycle $\mathcal{P}$ in G. Further, as all vertices of G occur in $\mathcal{P}$, this cycle must be Hamiltonian.

Hence, MSO is more expressive over incidence graphs than over the standard encoding of graphs. It is clear that MSO interpreted over incidence graphs is the same as considering the extension of MSO by quantification over sets of edges (rather than just sets of vertices) on the standard encoding. This logic is sometimes referred to as MSO_2 in the literature. A more general framework are MSO_2 *guarded logics*, that allow quantification only over tuples that occur together in some relation in the structure. On graphs, *guarded second-order logic* (GSO) is just MSO_2. As we will not be dealing with general structures in the rest of GSO this survey, we refrain from introducing guarded logics formally and refer to [2, 51] and references therein instead.

5.3.3 Coding Tree-Decompositions in Trees

The aim of the following sections is to show that model-checking and satisfiability testing for monadic second-order logic becomes tractable when restricted to graph classes of small tree-width. The proof of these results relies on a reduction from graph classes of bounded tree-width to classes of finite labelled trees. As a first step towards this we show how graphs of tree-width bounded by some constant k can be encoded in Σ_k-labelled finite trees for a suitable alphabet Σ_k

depending on k. We will also show that the class of graphs of tree-width k, for some $k \in \mathbb{N}$, is MSO-interpretable in the class of Σ_k-labelled trees.

A tree-decomposition $(T, (B_t)_{t \in V(T)})$ of a graph G is already a tree and we will take T as the underlying tree of the encoding. Thus, all we have to do is to define the labelling. Note that we cannot simply take the bags B_t as labels, as we need to work with a finite alphabet and there is no a priori bound on the number of vertices in the bags. Hence we have to encode the vertices in the bags using a finite number of labels. To simplify the presentation we will be using tree-decompositions of a special form.

leaf- **Definition 5.3.23** A leaf-decomposition *of a graph G is a tree-decomposition*
decomposition $\mathcal{T} := (T, (B_t)_{t \in V(T)})$ *of G such that all leaves of $V(T)$ contain exactly one vertex and every $v \in V(G)$ is contained in exactly one leaf of T.*

In other words, in leaf-decompositions there is a bijection ρ between the set of leaves of the decomposition and the set of vertices of the graph and the bag B_t of a leaf t contains exactly its image $\rho(t)$. It is easily seen that any tree-decomposition can be converted into a leaf-decomposition of the same width.

Lemma 5.3.24 *For every tree-decomposition $\mathcal{T}$ of a graph G there is a leaf-decomposition $\mathcal{T}'$ of G of the same width and this can be computed in linear time, given $\mathcal{T}$.*

To define the alphabet Σ_k, we will work with a slightly different form of tree-decompositions where the bags are no longer sets but ordered tuples of vertices. It will also be useful to require that all these tuples have the same length and that the tree underlying a tree-decomposition is a binary directed tree.[3]

Definition 5.3.25 *An* ordered tree-decomposition *of width k of a graph G is a pair $(T, (\bar{b}_t)_{t \in V(T)})$, where T is a directed binary tree and $\bar{b}_t \in V(G)^k$, so that $(T, (B_t)_{t \in V(T)})$ is a tree-decomposition of G, with $B_t := \{b_0, \ldots, b_k\}$ for $\bar{b}_t := b_0, \ldots, b_k$.*

An ordered leaf-decomposition is the ordered version of a leaf-decomposition.

Example 5.3.26 *Consider again the graph from Example 5.3.2. The following shows an ordered leaf-decomposition obtained from the tree-decomposition in*

---

[3] Note that, strictly speaking, to apply the results on MSO on finite trees we have to work with trees where an ordering on the children of a node is imposed. Clearly we can change all definitions here to work with such trees. But as this would make the notation even more complicated, we refrain from doing so.

Example 5.3.2 by first adding the necessary leaves containing just one vertex and then converting every bag into an ordered tuple of length 4.

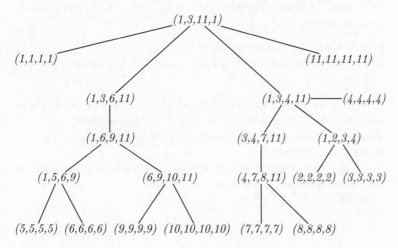

The graph G together with this leaf-decomposition induces the following Σ_3-labelled tree:

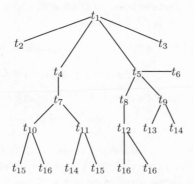

where, for instance, $\lambda(t_4) := \big(\text{eq}(t_4), \text{overlap}(t_4), \text{edge}(t_4)\big)$, *with*

- $\text{eq}(t_4) := \varnothing$,
- $\text{overlap}(t_4) := \{(0, 0), (0, 3), (1, 1), (3, 2)\}$, *and*
- $\text{edge}(t_4) := \{(0, 1), (1, 2), (1, 3), (2, 3)\} \cup \{(1, 0), (2, 1), (3, 1), (3, 2)\}$.

$\text{eq}(t_4) := \varnothing$, *as all positions of* $\overline{b}_{t_4}$ *correspond to different vertices in G. On the other hand,* $\text{eq}(t_{15}) := \{(i, j) : i, j \in \{0, \ldots, 3\}\}$, *as all entries of* $\overline{b}_{15}$ *refer to the same vertex 5.* ⊣

It is easily seen that every tree-decomposition of width k can be converted in linear time to an ordered tree-decomposition of width k. Combining this with Bodlaender's algorithm (Theorem 5.3.17) and Lemma 5.3.24 above yields the following lemma.

Lemma 5.3.27 *There is an algorithm that, given a graph G of tree-width $\leq k$, constructs an ordered leaf-decomposition of G of width* $\mathrm{tw}(G)$ *in time* $2^{\mathcal{O}(k^3)} \cdot |G|$.

Now let G be a graph and $\mathcal{L} := (T', (\overline{b}_t)_{t \in V(T')})$ be an ordered leaf-decomposition of G of width k. We code $\mathcal{L}$ in a labelled tree $\mathcal{T} := (T, \lambda)$, so that $\mathcal{L}$ and G can be reconstructed from $\mathcal{T}$, and this reconstruction can even be done by MSO formulas.

The tree T underlying $\mathcal{T}$ is the tree T' of $\mathcal{L}$. To define the alphabet and the labels of the nodes let $t \in V(T)$ and let $\overline{b}_t := b_0, \dots, b_k$.

$\lambda(t)$ We set

$$\lambda(t) := (\mathrm{eq}(t), \mathrm{overlap}(t), \mathrm{edge}(t))$$

where $\mathrm{eq}(t)$, $\mathrm{overlap}(t)$, $\mathrm{edge}(t)$ are defined as follows:

$\mathrm{eq}(t)$ • $\mathrm{eq}(t) := \{(i, j) : 0 \leq i, j \leq k \text{ and } b_i = b_j\}$.
          • If t is the root of T, then $\mathrm{overlap}(t) := \varnothing$. Otherwise let p be the predecessor
$\mathrm{overlap}(t)$ of t in T and let $\overline{b}_p := a_0, \dots, a_k$. We set

$$\mathrm{overlap}(t) := \{(i, j) : 0 \leq i, j \leq k \text{ and } b_i = a_j\}.$$

$\mathrm{edge}(t)$ • Finally, $\mathrm{edge}(t) := \{(i, j) : 0 \leq i, j \leq k \text{ and } \{b_i, b_j\} \in E(G)\}$.

Σ_k For every fixed k, the labels come from the finite alphabet

$$\Sigma_k := 2^{\{0, \dots, k\}^2} \times 2^{\{0, \dots, k\}^2} \times 2^{\{0, \dots, k\}^2}.$$

$\mathcal{T}(G, \mathcal{L})$ We write $\mathcal{T}(G, \mathcal{L})$ for the labelled tree encoding a leaf-decomposition $\mathcal{L}$ of a graph G. Note that the signature depends on the arity k of the ordered leaf-decomposition $\mathcal{L}$, i.e. on the bound on the tree-width of the class of graphs we are working with.

The individual parts of the labelling have the following meaning. Recall that we require all tuples $\overline{b}_t$ to be of the same length $k + 1$ and therefore they may contain duplicate entries. $\mathrm{eq}(t)$ identifies those entries in a tuple relating to the same vertex of the graph G. The label $\mathrm{overlap}(t)$ takes care of the same vertex appearing in tuples of neighbouring nodes of the tree. As we are working with directed trees, every node other than the root has a unique predecessor. Hence we can record in the overlap-label of the child which vertices in its bag occur at which positions of its predecessor. Finally, edge encodes the edge relation of G. As every edge is covered by a bag of the tree-decomposition, it suffices to record for each node $t \in V(T)$ the edges between elements of its bag $\overline{b}_t$.

The labels eq(t), overlap(t) and edge(t) satisfy some obvious consistency criteria, e.g. eq(t) is an equivalence relation for every t, eq(t) is consistent with edge(t) in the sense that if two positions i, i' refer to the same vertex, i.e. $(i, i') \in$ eq(t) and $(i, j) \in$ edge(t) then also $(i', j) \in$ edge(t), and likewise for eq(t) and overlap(t). We refrain from giving all necessary details. Note, though, that any Σ_k-labelled finite tree that satisfies these consistency criteria does encode a graph of tree-width at most k. Furthermore, the criteria as outlined above are easily seen to be definable in MSO, in fact even in first-order logic. Again we refrain from giving the exact formula as its definition is long and technical but absolutely straightforward. Let φ_{cons} be the MSO-sentence true in $\quad \varphi_{cons}$ a Σ_k-labelled tree if, and only if, it satisfies the consistency criteria, i.e. encodes a tree-decomposition of a graph of tree-width at most k.

Of course, to talk about formulas defining properties of Σ_k-labelled trees we first need to agree on how Σ_k-labelled trees are encoded as structures. For $k \in \mathbb{N}$ we define the signature $\qquad\qquad\qquad \sigma_k$

$$\sigma_k := \{E\} \cup \{\text{eq}_{i,j}, \text{edge}_{i,j}, \text{overlap}_{i,j} : 0 \leq i, j \leq k\},$$

where eq$_{i,j}$, overlap$_{i,j}$, and edge$_{i,j}$ are unary relation symbols. The intended meaning of eq$_{i,j}$ is that in a σ_k-structure A an element t is contained in eq$_{i,j}(A)$ if $(i, j) \in$ eq(t) in the corresponding tree. Likewise for overlap$_{i,j}$ and edge$_{i,j}$. σ_k-structures, then, encode Σ_k-labelled trees in the natural way. In the sequel, we will not distinguish notationally between a Σ_k-labelled tree T and the corresponding σ_k-structure A_T. In particular, we will write $T \models \varphi$, for an MSO-formula φ, instead of $A_T \models \varphi$.

Clearly, the information encoded in the Σ_k-labelling is sufficient to reconstruct the graph G from a tree $T(G, \mathcal{L})$, for some ordered leaf-decomposition of G of width k. Note that different leaf-decompositions of G may yield non-isomorphic trees. Hence, the encoding of a graph in a Σ_k-labelled tree is not unique but depends on the decomposition chosen. For our purpose this does not pose any problem, though.

The next step is to define an MSO-interpretation

$$\Gamma := (\varphi_{univ}(x), \varphi_{valid}, \varphi_E(x, y)) \qquad\qquad \Gamma$$

of the class $\mathcal{T}_k$ of graphs of tree-width at most k in the class $\mathcal{T}_{\Sigma_k}$ of Σ_k-labelled finite trees. To state the interpretation formally, we need to define the three formulas $\varphi_{univ}(x)$, φ_{valid}, and $\varphi_E(x, y)$. Recall that in a leaf-decomposition $\mathcal{L}$ there is a bijection between the leaves of T and the vertices of the graph that is being decomposed. Hence, we can take $\varphi_{univ}(x)$ to be the formula

$$\varphi_{univ}(x) := \forall y \neg E x y$$

saying that x is a leaf in $\mathcal{T}$.

Let G be a graph and $\mathcal{L} := (T, (\bar{b}_t)_{t \in V(T)})$ be an ordered leaf-decomposition of G of width k. Suppose we are given two leaves t_u, t_v of $\mathcal{L}$ containing u and v respectively and we want to decide whether there is an edge between u and v. Clearly, if $e := \{u, v\} \in E(G)$, then e must be covered by some bag, i.e. there are a node t in $\mathcal{L}$ with bag $\bar{b}_t := b_0 \ldots b_k$ and $i \neq j$ such that $b_i = u$ and $b_j = v$ and $(i, j) \in \text{edge}(t)$ in the tree $\mathcal{T} := \mathcal{T}(G, \mathcal{L})$. Further, u occurs in every bag on the path from t to t_u and likewise for v. Hence, to define $\varphi_E(x, y)$, where x, y are interpreted by leaves, we have to check whether there is such a node t and paths from x and y to t as before. For this, we need an auxiliary formula which we define next.

Recall that each position i in a bag $\bar{b}_t$ corresponds to a vertex in G. Hence, we can associate vertices with pairs (t, i). In general, a vertex can occur at different positions i and different nodes $t \in V(T)$. We can, however, identify any vertex v with the set

X_v $\qquad\qquad X_v := \{(t, i) : t \in V(T) \text{ and } v \text{ occurs at position } i \text{ in } \bar{b}_t\}.$

We call X_v the *equivalence set* of v. If $t \in V(T)$ and $0 \le i \le k$, we define the *equivalence set* of (t, i) as the equivalence set of b_i, where $\bar{b}_t := b_0, \ldots, b_k$.

Clearly, this identification of vertices with sets of pairs and the concept of equivalent sets extends to the labelled tree $\mathcal{T} := \mathcal{T}(G, \mathcal{L})$, as $\mathcal{T}$ and $\mathcal{L}$ share the same underlying tree.

To define sets X_v in MSO, we represent X_v by a tuple $\overline{X} := (X_0, \ldots, X_k)$ of sets $X_i \subseteq V(T)$, such that for all $0 \le i \le k$ and all $t \in V(T)$, $t \in X_i$ if, and only if, $(t, i) \in X_v$.

We are going to describe an MSO-formula $\psi(X_0, \ldots, X_k)$ that is satisfied by a tuple $\overline{X}$ if, and only if, $\overline{X}$ is the equivalence set of a pair (t, i), or equivalently of a vertex $v \in V(G)$. To simplify notation, we will say that a tuple $\overline{X}$ contains a pair (t, i) if $t \in X_i$. Consider the formulas

$$\psi_{eq}(X_0, \ldots, X_k) := \bigwedge_i \forall t \in X_i \left(\bigwedge_{j \neq i} \text{eq}_{i,j}(t) \to t \in X_j \right)$$

and

$$\psi_{overlap}(X_0, \ldots, X_k) := \forall s \forall t \bigwedge_{i,j} \left(E(s, t) \land t \in X_i \land \text{overlap}_{i,j}(t) \right) \to s \in X_j.$$

$\psi_{eq}(\overline{X})$ says of a tuple $\overline{X}$ that $\overline{X}$ is closed under the eq-labels and $\psi_{overlap}(\overline{X})$ says the same of the overlap-labels. Now let $\psi(\overline{X}) := \psi_{eq} \wedge \psi_{overlap}$. ψ is satisfied by a tuple $\overline{X}$ if whenever $\overline{X}$ contains at a pair (t, i), then it contains the complete equivalence set of (t, i). Now, consider the formula

$$\varphi_{vertex}(X_0, \ldots, X_k) := \psi(\overline{X}) \wedge \overline{X} \neq \varnothing \wedge \forall \overline{X}' \neq \varnothing \big(\overline{X}' \subsetneq \overline{X} \rightarrow \neg \psi(\overline{X}')\big)$$

φ_{vertex}

where "$\overline{X} \neq \varnothing$" defines that at least one X_i is non-empty and "$\overline{X}' \subsetneq \overline{X}$" is an abbreviation for a formula saying that $X_i' \subseteq X_i$, for all i, and for at least one i the inclusion is strict.

$\varphi_{vertex}(\overline{X})$ is true for a tuple if $\overline{X}$ is non-empty, closed under eq and overlap, but no proper non-empty subset of $\overline{X}$ is. Hence, $\overline{X}$ is the equivalence set of a single vertex $v \in V(G)$. The definition of $\varphi_{vertex}(\overline{X})$ is the main technical part of the MSO-interpretation $\Gamma := (\varphi_{univ}(x), \varphi_{valid}, \varphi_E(x, y))$.

We have already defined $\varphi_{univ}(x) := \forall y \neg Exy$. For φ_{valid}, recall from above the formula φ_{cons} true in a Σ_k-labelled tree $\mathcal{T}$ if, and only if, $\mathcal{T}$ encodes a tree-decomposition of a graph G of tree-width at most k. To define φ_{valid} we need a formula that not only requires $\mathcal{T}$ to encode a tree-decomposition of G but a leaf-decomposition.

To force the encoded tree-decomposition to be a leaf-decomposition, we further require the following two conditions.

1. For all leaves $t \in V(T)$ and all $i \neq j$, $(i, j) \in eq(t)$.
2. For all $t \in V(T)$ and all $0 \leq i \leq k$ the equivalence set of (t, i) contains exactly one leaf.

Both conditions can easily be defined by MSO-formulas φ_1 and φ_2, respectively, where in the definition of φ_2 we use the formula φ_{vertex} defined above.

Hence, the formula

$$\varphi_{valid} := \varphi_{cons} \wedge \varphi_1 \wedge \varphi_2$$

φ_{valid}

is true in a Σ_k-labelled tree $\mathcal{T}$ (or the corresponding σ_k-structure) if, and only if, $\mathcal{T}$ encodes a leaf-decomposition of width k.

Finally, we define the formula $\varphi_E(x, y)$ saying that there is an edge between x and y in the graph G encoded by a Σ_k-labelled tree $\mathcal{T} := (T, \lambda)$. Note that there is an edge in G between x and y if, and only if, there is a node $t \in V(T)$ and $0 \leq i \neq j \leq k$ such that $(i, j) \in edge(t)$ and x is the unique leaf in the equivalence set of (t, i) and y is the unique leaf in the equivalence set of (t, j). This is formalised by

$$\varphi_E(x, y) := \exists t \bigvee_{i \neq j} \left(\begin{array}{c} edge_{i,j}(t) \wedge \exists \overline{X} \exists \overline{Y} \varphi_{vertex}(\overline{X}) \wedge \varphi_{vertex}(\overline{Y}) \wedge \\ X_1(x) \wedge Y_1(y) \wedge X_i(t) \wedge Y_j(t) \end{array} \right).$$

This completes the definition of Γ. Now, the proof of the following lemma is immediate.

Lemma 5.3.28 *Let G be a graph of tree-width $\leq k$ and $\mathcal{L}$ be a leaf-decomposition of G of width k. Let $\mathcal{T} := \mathcal{T}(G, \mathcal{L})$ be the tree-encoding of $\mathcal{L}$ and G. Then $G \cong \Gamma(\mathcal{T})$.*

Further, by the interpretation lemma, for all MSO-formulas φ and all Σ_k-trees $\mathcal{T} \models \varphi_{valid}$,

$$\mathcal{T} \models \Gamma(\varphi) \quad \Longleftrightarrow \quad \Gamma(\mathcal{T}) \models \varphi.$$

5.3.4 Courcelle's Theorem

In this section and the next we consider computational problems for monadic second-order logic on graph classes of small tree-width. The algorithmic theory of MSO on graph classes of small tree-width has, essentially independently, been developed by Courcelle, Seese and various co-authors. We first consider the model-checking problem for MSO and present Courcelle's theorem. We then state a similar theorem by Arnborg, Lagergreen and Seese concerning the *evaluation problem* of MSO. In the next section, we consider the satisfiability problem and prove Seese's theorem.

Theorem 5.3.29 *(Courcelle* [13]*) The problem*

MC(MSO, tw)	
Input:	Graph G, $\varphi \in$ MSO
Parameter:	$|\varphi| + \text{tw}(G)$
Problem:	$G \models \varphi$?

is fixed parameter tractable and can be solved in time $f(|\varphi|) + 2^{p(\text{tw}(G))} \cdot |G|$, for a polynomial p and a computable function $f : \mathbb{N} \to \mathbb{N}$.

That is, the model-checking problem for a fixed formula $\varphi \in$ MSO can be solved in linear time on any class of graphs of bounded tree-width.

Proof. Let $\mathcal{C}$ be a class of bounded tree-width and let k be an upper bound for the tree-width of $\mathcal{C}$. Let $\varphi \in$ MSO be given.

On input $G \in \mathcal{C}$ we first compute an ordered leaf-decomposition $\mathcal{L}$ of G of width k. From this, we compute the tree $\mathcal{T} := \mathcal{T}(G, \mathcal{L})$. We then check whether $\mathcal{T} \models \Gamma(\varphi)$, where Γ is the MSO-interpretation of the previous section.

Correctness of the algorithm follows from Lemma 5.3.28. The time bounds follow from Lemma 5.3.24 and the fact that MSO model-checking is in linear time (for a fixed formula) on the class of trees (see e.g. [63, Chapter 7] or [46, Chapter 10]). □

We will see a different proof of this theorem using logical types later when we prove Lemma 5.7.12. The result immediately implies that parametrized problems such as the independence set or dominating set problem or problems such as 3-colourability and Hamiltonicity are solvable in linear time on classes of graphs of bounded tree-width.

Without proof we state the following extension of Courcelle's theorem which essentially follows from [4]. The proof uses the same methods as described above and the corresponding result for trees.

Theorem 5.3.30 (*Arnborg, Lagergreen, Seese* [4]) *The problem*

Input:	Graph G, $\varphi(X) \in$ MSO, $k \in \mathbb{N}$.
Parameter:	$|\varphi| + \text{tw}(G)$.
Problem:	Determine whether there is a set $S \subseteq V(G)$ such that $G \models \varphi(S)$ and $|S| \leq k$ and compute one if it exists.

is fixed-parameter tractable and can be solved by an algorithm with running time $f(|\varphi|) + 2^{p(\text{tw}(G))} \cdot |G|$, for a polynomial p and a computable function $f : \mathbb{N} \to \mathbb{N}$.

Recall that by the results discussed in Section 5.3.2 the previous results also hold for MSO on incidence graphs, i.e. MSO_2 where quantification over sets of edges is allowed also.

Corollary 5.3.31 *The results in Theorem 5.3.29 and 5.3.30 extend to MSO_2.*

5.3.5 Seese's Theorem

We close this section with another application of the interpretation defined in Section 5.3.3. Recall that MSO_2 has set quantification over sets of vertices as well as sets of edges and corresponds to MSO interpreted over the incidence encoding of graphs.

Theorem 5.3.32 (*Seese* [81]) *Let $k \in \mathbb{N}$ be fixed. The MSO_2-theory of the class of graphs of tree-width at most k is decidable.*

Proof. Let $\Gamma := (\varphi_{univ}, \varphi_{valid}, \varphi_E)$ be the interpretation defined in Section 5.3.3. On input φ we first construct the formula $\varphi^* := \Gamma(\varphi)$. Using the decidability of the MSO-theory of finite labelled trees, we then test whether there is a Σ_k-labelled tree $\mathcal{T}$ such that $\mathcal{T} \models \varphi_{valid} \wedge \varphi^*$.

If there is such a tree $\mathcal{T}$, then, as $\mathcal{T} \models \varphi_{valid}$, there is a graph G of tree-width at most k encoded by $\mathcal{T}$ which satisfies φ. Otherwise, φ is not satisfiable by any graph of tree-width at most k.   □

Again without proof, we remark that the following variant of Seese's theorem is also true.

Theorem 5.3.33 (*Adler, Grohe, Kreutzer* [1]) *For every k it is decidable whether a given* MSO-*formula is satisfied by a graph of tree-width exactly k.*

We remark that there is a kind of converse to Seese's theorem which we will prove in Section 5.6 below.

Theorem 5.3.34 (*Seese* [81]) *If $\mathcal{C}$ is a class of graphs with a decidable* MSO$_2$-*theory, then $\mathcal{C}$ has bounded tree-width.*

The proof of this theorem relies on a result proved by Robertson and Seymour as part of their proof of the graph minor theorem. We will present the graph theory needed for this in Section 5.5 and a proof of Theorem 5.3.34 in Section 5.6.

5.4 From Trees to Cliques

In the previous section we considered graphs that are sufficiently tree-like so that efficient model-checking algorithms for monadic second-order logic can be devised following the tree-structure of the decomposition. On a technical level these results rely on Feferman-Vaught style results allowing to infer the truth of an MSO sentence in a graph from the MSO types of the smaller subgraphs it can be decomposed into. In this section we will see a different property of graphs that also allows for efficient MSO model-checking. It is not based on the idea of decomposing the graph into smaller parts of lower complexity, but instead it is based on the idea of the graphs being *uniform* in some way, i.e. not having too many types of its vertices.

As a first example let us consider the class $\{K_n : n \in \mathbb{N}\}$ of cliques. Obviously, these graphs have as many edges as possible and cannot be decomposed in any meaningful way into parts of lower complexity. However, model-checking for first-order logic or monadic second-order logic is simple, as all vertices look the same. In a way, a clique is no more complex than a set: the edges do not impose any meaningful structure on the graph. This intuition is generalised by the notion of *clique-width* of a graph. It was originally defined in terms of graph grammars by Courcelle, Engelfriet and Rozenberg [17]. Independently, Wanke introduced k-NLC graphs, a notion that is equivalent to Courcelle et al.'s definition up to a factor of 2. The term clique-width was introduced in [19]. Clique-decompositions (or k-expressions as they are called) are useful for the design of algorithms, as they again provide a tree-structure along which algorithms can work. However, until recently algorithms using clique-decompositions had to be given the decomposition as input, as no fixed-parameter algorithms were known to compute the decomposition.

In 2006, Oum and Seymour [71] introduced the notion of *rank-width* and corresponding *rank-decompositions*, a notion that is broadly equivalent to clique-width in the sense that for every class of graphs, one is bounded if, and only if, the other is bounded. Rank-decompositions can be computed by fpt-algorithms parametrized by the width and from a rank-decomposition a clique-decomposition can be generated. In this way, the requirement of algorithms being given the decomposition as input has been removed. But rank-decompositions are also in many other ways the more elegant notion.

We first recall the definition of clique-width in Section 5.4.1. In Section 5.4.2, we then introduce general rank-decompositions of submodular functions, of which the rank-width of a graph is a special case. As a side effect, we also obtain the notion of branch-width, which is another elegant characterisation of tree-width. Model-checking algorithms for MSO on graph classes of bounded rank-width are presented in Section 5.4.3, where we also consider the satisfiability problem for MSO and a conjecture by Seese.

5.4.1 Clique-Width

Definition 5.4.1 (k-**expression**) *Let* $k \in \mathbb{N}$ *be fixed. The set of* k-expressions *is inductively defined as follows:*   *k-expression*

   (i) $\mathbf{i}$ *is a* k-expression *for all* $i \in [k]$.
   (ii) *If* $i \neq j \in [k]$ *and* φ *is a* k-expression, *then so are* $\mathrm{edge}_{i-j}(\varphi)$ *and* $\mathrm{rename}_{i \to j}(\varphi)$.
   (iii) *If* φ_1, φ_2 *are* k-expressions, *then so is* $(\varphi_1 \oplus \varphi_2)$.

A k-expression φ generates a graph $G(\varphi)$ coloured by colours from $[k]$ as follows: The k-expression $\mathbf{i}$ generates a graph with one vertex coloured by the $\mathbf{i}$ colour i and no edges.

The expression edge_{i-j} is used to add edges. If φ is a k-expression generat- edge_{i-j} ing the coloured graph $G := G(\varphi)$ then $\mathrm{edge}_{i-j}(\varphi)$ defines the graph H with $V(H) := V(G)$ and

$$E(H) := E(G) \cup \big\{ \{u, v\} : u \text{ has colour } i \text{ and } v \text{ has colour } j \big\}.$$

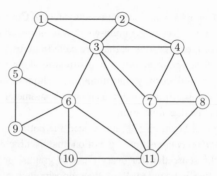

Figure 5.4 Graph from Example 5.3.2

Hence, $\text{edge}_{i-j}(\varphi)$ adds edges between all vertices with colour i and all vertices with colour j.

$\text{rename}_{i\to j}(\varphi)$ The operation $\text{rename}_{i\to j}(\varphi)$ recolours the graph. Given the graph G generated by φ, the k-expression $\text{rename}_{i\to j}(\varphi)$ generates the graph obtained from G by giving all vertices which have colour i in G the colour j in H. All other vertices keep their colour.

Finally, if φ_1, φ_2 are k-expressions generating coloured graphs G_1, G_2 respectively, then $(\varphi_1 \oplus \varphi_2)$ defines the disjoint union of G_1 and G_2.

We illustrate the definition by an example.

Example 5.4.2 *Consider again the graph from Example 5.3.2 depicted in Figure 5.3. For convenience, the graph is repeated below. We will show how this graph can be obtained by a 6-expression.*

Consider the expression φ_0 in Figure 5.5, which generates the graph in Figure 5.6 a). The labels in the graph represent the colours. Here we use obvious abbreviations such as $\text{edge}_{i-j,s-t}$ to create edges between i and j as well as edges between s and t in one step.

The vertices generated so far correspond to the vertices 5, 6, 9, 10 of the graph in Figure 5.4. Note that we have already created all edges incident to vertex 9. Hence, in the construction of the rest of the graph, the vertex 9 (having colour 2) does not have to be considered any more. We will use the colour 0 to mark vertices that will not be considered in further steps of the k-expression. Let $\varphi_1 := \text{rename}_{2\to 0}(\varphi_0)$ be the 6-expression that generates the graph in Figure 5.6 a), but where the vertex with colour 2 now has colour 0.

The next step is to generate the vertex 11 of the graph. This is done by the expression $\varphi_2 := \text{rename}_{5\to 0}\Big(\text{edge}_{1-5,1-4}\big(\mathbf{1} \oplus \varphi_1\big)\Big)$. We proceed by adding the

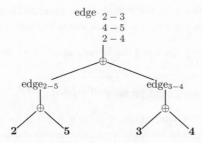

Figure 5.5 The 6-expression φ_0 generating the graph in Fig. 5.6 a)

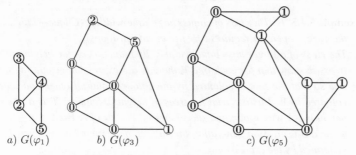

a) $G(\varphi_1)$ *b)* $G(\varphi_3)$ *c)* $G(\varphi_5)$

Figure 5.6 Graphs generated by the 6-expressions in Example 5.4.2

vertices 1 *and* 3 *and the appropriate edges. Let*

$$\varphi_3 := \mathrm{rename}_{3\to 0,4\to 0}\mathrm{edge}_{2-3,4-5,1-5}\Big(\varphi_2 \oplus \big(\mathrm{edge}_{2-5}(\mathbf{2} \oplus \mathbf{5})\big)\Big)$$

This generates the graph depicted in Figure 5.6 b). The next step is to add the vertices 7 *and* 8. *Let*

$$\varphi_4 := \mathrm{rename}_{1\to 0}\mathrm{edge}_{1-3,1-4,3-5}\big(\varphi_3 \oplus \mathrm{edge}_{3-4}(\mathbf{3} \oplus \mathbf{4})\big)$$

Finally, we add the vertex 2 *and rename the colour of the vertex* 2 *to* 0, *i.e. essentially remove the colour, and rename all other colours to* 1.

$$\varphi_5 := \mathrm{rename}_{2\to 0,5\to 1,3\to 1,4\to 1}\mathrm{edge}_{1-2,1-5}(\mathbf{1} \oplus \varphi_4)$$

This generates the graph in Figure 5.6 c).

Finally, we add the vertex 4 *and edges to all other vertices marked by the colour* 1.

The complete expression generating the graph is therefore $\mathrm{edge}_{1-2}(\mathbf{2} \oplus \varphi_5)$.

⊣

214   *Stephan Kreutzer*

It is easily seen that every finite graph can be generated by a k-expression for some $k \in \mathbb{N}$. Just choose a colour for each vertex and add edges accordingly.

Lemma 5.4.3 *Every finite graph can be generated[4] by a k-expression for some $k \in \mathbb{N}$.*

Hence, the following concepts are well defined.

clique-width **Definition 5.4.4** *The* clique-width $\mathrm{cw}(G)$ *of a graph G is defined as the least $k \in \mathbb{N}$ such that G can be generated by a k-expression. A class $\mathcal{C}$ of graphs has bounded clique-width if there is a $k \in \mathbb{N}$ such that $\mathrm{cw}(G) \leq k$ for all $G \in \mathcal{C}$.*

We give a few more examples.

Example 5.4.5 *1. The class of cliques has clique-width* 2. *(Clique-width* 2, *as the* $\mathrm{edge}_{i,j}$ *operator requires $i \neq j$ to avoid self-loops).*

2. The class of all trees has clique-width 3. *By induction on the height of the trees we show that for each tree T there is a 3-expression generating this tree so that the root is coloured by the colour* **1** *and all other nodes are coloured by* **0**. *This is trivial for trees of height* 0. *Suppose T is a tree of height $n + 1$ with root r and successors $v_1, \ldots, v_k$ of r. For $1 \leq i \leq k$ let φ_i be a 3-expression generating the subtree of T rooted at v_i. Then T is generated by the expression*

$$\mathrm{rename}_{2 \to 1}\mathrm{rename}_{1 \to 0}\mathrm{edge}_{2-1}(\mathbf{2} \oplus \varphi_1 \oplus \cdots \oplus \varphi_k).$$

3. It can be shown that the clique-width of the $(n \times n)$-grid is $\Omega(n)$. (This follows, for instance, from Theorem 5.4.7 below). ⊣

The next theorem due to Wanke and also Courcelle and Olariu relates clique-width to tree-width.

Theorem 5.4.6 ([91, 19]) *Every graph of tree-width at most k has clique-width at most $2^{k+1} + 1$.*

As the examples above show, there is no hope to bound the tree-width of a graph in terms of its clique-width. Hence, clique-width is more general than tree-width in the sense that more graph classes have bounded clique-width than bounded tree-width. Gurski and Wanke [55] established the following relation between clique-width and tree-width in terms of complete bipartite subgraphs.

---

[4] By "generating" we always mean up to isomorphism. That is, a graph G is generated by an expression φ if φ defines a graph isomorphic to G.

Theorem 5.4.7 (*Gurski, Wanke* [55]) *Let G be a graph of clique-width[5] k such that for some n > 1 the complete bipartite graph $K_{n,n}$ is not a subgraph of G. Then* $\mathrm{tw}(G) \leq 3k(n-1) - 1$.

Another interesting relation between clique-width and tree-width follows from a connection, due to Oum [70], between the branch-width of a graph and the rank-width of its incidence graph which we will present at the end of Section 5.4.2.

As seen in the previous section, the notion of tree-width is preserved by taking subgraphs, induced subgraphs, minors, and other transformations. Clique-width is less robust. It is easily seen that clique-width is preserved under taking induced subgraphs. But it is not preserved under taking arbitrary subgraphs and hence not preserved under taking minors. For instance, cliques have clique-width 2 but every graph is a subgraph of a clique and we know that there are graphs of arbitrarily high clique-width.

Proposition 5.4.8 *(i) If G is a graph and H is an induced subgraph of G, then* $\mathrm{cw}(H) \leq \mathrm{cw}(G)$.

(ii) Clique-width is not preserved under taking subgraphs and hence not preserved under taking minors. That is, there are graphs G and $H \subseteq G$ with $\mathrm{cw}(H) > \mathrm{cw}(G)$ *and the difference can be arbitrarily large.*

We close this section with a negative result concerning the complexity of deciding clique-width and related measures. Gurski and Wanke showed that deciding the NLC-width of a graph is NP-complete. For clique-width, this was shown by Fellows, Rosamond, Rotics and Szeider.

Theorem 5.4.9 *1. Given a graph G and an integer k, the problem to decide whether G has NLC-width at most k is NP-complete (see [56]).*

2. Given a graph G and an integer k, the problem to decide whether G has clique-width at most k is NP-complete (see [42]).

However, as we will see in the next section, there are FPT-algorithms, parametrized by the clique-width, to compute an approximate clique-decomposition of a given graph.

Finally, we mention a result by Espelage, Gurski and Wanke [41], that the clique-width of a graph can be computed in linear time on graph classes of bounded tree-width.

---

[5] In [91] Wanke introduced the notion of k-node label controlled graphs (k-NLC). They are defined by similar operations as in k-expressions and for every graph G we have $\mathrm{cw}(G) \leq \mathrm{nlc}(G) \leq 2 \cdot \mathrm{cw}(G)$, where $\mathrm{nlc}(G)$ denotes the NLC-width. The result in [55] is actually stated and proved in terms of NLC-width.

5.4.2 Rank-Width

In this section we consider an alternative characterisation of graph classes of bounded clique-width – the *rank-width* of a graph. Rank-width is a special case of abstract *branch-decompositions* of connectivity functions which we present first. Another special case of this abstract notion is the *branch-width of graphs*, a notion that is equivalent up to a small constant factor to tree-width.

Branch-Decompositions of Connectivity Functions

Let M be a finite non-empty set and $f : 2^M \to \mathbb{R}$ be a function. A *branch-* *decomposition* of the pair (M, f) is a pair (T, β) consisting of a binary tree *T* and a bijection $\beta : L(T) \to M$ from the set $L(T)$ of leaves of T to M. We inductively define a map $\beta^* : V(T) \to 2^M$ by setting

abstract branch-decomposition

$$\beta^*(t) := \begin{cases} \{\beta(t)\} & \text{if } t \text{ is a leaf} \\ \beta^*(t_1) \cup \beta^*(t_2) & \text{if } t \text{ is an inner node with successors } t_1 \cup t_2. \end{cases}$$

abstract branch-width

The *width* of (T, β) is defined as $\max\{f(\beta^*(t)) : t \in V(T)\}$ and the *branch-width* of (M, f) is defined as the minimal width of any of its branch-decompositions. If M is empty, we define the branch-width of M to be $f(\varnothing)$. Note that in this case, (M, f) does not have a branch-decomposition, as a tree, being connected, cannot be empty.

Of particular interest are branch-decompositions of connectivity functions f which are integer valued, symmetric and submodular. A function $f : 2^M \to \mathbb{R}$ is *symmetric* if $f(A) = f(M \setminus A)$ for all $A \subseteq M$ and it is *submodular* if $f(A) + f(B) \geq f(A \cap B) + f(A \cup B)$ for all $A, B \subseteq M$. Submodular and symmetric connectivity functions are algorithmically particularly well-behaved. Note that if f is symmetric we can take the tree T of a branch-decomposition of (M, f) to be undirected and cubic (i.e. every vertex has degree 1 or 3). We will occasionally do so, for instance in Figure 5.7 below.

symmetric submodular

In [71], Oum and Seymour showed that optimal branch-decompositions of submodular, symmetric, and integer valued connectivity functions can be approximated up to a factor 3 by an fpt-algorithm. Before we can state the result we need to define how the input to such an algorithm is represented. Let $\mathcal{M}$ be a class of pairs (M, f), where $f : 2^M \to \mathbb{N}$ is symmetric and submodular. $\mathcal{M}$ is a *tractable class of connectivity functions* if there is a representation of the pairs $(M, f) \in \mathcal{M}$ such that, given the representation of a pair (M, f), the underlying set M and the values $f(A)$ can be computed in polynomial time for all $A \subseteq M$.

tractable class

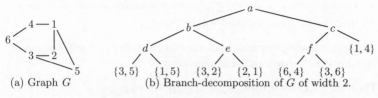

(a) Graph G (b) Branch-decomposition of G of width 2.

Figure 5.7 Branch-decomposition of width 2

We are primarily interested in certain connectivity functions naturally associated with graphs and in this case the graph itself will be the representation.

Theorem 5.4.10 (*Oum, Seymour* [71]) *Let $\mathcal{M}$ be a tractable class of connectivity functions. Then there is an fpt-algorithm that, on input (the representation of) $(\mathcal{M}, f)$ and a parameter k, computes a branch-decomposition of $(\mathcal{M}, f)$ of width at most $3k$ provided that the branch-width of $(\mathcal{M}, f)$ is at most k. If the branch-width of $(\mathcal{M}, f)$ is greater than k, then the algorithm may halt without output or still compute a branch-decomposition of $(\mathcal{M}, f)$ of width $\leq 3k$.*

As a first example of abstract branch-decompositions we consider the branch-width of graphs.

Branch-Width of Graphs

Let G be a graph. The *boundary* ∂F of a set $F \subseteq E(G)$ is defined as the set of vertices incident to an edge in F and also an edge in $E(G) \setminus F$.   *boundary, ∂F*

We define a function $b_G : 2^{E(G)} \to \mathbb{N}$ by $b_G(F) := |\partial F|$ for all $F \subseteq E(G)$. The function b_G is symmetric and submodular. A *branch-decomposition* of G *branch-* is a branch-decomposition of $(E(G), b_G)$ and the *branch-width* bw(G) of G is   *decomposition* defined as the branch-width of $(E(G), b_G)$.   *branch-width*

Example 5.4.11 *Figure 5.7 shows a graph and its branch-decomposition of width 2. For example, $\beta^*(d) = \{\{1,5\}, \{3,5\}\}$ and $\partial \beta^*(d) = \{1,3\}$, as the vertex 5 has no edge to a vertex other than $1, 3$. Similarly, $\partial \beta^*(b) = \partial \beta^*(e) = \partial \beta^*(e) = \{1,3\}$ and $\partial \beta^*(f) = \{3,4\}$.* ⊣

Example 5.4.12 (*Robertson, Seymour* [75]) *1. For every $n \geq 3$, the n-clique K_n has branch-width $\frac{2}{3} \cdot n$.*
2. For all $n \geq 2$, the $n \times n$-grid has branch-width n.

3. *A graph has branch-width 0 if, and only if, it has maximal degree at most 1.*
4. *Trees and cycles have branch-width at most 2.*   ⊣

As the following theorem shows, the branch-width of a graph is equivalent to its tree-width up to a small constant factor.

Theorem 5.4.13 (*Robertson, Seymour* [75]) *For all graphs G*

$$\mathrm{bw}(G) \quad \leq \quad \mathrm{tw}(G) + 1 \quad \leq \quad \max\{2, \frac{3}{2}\,\mathrm{bw}(G)\}.$$

Proof. To show $\mathrm{bw}(G) \leq \mathrm{tw}(G) + 1$, let $\mathcal{T} := (T, (B_t)_{t \in V(T)})$ be a tree-decomposition of G of width $k := \mathrm{tw}(G)$, such that T is a binary tree and every edge of G is covered by exactly one leaf of T. Clearly, given a tree-decomposition of G we can easily find one of the same width with this additional property. We define a branch-decomposition $\mathcal{B} := (T', \beta)$ of G as follows: $T' = T$ and for a leaf $t \in L(T)$ of T we set $\beta(t) := e$, where e is the (unique) edge covered by B_t. We define $\beta^* : V(T) \to 2^{E(G)}$ as before. It is easily seen that for all $t \in V(T)$, $\partial \beta^*(t) \subseteq B_t$ and hence the width of $\mathcal{B}$ is at most $k + 1$.

Conversely, let $\mathcal{B} := (T, \beta)$ be a branch-decomposition of G of width $\mathrm{bw}(G)$. For each $t \in V(T)$ we define $B_t \subseteq V(G)$ as follows. If t is a leaf of T define $B_t := \beta(t)$. Now let t be an inner node with children t_1, t_2. For $i = 1, 2$ let $F_i := \beta^*(t_i)$ and let $F_3 := \big(E(G) \setminus \beta^*(t)\big) = \big(E(G) \setminus (F_1 \cup F_2)\big)$. We define $B_t := \partial F_1 \cup \partial F_2 \cup \partial F_3$.

By construction, $|F_i| \leq \mathrm{bw}(G)$. We claim that for all $v \in V(G)$, if v occurs in some ∂F_i then it also occurs in ∂F_j for some $j \neq i$. For, if $v \in \partial F_i$ then there must be edges $e \in F_i$ and $e' \in E(G) \setminus F_i$ with $v \in e$ and $v \in e'$. Hence, $e' \in F_j$ for some $j \neq i$ and therefore $v \in \partial F_j$. If follows that $|B_t| \leq \max\{2, \frac{3}{2}\,\mathrm{bw}(G)\}$.

Now let $\mathcal{T} := (T, (B_t)_{t \in V(T)})$. It is easily verified that $\mathcal{T}$ is indeed a tree-decomposition of G.[6] Hence, we obtain a tree-decomposition of G of width $\leq \max\{2, \frac{3}{2}\,\mathrm{bw}(G)\} - 1$.   □

In principle one can use the general algorithm from Theorem 5.4.10 to compute approximate branch-decompositions of graphs. However, as for the case of tree-width, better algorithms are known.

Theorem 5.4.14 (*Bodlaender, Thilikos* [7]) *There is an algorithm that, given a graph G and $k \in \mathbb{N}$, computes a branch-decomposition of G of width at*

---

[6] At least if G has no isolated vertices. If it does, add a bag for each isolated vertex.

most k, if it exists, in time $f(k) \cdot |G|$, for some computable function $f : \mathbb{N} \to \mathbb{N}$.

Clique- and Rank-Width

We now turn back to the original goal of giving a different characterisation of clique-width of a graph in terms of its *rank-width*. Recall that the branch-width of a graph is based on a decomposition of its edge set. For rank-width we decompose its vertex set.

Let G be a graph. For $U, W \subseteq V(G)$ we define a $|U| \times |W|$-matrix $M_G(U, W)$ with entries $m_{u,w}$ for $u \in U$ and $w \in W$, where $\qquad M_G(U, W)$

$$
m_{u,w} := \begin{cases} 1 & \text{if } \{u, w\} \in E(G) \\ 0 & \text{otherwise.} \end{cases}
$$

Note that $M_G(V(G), V(G))$ is the adjacency matrix of G. For all $U, W \subseteq V(G)$ let rk $\big(M_G(U, W)\big)$ be its row rank when viewed as a matrix over GF(2). This rk $\big(M_G(U, W)\big)$ induces the following connectivity function $r_G : 2^{V(G)} \to \mathbb{N}$ defined as

$$
r_G(U) := \text{rk}\,\big(M_G(U, V(G) \setminus U)\big)
$$

for $U \subseteq V(G)$. Obviously, r_G is symmetric, as the row and column rank of the matrix coincide. It is left as an exercise to show that it is also submodular.

Definition 5.4.15 *A* rank-decomposition *of a graph G is a branch-decomposition of the pair $(V(G), r_G)$. The* rank-width *of G, in terms* rw(G), *is* rank-width, *the minimal width of any of its rank-decompositions.* rw(G)

Example 5.4.16 *Consider again the graph G from Example 5.3.2 depicted in Figure 5.3. The following is a rank-decomposition of G of width 3.*

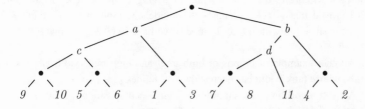

The relevant matrices determining the width of the decomposition are the matrices $M_a, \ldots, M_d$ at the nodes $a, \ldots, d$.

$$M_c := M_G(\{5, 6, 9, 10\}, \{1, 2, 3, 4, 7, 8, 11\}) = \begin{pmatrix} 1 & 0 & 0 & 0 & 0 & 0 & 0 \\ 0 & 0 & 1 & 0 & 0 & 0 & 1 \\ 0 & 0 & 0 & 0 & 0 & 0 & 0 \\ 0 & 0 & 0 & 0 & 0 & 0 & 1 \end{pmatrix}$$

$$M_d := M_G(\{7, 8, 11\}, \{1, 2, 3, 4, 5, 6, 9, 10\}) = \begin{pmatrix} 0 & 0 & 1 & 1 & 0 & 0 & 0 & 0 \\ 0 & 0 & 0 & 1 & 0 & 0 & 0 & 0 \\ 0 & 0 & 1 & 0 & 0 & 1 & 0 & 1 \end{pmatrix}$$

$$M_a := M_G(\{1, 3, 5, 6, 9, 10\}, \{2, 4, 7, 8, 11\}) = \begin{pmatrix} 1 & 0 & 0 & 0 & 0 \\ 1 & 1 & 1 & 0 & 1 \\ 0 & 0 & 0 & 0 & 0 \\ 0 & 0 & 0 & 0 & 1 \\ 0 & 0 & 0 & 0 & 0 \\ 0 & 0 & 0 & 0 & 1 \end{pmatrix}$$

$$M_b := M_G(\{2, 4, 7, 8, 11\}, \{1, 3, 5, 6, 9, 10\}) = \begin{pmatrix} 1 & 1 & 0 & 0 & 0 & 0 \\ 0 & 1 & 0 & 0 & 0 & 0 \\ 0 & 1 & 0 & 0 & 0 & 0 \\ 0 & 0 & 0 & 0 & 0 & 0 \\ 0 & 1 & 0 & 1 & 0 & 1 \end{pmatrix}$$

Obviously, $\mathrm{rk}(M_a) = \mathrm{rk}(M_b) = \mathrm{rk}(M_c) = 3$ and this is the maximal rank occurring in the decomposition. Hence, the decomposition has width 3. ⊣

It is not too hard to see that the rank-width of a graph can be bounded in terms of its branch-width and hence its tree-width. The following theorem due to Oum gives an exact bound.

Theorem 5.4.17 (*Oum* [70]) $\mathrm{rw}(G) \leq \max\{1, \mathrm{bw}(G)\}$ *for all graphs G.*

It is easily seen that the rank of width a complete graph is 1 (all entries in all matrices are 1). Hence, there can be an arbitrarily large difference between the rank-width and the branch-width of a graph. On the other hand, Oum [70] proved that if $I(K_n)$ denotes the incidence graph of the n-clique K_n, then for all $n \geq 3$ with $n = 0, 1 \mod 3$ we have $\mathrm{rw}(I(K_n)) = \mathrm{bw}(I(K_n)) = \lceil \frac{2}{3} \cdot n \rceil$.

Another example of graphs of high tree- and high rank-width are $n \times n$-grids, whose rank-width has been shown by Jelínek [58] to be n.

An fpt-algorithm for computing rank-decompositions follows from Theorem 5.4.10 but more efficient algorithms are known.

Theorem 5.4.18 (*Hlineny, Oum* [40]) *There is an algorithm that, given a graph G and $k \in \mathbb{N}$, computes a rank-decomposition of G of width at most k, provided* $\mathrm{rw}(G) \leq k$, *in time* $f(k) \cdot |G|^3$, *for some computable function* $f : \mathbb{N} \to \mathbb{N}$.

Oum and Seymour [71] established the following connection between rank-width and clique-width:

$$\mathrm{rw}(G) \leq \mathrm{cw}(G) \leq 2^{\mathrm{rw}(G)+1} - 1.$$

In particular, a class of graphs has bounded clique-width if, and only if, it has bounded rank-width (see [71]). Together with Theorem 5.4.18 this yields a parameterized algorithm for computing approximate clique-decompositions of graphs.

We have already seen that clique-width and tree-width and hence branch-width of graphs can differ arbitrarily and this clearly extends to rank-width. However, Oum [70] established the following relation between the branch-width of a graph and the rank-width of the incidence graph.

$$\mathrm{bw}(G) - 1 \leq \mathrm{rw}(I(G)) \leq \mathrm{bw}(G)$$

5.4.3 Monadic Second-Order Logic and Bounded Clique-Width

In this section we aim at extending Courcelle's and Seese's theorems from tree-width to clique-width. As in Section 5.3, we will do so by a reduction to MSO model-checking and satisfiability on trees. In particular, we show next that for each k the class of graphs of clique-width k can be interpreted in the class of coloured trees for a suitable set of colours depending on k. The idea is simple: the class of graphs of clique-width k is the class of graphs generated by k-expressions whose syntax trees will be the class of trees we are looking for. Hence, let

$$\Sigma_k := \{\mathbf{0}, \dots, \mathbf{k-1}, \oplus, \mathrm{edge}_{i,j}, \mathrm{rename}_{i \to j} : 0 \leq i \neq j < k\}$$

be the symbols used in k-expressions and let $\mathcal{T}_{\Sigma_k}$ be the class of all Σ_k-labelled directed trees. Obviously, not every Σ_k-labelled tree is the syntax tree of a k-expression. However, every Σ_k-labelled directed tree such that the symbol $\oplus$ occurs precisely at the nodes with two successors, no node has more than two successors and the leaves are precisely the nodes labelled by a symbol from $\{\mathbf{0}, \dots, \mathbf{k-1}\}$ are syntax trees of k-expressions. These conditions are easily expressed by an MSO-sentence φ_{valid}. Hence, for all $T \in \mathcal{T}_{\Sigma_k}$, $T \models \varphi_{valid}$ if, and only if, T is the syntax tree of a k-expression. The formula φ_{valid} is one part of an interpretation $\Gamma_k := \left(\varphi_{univ}, \varphi_{valid}, \varphi_E(x, y) \right)$ from Σ_k-labelled trees to graphs of clique-width at most k.

The formula $\varphi_{univ}(x)$ defining the universe of a graph generated by a k-expression coded in a tree T is trivial: $\varphi_{univ}(x)$ just defines the set of leaves.

Finally, we have to define the formula $\varphi_E(x, y)$ such that for all $T \in \mathcal{T}_\Sigma$ with $T \models \varphi_{valid}$ and all leaves $u, v \in V(T)$ we have $T \models \varphi_E(u, v)$ if, and only if, there is an edge between u and v in the graph G generated by T. Note that such an edge exists if, and only if, there is a common ancestor t of u and v in T labelled by edge_{i-j}, for some $0 \le i \ne j < k$, so that at the node t, one of u, v has colour i and the other the colour j. To check this, we only need to look at the unique path from t to u (and v respectively) and keep track of how the colour of u (resp. v) changes along this path. This can easily be formalised in MSO by a formula $\varphi_E(x, y)$ as required. Hence, the triple $\Gamma_k := (\varphi_{univ}(x), \varphi_{valid}, \varphi_E(x, y))$ is an interpretation from σ_k-structures to graphs, where $\sigma_k := \{E\} \cup \Sigma_k$ is the signature of Σ_k-labelled trees.

The interpretation is the key to tractability results for MSO model-checking and satisfiability. We consider model-checking first and prove the following extension of Courcelle's theorem. It was first proved by Courcelle in terms of certain graph grammars (see [12, 14]) and then by Courcelle, Makowski and Rotics for graph classes of bounded clique-width.

Theorem 5.4.19 (*Courcelle, Makowski, Rotics* [18]) *Let $\mathcal{C}$ be a class of graphs of bounded clique-width. Then the model-checking problem for MSO on $\mathcal{C}$ is fixed-parameter tractable.*

Proof. Let $\varphi \in$ MSO be fixed and let k be an upper bound for the clique-width of the graphs in $\mathcal{C}$. Given a graph G we first compute a k-expression ϑ generating G. This can be done in polynomial time (see Section 5.4.2). Let T be the Σ_k-labelled syntax tree of ϑ. We can now test whether $T \models \Gamma_k(\varphi)$. $\square$

We now consider the satisfiability problem for monadic second-order logic.

Theorem 5.4.20 *For every k, the MSO-theory of the class CW_k of graphs of clique-width at most k is decidable.*

Proof. Let $\varphi \in$ MSO[$\{E\}$] be given. By the interpretation lemma, φ is valid in CW_k if, and only if, $\Gamma_k(\varphi) \in$ MSO[σ_k] is valid in the class $\{T \in \mathcal{T}_\Sigma : T \models \varphi_{valid}\}$ if, and only if, $\Gamma_k(\varphi) \wedge \varphi_{valid}$ is valid in the class of finite Σ_k-trees. The latter is well known to be decidable [31, 85]. $\square$

Seese conjectured a kind of converse to the theorem, the famous Seese conjecture [81].

Conjecture 5.4.21 (Seese's conjecture) *Every class $\mathcal{C}$ of structures with decidable MSO_1-theory has bounded clique-width.*

This conjecture can be rephrased in terms of MSO-interpretations using the following result due to Engelfriet and V. van Oostrom and also Courcelle and Engelfriet.

Lemma 5.4.22 ([39, 16]) *A class of graphs has bounded clique-width if, and only if, it is interpretable in the class of coloured trees for some suitable set of colours.*

Note that these papers use so-called MSO-transductions instead of interpretations. An MSO-transduction is essentially the same as an interpretation except that the formulas are allowed to have free second-order variables, the parameters. A graph is then interpretable in a tree if there is an interpretation of the parameters by sets of tree-nodes satisfying the formulas in the MSO-transduction. Hence, the parameters play exactly the same role as the colours of the trees we use here. As the colours/parameters in our context are the symbols of k-expressions, we prefer to have them as labels of the syntax trees rather than as free variables in the interpretation.

Using the previous lemma we can rephrase Seese's conjecture as follows:

Conjecture 5.4.23 (Seese's conjecture) *Every class $\mathcal{C}$ of structures with decidable MSO_1-theory is MSO-interpretable in the class of coloured trees for some set of colours.*

In [20], Courcelle and Oum prove the following weakening of the conjecture. Let C_2MSO be the extension of MSO by atoms $\text{EVEN}(X)$, where X is a monadic second-order variable, stating that the interpretation of X has even cardinality. Hence, C_2MSO extends MSO by counting modulo 2.

Theorem 5.4.24 (*Courcelle, Oum* [20]) *Every class of graphs with a decidable C_2MSO theory has bounded clique-width, i.e. is interpretable in a class of coloured trees.*

Note that the theorem is weaker than Seese's conjecture as there are less classes of graphs whose C_2MSO theory is decidable than there are classes of graphs with a decidable MSO-theory.

5.4.4 MSO Model-Checking Beyond Tree- and Clique-Width

In the previous section we showed that the model-checking problem for monadic second-order logic is fixed-parameter tractable on classes of graphs with bounded tree- or clique-width. There is not much hope for extending these results to other or larger classes of graphs such as planar graphs or graphs of bounded degree. This follows immediately from the following theorem by Garey, Johnson and Stockmeyer and the fact that 3-colourability is MSO-definable.

Theorem 5.4.25 (*Garey, Johnson, Stockmeyer* [49]) *3-colourability is* NP-*complete on the class of planar graphs of degree at most* 4.

We will see much stronger intractability results for MSO$_2$ in Section 5.8 below. However, first-order logic is tractable on many more classes of graphs. For instance, Seese [82] showed that first-order logic admits linear time model-checking (for a fixed formula) on any class of graphs of bounded degree. The same complexity bound was later obtained by Frick and Grohe [47] for planar graphs and classes of graphs of bounded *local tree-width*, a notion that properly extends both planarity and bounded degree (see below).

The most general results in this respect are fixed-parameter algorithms for first-order model-checking on H-minor free graphs and an extension thereof, called locally excluded minors. These results make heavy use of concepts and results developed by Robertson and Seymour in their celebrated proof of the graph minor theorem. In the next section, we will therefore give a brief overview of the relevant concepts of the graph minor theory used in the proofs. One such theorem, the excluded grid theorem, will be used later to prove the converse of Seese's theorem mentioned above. This will be the topic of Section 5.6. We return to first-order model-checking in Section 5.7.

5.5 Graph Minors

In this section we present relevant terminology and results from graph minor theory used later in the paper. Most of the results were developed in Robertson and Seymour's celebrated proof of the graph minor theorem (Theorem 5.5.2 below) presented in a series [76] of 23 papers, with additions and improvements by other authors.

5.5.1 Minors and Minor Ideals

G/e Let G be a graph and $e := \{v, w\} \in E(G)$ be an edge. The graph G/e obtained
contraction   from G by *contracting* the edge e is the graph obtained from G by removing e, identifying its two endpoints, and possibly removing parallel edges. Formally,

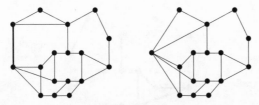

Figure 5.8 Contracting an edge

G/e is defined by

$$V(G/e) := V(G) \setminus \{v, w\} \cup \{x_e\},$$

where x_e is a new vertex, and

$$E(G/e) := \begin{matrix} \left(E(G) \setminus \left\{\{u, u'\} : \{u, u'\} \cap e \neq \varnothing \right\} \right) \cup \\ \left\{\{u, x_e\} : u \in V(G/e) \text{ and} \{u, v\} \in E(G) \text{ or } \{u, w\} \in E(G) \right\}. \end{matrix}$$

Figure 5.8 illustrates edge contraction.

A graph H is a *minor* of a graph G if H can be obtained from G by deleting   *minor*
vertices and edges and contracting edges. We write $H \preccurlyeq G$ to denote that H is
isomorphic to a minor of G.

An alternative definition of minors is in terms of minor maps. A *minor map*
from H to G is a function μ that associates with every vertex $v \in V(H)$ a
connected subgraph $\mu(v) \subseteq G$ and with every edge $e \in E(H)$ an edge $\mu(e) \in$
$E(G)$ such that

- if $u, v \in V(H)$ and $u \neq v$ then $\mu(v)$ and $\mu(u)$ are vertex disjoint and
- if $e := \{u, v\} \in E(H)$ then $\mu(e) := \{u', v'\}$ for some $u' \in V(\mu(u))$ and $v' \in V(\mu(v))$.

The subgraph $G_\mu \subseteq G$ with

$$V(G_\mu) := \bigcup \left\{ V(\mu(v)) : v \in V(H) \right\}$$

and

$$E(G_\mu) := \bigcup \left\{ E(\mu(v)) : v \in V(H) \right\} \cup \left\{ \mu(e) : e \in E(H) \right\}$$

is called a *model* or *image* of H in G. In graph theory literature, the term model   *model, image*
is commonly used. We prefer the name image here to avoid confusion with
logical models. Figure 5.9 illustrates an image of K_5 in a graph G.

It is easily seen that we can always choose an image of H in G so that each
vertex is represented by a tree in G.

Figure 5.9 Image of K_5 in a graph G

subdivision   Let G, H be graphs. G is a *subdivision* of H if H can be obtained from G by replacing some edges in G by paths which are pairwise internally vertex disjoint, i.e. H can be constructed from G by repeatedly *subdividing* edges. If a subgraph of G is isomorphic to a subdivision of H, then H is called a *topological* *topological minor* of G. Now suppose $H \preccurlyeq G$ and H has maximal degree 3. *minor* Let μ be a minor map from H into G so that the image of all vertices of H are trees in G. Then each of these trees has at most 3 leaves and hence at most one vertex of degree more than two. It follows that every graph H of maximal degree ≤ 3 that is a minor of G also is a topological minor of G.

Lemma 5.5.1 *Let H, G be graphs. If $\Delta(H) \leq 3$ and $H \preccurlyeq G$, then H is a topological minor of G.*

If $H \npreccurlyeq G$, we say that H is a *forbidden* minor of G, or that G *excludes* *Excl(H)* H. For any graph H let $Excl(H) := \{G : H \npreccurlyeq G\}$ be the class of graphs not *Excl(H)* containing H as a minor. Analogously, if $\mathcal{H}$ is a set of graphs, then $Excl(\mathcal{H}) := \bigcap \{Excl(H) : H \in \mathcal{H}\}$ is the class of graphs not containing any member of $\mathcal{H}$ as a minor.

minor ideal   A class $\mathcal{C}$ of graphs is a *minor ideal* if for all $G \in \mathcal{C}$ and $H \preccurlyeq G$ also $H \in \mathcal{C}$. *proper* It is *proper* if it is not the class of all graphs.
minor ideal   A class $\mathcal{C}$ is *characterised* by a class $\mathcal{F}$ of graphs if $\mathcal{C} = Excl(\mathcal{F})$. Note that any minor ideal $\mathcal{C}$ can be characterised by a class of excluded minors, e.g. $\mathcal{C} = Excl(\text{GRAPHS} \setminus \mathcal{C})$. As the main result of their fundamental work on graph minors, Robertson and Seymour proved that any minor ideal can in fact be characterised by a *finite* set of forbidden minors.

Theorem 5.5.2 (*Robertson, Seymour* [80]) *For every minor ideal $\mathcal{C}$ there is a finite set $\mathcal{F}$ of graphs such that $\mathcal{C} = Excl(\mathcal{F})$.*

There are many natural examples of minor ideals.

- Every cycle can be contracted to a triangle. Hence, $Excl(K_3)$ is the class of acyclic graphs.
- Kuratowski's theorem [61] (or rather a variant established by Wagner [90]) implies that planar graphs are characterised by excluding $K_{3,3}$ and K_5.
- *Series-parallel* graphs and *outerplanar* graphs exclude K_4. It can be shown that $Excl(K_4)$ is the class of subgraphs of series-parallel graphs and the class of outerplanar graphs is characterised by $Excl(K_4, K_{2,3})$. (See e.g. [30, Exercises 7.32 and 4.20].)
- The class of graphs *not* having k vertex disjoint cycles, for any fixed $k \in \mathbb{N}$. For $k \in \mathbb{N}$ let T_k be the graph consisting of k disjoint copies of a triangle. Clearly, every graph containing k vertex disjoint cycles contains T_k as a minor. Conversely, every graph containing T_k as a minor also contains k vertex disjoint cycles. Hence the class C_k of graphs *not* having k disjoint cycles is characterised by T_k.

It is easily seen that for each $k \in \mathbb{N}$ the class $\mathcal{T}_k$ of graphs of tree-width at most k and the class $\mathcal{B}_k$ of graphs of branch-width at most k are minor ideals and so is the class of graphs of genus at most k. Finally, let us mention another famous example of a minor ideal: the class of *knotlessly* embeddable graphs.

On the other hand, the class of graphs of clique-width at most k is not minor closed and hence not a minor ideal. Also, the class of graphs of crossing number $k \geq 1$ is *not* minor closed.

Robertson and Seymour also proved that for any fixed graph H, testing if a graph G contains H as a minor can be done in cubic time (we will say more about this later in this section). Hence, combining this minor test with Theorem 5.5.2 implies that every minor-ideal can be decided in cubic time.

Corollary 5.5.3 *Every minor ideal can be decided in cubic time.*

The various concepts and results developed in the course of the proof of Theorem 5.5.2 have sparked of a rich algorithmic theory of graphs based on structural restrictions of instances. We have already hinted at the algorithmic theory of graphs of bounded tree-width. However, the algorithmic applications of the graph minor theory developed by Robertson and Seymour extend far beyond tree-like graphs. In the following two sections we present some of the results and methods with implications for algorithms and model-checking on graphs.

However, the following can only give a glimpse into the deep results underlying the proof of the graph minor theorem – we will not even be able to state the relevant results in full detail let alone attempt to prove them. While we are trying to give an intuitive account of the results and proof methods, we will necessarily have to be brief and the presentation may not always reflect the actual proofs.

5.5.2 Disjoint Paths and the Trinity Lemma

Let us try to prove Theorem 5.5.2. Clearly, the statement of the theorem is equivalent to the statement that in every infinite class of finite graphs one graph is a minor of another. Let $\mathcal{C} := \{H, G_1, G_2, \dots\}$ be an infinite class of finite graphs. If H is a minor of some G_i, then the claim is trivially satisfied by H. Hence, the only interesting case is when no $G_i \in \mathcal{C}$ contains H as a minor. For this reason, much of the theory developed by Robertson and Seymour deals with graphs not containing another fixed graph H as a minor. We refer to such *H-minor free* graphs as *H-minor free*. Clearly, if G is H-minor free, then G also excludes a clique K_k as a minor, for instance taking $k := |V(H)|$. Let us fix k for the rest of the section.

The key to studying the structure of K_k-minor free graphs is the following theorem, proved by Robertson and Seymour in [74]. Recall from Section 5.2 that $G_{k \times k}$ denotes the $k \times k$-grid.

Theorem 5.5.4 (Excluded Grid Theorem [74]) *There is a computable function $f : \mathbb{N} \to \mathbb{N}$ such that every graph of tree-width at least $f(k)$ contains $G_{k \times k}$ as a minor.*

We refer to [30] for a proof of this theorem. As every planar graph is a minor of a suitably large grid, the theorem implies – is equivalent, in fact – to the following statement.

Corollary 5.5.5 *For all H, the class $Excl(H)$ of H-minor free graphs has bounded tree-width if, and only if, H is planar.*

The function f in the original proof of Theorem 5.5.4 was huge. In [73], Robertson, Seymour and Thomas significantly improved the bounds on f to 20^{2k^5}. However, no matching lower bounds have been established and it is conjectured that the actual bound may be as small as polynomial in k. For planar graphs G a much better bound can be obtained.

Theorem 5.5.6 (*Robertson, Seymour, Thomas* [73]) *Every planar graph with no $k \times k$-grid minor has tree-width $\leq 6k - 5$.*

For branch-width a slightly tighter bound has been established: every planar graph of branch-width at least $4k - 3$ contains a $k \times k$-grid minor (see [73]). Whereas it is still open whether optimal tree-decompositions of planar graphs can be computed in polynomial time, in [84] Seymour and Thomas proved

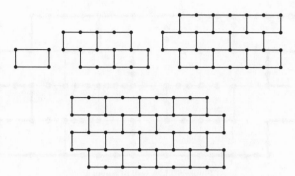

Figure 5.10 Elementary walls of height 1–4

that optimal branch-decompositions of planar graphs can be computed in time $\mathcal{O}(n^4)$. This has later been improved to $\mathcal{O}(n^3)$ by Gu and Tamaki [54]. It should be noted that these algorithms do not contain any large hidden constants and perform reasonably well in practise. Optimal branch-decompositions of planar graphs with up to 50.000 edges have been computed by actual implementations of the algorithms (see e.g. [5]).

To give an application of the grid-theorem on planar graphs, we note that it implies an $2^{\mathcal{O}(\sqrt{k})} \cdot n^c$ algorithm, for some $c \in \mathbb{N}$, for deciding whether a planar graph has a path of length k. For this, use an $\mathcal{O}(n^3)$ algorithm for testing whether a given planar graph G has branch-width at most $4\sqrt{k} - 3$. If so, then one can compute a suitable branch-decomposition and use dynamic programming to decide whether a path of length k exists. Otherwise, the planar grid theorem tells us that the graph contains a $\sqrt{k} \times \sqrt{k}$ grid as a minor and hence a path of length at least k following the grid structure. A similar algorithmic idea has found numerous applications, for instance on H-minor free graphs, in the form of *bidimensionality theory*. See e.g. [25, 32, 27, 33, 24, 26, 28] and references therein.

For the rest of this section we will work with a somewhat simpler structure than grids, called *walls*. *wall*

An *elementary wall* is a graph as displayed in Figure 5.10. A wall of height h is a subdivision of an elementary wall of height h. See Figure 5.11 for a wall of height 4. The induced cycles of a wall, i.e. the cycles of length 6 in an elementary wall or their subdivisions in general walls, are called the *bricks* of *brick* the wall. We assign coordinates $(i, j) = (row, col)$ to the bricks of a wall. The brick in the lower left corner is assigned $(1, 1)$, its neighbour to the right $(1, 2)$, the brick just above it $(2, 1)$ and so on. The *central brick* of H is the brick with *central brick*

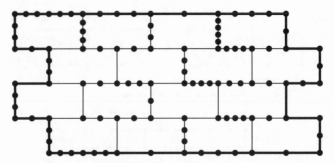

Figure 5.11 A wall of height 4

central coordinates ($\lceil h/2 \rceil$, $\lceil h/2 \rceil$). A *central vertex* of a wall is a vertex contained in
vertex the central brick but not in its neighbours to the left or right.
perimeter   The outermost (non-induced) cycle of a wall W is called its *perimeter*.
   Clearly, every large grid contains a large wall as a subgraph and conversely
every large wall contains a large grid as a minor. The main advantage of working
with walls rather than grids is that if G contains an elementary wall as a minor
then, by Lemma 5.5.1, it contains a wall of the same height as a subgraph.
   Let us come back to the analysis of the structure of graphs. Let t be a bound
on the tree-width we want to consider. If G has tree-width at most t, then it is
sufficiently tree-like and its structure is well understood. So suppose G has large
tree-width. By the Excluded Grid Theorem 5.5.4, we know that G contains a
large wall W as a subgraph. We can use W as a drawing board on which we
draw the rest of the graph G. Clearly, as G is not required to be planar, this
"drawing" will not necessarily be plane, i.e. edges may cross. In particular,
edges or paths may span over different bricks of the wall. This is called a
"crossing". More formally, a crossing consists of two pairwise vertex-disjoint
paths with endpoints v_1, v_3 and v_2, v_4 such that v_1, v_2, v_3, v_4 occur clockwise
in this order on some cycle of the grid. Figure 5.12 illustrates the concept of
crossings.
   Crossings are important for our purpose. For, if G contains many crossings
which, in addition, are sufficiently far apart from each other on the wall used to
draw G, then we can use the crossings to find a large clique minor of the graph.
To see this, take a large clique and draw it "flat" on the wall W. Necessarily
(unless your clique has less than five vertices) some of the edges in the clique
will cross each other. However, if the wall W is large enough and there are
sufficiently many crossings far apart from each other, then we can replace the
edges of the clique by disjoint paths in G so that edges that cross are replaced

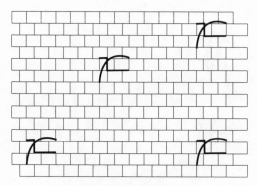

Figure 5.12 Crossings in a graph

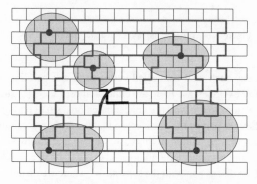

Figure 5.13 A K_5-minor in a wall with one crossing

by disjoint paths that cross each other using a "crossing" in the drawing of G. The following Figure 5.13 illustrates this with K_5 and one crossing. The grey areas are (essentially) the parts that are being contracted for each vertex in the clique.

Hence, if W is large enough and there are many crossings pairwise far apart in W, then G contains a large clique minor. So, how does a graph G drawn on a large wall look like if it does not contain a large clique minor?

As explained before, all but a small number of crossings must be grouped together in a bounded number of small parts of the wall. These regions with many crossings are called *vortices*. Further, there can be some vertices which *vortex* are very well connected to the rest of the graph, i.e. a set X of vertices that have edges to arbitrary vertices in the graph, where edges can be replaced by paths of arbitrary length. The vertices in X are called *apices* (see Figure 5.14). *apex*

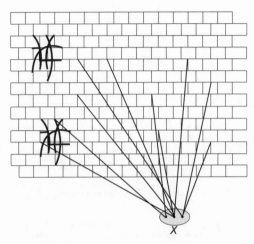

Figure 5.14 *Vortices* and *apices* in a graph drawn on a wall

However, any such well-connected vertex in X can be used as a crossing and hence, if G excludes K_k, there are either at most $|X| < \binom{k}{2}$ such elements, or their connections to the wall are concentrated on a small part of the wall W (and hence they are part of the vortices) so that the crossings cannot be used to route the edges of a K_k-minor. In this case, we will find a subwall of W which is still "large" and is connected only to a subset of X of size $< \binom{k}{2}$. Hence, we can continue the discussion with the subwall W' where we do not have vortices and only a bounded number of apices.

Besides the apices, there can be other parts of the graph with direct connections to the interior of the wall,[7] which do not induce any further crossings. We call these *extensions*. Essentially, an extension is a subgraph D of G that is connected to the wall only within a brick and only with at most 3 vertices. This is important as with three vertices the extensions cannot induce further crossings in the wall.

Furthermore, we can assume that the tree-width of any such extension is bounded, as otherwise we could forget about the rest of the graph and do the same analysis within the extension, either producing a large clique minor or a large wall with vortices, apices and extensions. Note, though, that the apices may have connections to the extensions. See Figure 5.15 for an illustration.

---

[7] There may also be parts of the graph connected to the wall only through its perimeter. These parts are not relevant here but we come back to this in the next section.

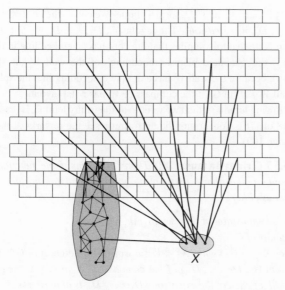

Figure 5.15 Apices, extensions and connections within the subwall W'

The discussion so far presents the main ideas in the proof of the next lemma, one of the important results in the Graph Minor Series. To state it precisely, we need some further notation.

For a subgraph D of a graph G, we let $\partial^G D$ be the set of all vertices of D that are incident with an edge in $E(G) \setminus E(D)$. In the following, let W be a wall of height at least 2 in a graph G and let P be the perimeter of W, i.e. the boundary cycle of W. Let K' be the unique connected component of $G \setminus P$ that contains $W \setminus P$. The graph $K = K' \cup P$ is called the *compass* of W in G. A *layout of K (with respect to the wall W in G)* is a family $(C, D_1, \ldots, D_m)$ of connected subgraphs of K such that:

1. $K = C \cup D_1 \cup \ldots \cup D_m$,
2. $W \subseteq C$ and there is no separation (X, Y) of C of order ≤ 3 with $V(W) \subseteq X$ and $Y \setminus X \neq \varnothing$,
3. $\partial^G D_i \subseteq V(C)$ for all $i \in \{1, \ldots, m\}$,
4. $|\partial^G D_i| \leq 3$ for all $i \in \{1, \ldots, m\}$,
5. $\partial^G D_i \neq \partial^G D_j$ for all $i \neq j \in \{1, \ldots, m\}$.

We let $\overline{C}$ be the graph obtained from C by adding new vertices $d_1, \ldots, d_m$ and, for $1 \leq i \leq m$, edges between d_i to the vertices in $\partial^G D_i$ and edges between all vertices in $\partial^G D_i$. Hence, for each $i \in \{1, \ldots, m\}$, the vertex d_i together with the (at most 3) vertices in $\partial^G D_i$ form a clique. We call $\overline{C}$ the *core* of the layout and $D_1, \ldots, D_m$ its *extensions*. The layout $(C, D_1, \ldots, D_m)$ is *flat* if its core $\overline{C}$ is planar. Note that this implies that the core has an embedding in the plane that extends the "standard planar embedding" of the wall W (as shown in Figure 5.10), because the wall W has a unique embedding into the sphere. We call the wall W *flat* (in G) if the compass of W has a flat layout.

The following lemma, which we refer to as the trinity lemma, is (essentially) Lemma 9.8 of [78]. Concerning the uniformity, see the remarks at the end of [78] (on page 109).

Lemma 5.5.7 (Trinity Lemma [78]) *There are computable functions $f, g :$ $\mathbb{N}^2 \to \mathbb{N}$ and an algorithm A that, given a graph G and non-negative integers k, h, computes either*

1. *a tree-decomposition of G of width $f(k, h)$,*
2. *a K_k-minor of G, or*
3. *a subset $X \subseteq V(G)$ with $|X| < \binom{k}{2}$, a wall W of height h in $G \setminus X$, and a flat layout $(C, D_1, \ldots, D_m)$ of the compass of W in $G \setminus X$ such that the tree-width of each of the extensions $D_1, \ldots, D_m$ is at most $f(k, h)$.*

Furthermore, the running time of the algorithm is bounded by $g(k, h) \cdot |V(G)|^2$.

Using the trinity lemma, we can now sketch the proof of the following theorem due to Robertson and Seymour [78].

Theorem 5.5.8 (*Robertson, Seymour* [78]) *The following problem is fixed-parameter tractable with a cubic fpt algorithm.*

p-DISJOINT-PATHS	
Input:	Graph $G, s_1, \ldots, s_k, t_1, \ldots, t_k \in V(G)$.
Parameter:	k.
Problem:	Are there k vertex disjoint paths connecting s_i and t_i, $1 \le i \le k$?

The idea of the algorithm is as follows. Apply the trinity lemma on G for suitable values of k and h. If G has tree-width $\le f(k, h)$, then the disjoint paths problem can be solved by standard techniques using dynamic programming (or by formalising the problem in MSO and using Courcelle's theorem). Otherwise, if G contains a large clique minor (say at least K_{3k}), then we can do the following. To simplify the presentation, let us assume that G actually contains the $3k$-clique as a subgraph. If there are $2k$ vertex disjoint paths connecting $\{s_1, \ldots, s_k, t_1, \ldots, t_k\}$ to the clique, then these paths together with the edges of the clique yield the k vertex-disjoint paths connecting s_i, t_i as desired.

Otherwise, by Menger's theorem, there is a separator $X \subseteq V(G)$ of size at most $2k$ separating the clique and (part of) the $\{s_i, t_i : 1 \leq i \leq k\}$. But now, the problem can be reduced to a constant number of disjoint paths problems on smaller subgraphs, trying to connect s_i, t_i with all possible combinations of elements in the separator.

If G does not contain the clique as a subgraph but as a minor, then the argument becomes considerably more complicated, but can still be done. Hence, the case where G contains a large enough clique minor can be solved efficiently.

Finally, consider the third case of the trinity lemma, where G contains a large wall W and we are given a flat layout of W, its extensions and the apices. This is the tricky bit. However, one can show that if W is large enough, then it must contain a subwall W', which is still large, does not contain any of the s_i's or t_i's and is "homogeneous" with respect to the apices. Informally, homogeneous means that every type of a small part of the wall with respect to the apices is realised sufficiently often all over the subwall W'. In [78], Robertson and Seymour show how such a homogeneous subwall can be constructed efficiently. To simplify the presentation, assume that W' has actually no direct connection to the apices (other than those using vertices of $W \setminus W'$). Now suppose there are k vertex-disjoint paths connecting s_i and t_i, $1 \leq i \leq k$. Some of these paths may use parts of W'. As none of the endpoints s_i, t_i is in W', the paths merely cross W', although they may do so in a rather irregular and complicated way. However, it can be shown that if W' is homogeneous and large enough, then any such set of paths can be rerouted so as to avoid a central vertex v of the wall (recall from above that the central vertices are those in the middle of the wall). This implies, that k vertex-disjoint paths connecting s_i, t_i exist in G if, and only if, such paths exist in $G - v$. Hence, we can remove the central vertex v and start the whole procedure again on the smaller graph.

It seems intuitively obvious that on a very large wall, everything that can be routed through the wall can be routed without using the central vertex. A formal proof of this is extremely complicated and uses a major part of the deep structure theory developed in the graph minor series.

As mentioned above, the solution to the disjoint paths problem was given by Robertson and Seymour in [78]. In fact, they solve the following more general problem. A *rooted graph* $(G, v_1, \ldots, v_k)$ is a graph G together with vertices $v_i \in V(G)$. A rooted graph $(H, t_1, \ldots, t_k)$ is a minor of $(G, v_1, \ldots, v_k)$, if there is a minor map μ from H to G such that $v_i \in \mu(t_i)$ for all $1 \leq i \leq k$.

Theorem 5.5.9 (*Robertson, Seymour* [78]) *The following problem is fixed-parameter tractable with a cubic fpt algorithm.*

---

p-ROOTED-MINOR
    Input: Rooted graphs $(G, v_1, \ldots, v_k)$, $(H, t_1, \ldots, t_k)$.
    Parameter: k.
    Problem: Is $(H, t_1, \ldots, t_k)$ a minor of $(G, v_1, \ldots, v_k)$?

Clearly, this implies Theorem 5.5.8 and also Corollary 5.5.3. This is a truly remarkable consequence of the proof of the graph minor theorem. Note, however, that the statement is purely existential. For every minor ideal there is a finite set of excluded minors and for each member H of the set we can decide in cubic time, whether a graph G contains H as a minor. The theory does not yield an algorithm to compute a set of excluded minors and hence it only states the existence of a polynomial time membership test but not an actual algorithm. We come back to this in Section 5.5.4 where we consider ways in which to overcome this non-constructive element in the theory.

5.5.3 The Structure of H-Minor Free Graphs

The proof of the graph minor theorem relies on a structure theory for graphs G excluding a fixed graph H as a minor. We have already seen some of the results developed in the proof. In this section we focus on describing the structure of graphs in terms of simple building blocks into which they can be decomposed.

The key to the decomposition theorem we are going to describe is once again the grid theorem, or in this case the trinity lemma as described in the previous section. Clearly, as G excludes a fixed graph H as a minor, it is obvious that, if we choose the values for k and h correctly, of the three cases of the trinity lemma, the second is impossible: if G excludes H it cannot contain a large clique minor. Further, if G has small tree-width, then it can be decomposed into subgraphs of constant size. Hence, we primarily have to deal with the third case, where G has large tree-width but does not contain a large clique minor.

Recall our exploration of the trinity lemma in the previous section. Let us assume that G is highly connected. If not, we first decompose it into parts that are highly connected. We will come back to this later.

As G has high tree-width it must contain a large wall as a subdivision. This wall may contain "crossings", in particular there may be a bounded number of apices and vortices. As explained before, apart from the vortices and the apices, the rest of the graph, the extensions, must fit nicely into the planar structure of the wall, i.e. they fit into the individual bricks. So far, however, we only have discussed the interior of the wall. There may be more to the graph, which is connected to the wall only through the perimeter. These connections cannot be too wild, though, as otherwise we would again find a large clique minor.

We can now subdivide the exterior cycle of the wall into a bounded number of regions and glue some of them together. In this way we obtain a graph that can be embedded into a surface of bounded genus: any such surface can be obtained from a convex polygon in the plane by gluing some edges together. Hence, after removing a bounded number of apices and vortices we obtain a graph that can be embedded into a surface of bounded genus. We say that G has *almost bounded genus*. Recall that we assumed that G is highly connected. If it is not, then we can decompose it into pieces with this property. This realisation is the main structural theorem in Robertson and Seymour's proof of the graph minor theorem: *if C is a class of graphs excluding a fixed minor H, then every graph $G \in C$ can be decomposed into graphs that have almost bounded genus.*

We still have to make precise what we mean by "decomposing a graph". Intuitively, we recursively find a small separator in the graph and split the graph along the separator until the remaining graph is highly connected, and hence no such separators can be found. However, by doing so some information is lost. Let G be a graph and X be a small separator. We want to decompose the graph into subgraphs each containing X and a component of $G - X$. Clearly, in a graph obtained from X and a component C of $G \setminus X$, we lose the connections between elements of X through the other components of $G \setminus X$. In particular, elements of X which are far apart in $X \cup C$ can be close together in other components and hence in G. This loss of information in the decomposition process needs to be avoided. A rather drastic approach, which we take here, is to add all possible edges between elements of the separator X, i.e. to turn X into a clique.

Let $T := (T, (B_t)_{t \in V(T)})$ be a tree-decomposition of a graph G and let $t \in V(T)$ be a node with neighbours $t_1, \ldots, t_k$. The *torso* $[B_t]$ of the bag B_t is *torso, $[B_t]$* defined as $G[B_t] \cup \bigcup_{i=1}^{k} K[B_{t_i}]$, where $K[B_{t_i}]$ is the complete graph on the vertex set B_{t_i}. The tree-decomposition T of G is *over a class C* of graphs if the *tree-decompo-* torsi of all bags in T belong to C. *sition over C*

Example 5.5.10 *Figure 5.5.10 shows a tree-decomposition of a graph over the class of triangles. Part b) shows the tree-decomposition and Part c) the corresponding torsi.*

A graph G is called *decomposable* over a class C if it has a tree-decomposition over C. For every class C we denote the class of graphs

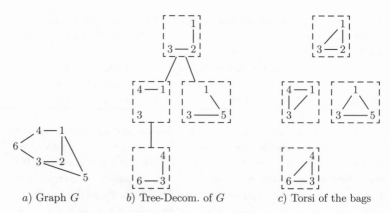

a) Graph G b) Tree-Decom. of G c) Torsi of the bags

Figure 5.16 Tree-Decomposition over the class of triangles

$\mathfrak{D}(\mathcal{C})$ decomposable over $\mathcal{C}$ by $\mathfrak{D}(\mathcal{C})$. It is not hard to see that if $\mathcal{C}$ is minor closed then so is $\mathfrak{D}(\mathcal{C})$.

Example 5.5.11 *Let C_{k+1} be the class of graphs of order at most $k + 1$ and let $\mathcal{T}_k$ be the class of graphs of tree-width at most k. Then $\mathcal{T}_k = \mathfrak{D}(C_{k+1})$.*   ⊣

Robertson and Seymour's structure theorem for classes of graphs excluding a minor can now be reformulated as follows.

Theorem 5.5.12 (*Robertson, Seymour* [79]) *For every minor ideal $\mathcal{D}$ there is a class $\mathcal{C}$ of graphs of almost bounded genus such that $\mathcal{D} \subseteq \mathfrak{D}(\mathcal{C})$.*

We will not make the notion of "almost bounded genus" precise here and instead refer to [79] or to [30, Chapter 12] which contains a more elaborate introduction to the theory. For the applications we have in mind, we do not have to work with almost bounded genus graphs, vortices and apices directly but can use a simpler version of the structure theorem. This relies on the following lemma, proved by Grohe in [52].

local tree-width   The *local tree-width* is the function ltw : GRAPHS × ℕ → ℕ defined as

$$\text{ltw}(G, r) := \max \left\{ \text{tw}\left(G[N_r(v)]\right) : v \in V(G) \right\},$$

where $N_r(v)$ is the r neighbourhood of v, i.e. the set of vertices of distance at most r from v. That is, the local tree-width of a graph assigns to every radius $r \in \mathbb{N}$ the maximal tree-width of an r-neighbourhood in the graph G. See Section 5.7.3 for more on local tree-width.

Lemma 5.5.13 (*Grohe* [52]) *Let S be a surface. Then the class of all minors of graphs almost embeddable into S has linear local tree-width.*

For all $\lambda, \mu \geq 1$ define

$$\mathcal{L}(\lambda) := \{G : \mathrm{ltw}(H, r) \leq \lambda \cdot r \text{ for all } H \preccurlyeq G\}$$

and

$$\mathcal{L}(\lambda, \mu) := \{G : \text{ there is } X \subseteq V(G), |X| \leq \mu \text{ s.th. } G \setminus X \in L(\lambda)\}.$$

Then, the previous lemma implies the following simpler structure theorem that will be used in later sections.

Theorem 5.5.14 *For every minor ideal $\mathcal{D}$ there exist $\lambda, \mu \geq 1$ such that $\mathcal{D} \subseteq \mathfrak{D}(\mathcal{L}(\lambda, \mu))$.*

Furthermore, Grohe proves the existence of an algorithm for computing the decompositions over $\mathcal{L}(\lambda, \mu)$, based on the following lemma.

Lemma 5.5.15 (*Grohe* [52]) *Let $\mathcal{C}$ be a minor closed class of graphs. Then there is a polynomial-time algorithm that, given a graph G, either computes a tree-decomposition of G over $\mathcal{C}$ or rejects G, if no such decomposition exists.*

Taking $\mathcal{C}$ to be $\mathcal{L}(\lambda, \mu)$, the lemma implies the existence of an algorithm for computing tree-decompositions over $\mathcal{L}(\lambda, \mu)$. However, the algorithm outlined in [52] uses non-constructive elements of the graph minor theory and hence, while proving the existence of an algorithm, does not actually state one.

In [29], Demaine, Hajiaghayi and Kawarabayashi proved that the decompositions as guaranteed by Theorem 5.5.12 can be computed in polynomial time for every fixed class of graphs excluding at least one minor H.

Theorem 5.5.16 (*Demaine, Hajiaghayi, Kawarabayashi* [29]) *For every fixed H, there is a polynomial-time algorithm for computing the decompositions of H-minor free graphs as stated in Theorem 5.5.12.*

From this, for each fixed H, a polynomial time algorithm which computes a tree-decomposition of an H-minor free graph G over $\mathcal{L}(\lambda, \mu)$, for suitable values of λ, μ, can easily be derived.

One may wonder why we only considered classes $\mathcal{L}(\lambda)$ of *linear* local tree-width instead of classes of graphs where the local tree-width is bounded by a polynomial $p(r)$ or even worse. In [24], Demaine and Hajiaghayi showed that minor closed classes of bounded local tree-width always have linear local tree-width. Hence, there is no need to consider non-linear local tree-width here, as all classes $\mathcal{L}(\lambda)$ are minor-closed.

5.5.4 Computing Excluded Minor Characterisations

Recall from Section 5.5.1 that every minor ideal can be characterised by a finite set of excluded minors (Theorem 5.5.2) and that for each fixed H it is decidable in cubic time whether a graph G contains H as a minor (Theorem 5.5.9). As a consequence we obtain Corollary 5.5.3 stating that every minor ideal can be decided in cubic time. Note that the result contains a non-constructive element as it does not give a way to compute the excluded minors for a minor ideal. For instance, while we know that the class of knotlessly embeddable graphs can be decided in cubic time, no algorithm for doing so is actually known.

This naturally raises the question whether this non-constructive element can be removed from the proof, i.e. whether characterisations of minor ideals in terms of their excluded minors can be computed. Clearly, to state this precisely, we have to specify how we want to represent a minor ideal as an input to an algorithms and also what exactly we want to understand by a characterisation of a minor ideal in terms of excluded minors.

obstruction   Let C be a minor ideal. A graph H is an *obstruction* for C if H is an excluded minor of C but for all $H' \preccurlyeq H$ with $H' \not\cong H$ we have $H' \in C$. Hence, obstructions are minimal excluded minors. We denote the set of obstruc-
$\mathcal{O}(C)$ tions of C by $\mathcal{O}(C)$. It is easily seen that for all minor ideals C, $\mathcal{O}(C)$ is unique up to isomorphism and it is finite by the Graph Minor Theorem. We will therefore take $\mathcal{O}(C)$ as the characterisation of minor ideals we want to compute.

This leaves us with the question how to specify a minor ideal as an input for algorithms. A natural choice is to provide a Turing-machine deciding the ideal and use this as input. However, Fellows and Langston [44] observed that there is no algorithm which, given a Turing-machine deciding a minor ideal C, computes the set $\mathcal{O}(C)$. Later, Courcelle, Fellows and Langston [15] showed that there is no algorithm which, given an MSO-sentence defining a minor ideal C, computes $\mathcal{O}(C)$.

On the other hand, it is known that obstructions can be computed for a number of natural minor ideals. For instance, for all $k \geq 1$ the obstructions can be computed for the class $\mathcal{T}_k$ of all graphs of tree-width $\leq k$ (see [62]), for the class $\mathcal{B}_k$ of all graphs of branch-width $\leq k$ (see [50]) and for the class $\mathcal{G}_k$ of graphs of genus $\leq k$ (this follows from [83] or a combination of [87] and [43]).

Fellows and Langston were the first to study algorithmic issues related to the graph minor theorem and ways to overcome its non-constructiveness. In [43], they propose a general method for computing obstruction sets based on a generalisation of the Myhill-Nerode theorem of formal language theory to "graph languages". Adler, Courcelle, Grohe and Kreutzer[8] present a similar method for computing obstruction sets based on definability in monadic second-order logic (see [1]). We will give a brief presentation of this method and illustrate it by an example. For all minor ideals C and D, their union $C \cup D$ is minor closed and hence a minor ideal. We will show below that the set of obstructions for $C \cup D$ can be computed from $\mathcal{O}(C)$ and $\mathcal{O}(D)$.[9] The proof of this result also contains a nice application of the Trinity Lemma 5.5.7.

We first establish some lemmas which are all easily proved using well-known results from automata theory and the connection between monadic second-order logic on trees and tree-automata (see e.g. [86, 10]).

Lemma 5.5.17 *There is an algorithm which, given a formula $\varphi \in$ MSO defining a minor ideal C, computes a formula $\psi \in$ MSO defining $\mathcal{O}(C)$.*

Proof. A graph H is an obstruction for C if $H \notin C$ but $H - v \in C$, $H - e \in C$ and $H/e \in C$ for all $v \in V(H)$ and $e \in E(H)$. Given the formula φ defining C, this can be easily be formalised in MSO. $\square$

The next lemma is based on a pumping lemma for tree-automata (see [10]).

Lemma 5.5.18 *There is an algorithm which, given a formula $\varphi \in$ MSO so that the class $\mathrm{Mod}(\varphi) := \{H : H \models \varphi\}$ is finite (up to isomorphism) and a $k \in \mathbb{N}$ such that $\mathrm{tw}(H) \leq k$ for all $H \in \mathrm{Mod}(\varphi)$, computes $\mathrm{Mod}(\varphi)$.*

Proof (sketch). Suppose φ has only finitely many models each of tree-width $\leq k$. As we are given k explicitly, we can use the interpretation defined in Section 5.3.3 to encode the models of φ as coloured trees over a suitable alphabet and reduce the problem of computing the models of φ to the problem of computing the corresponding tree-encodings. An upper bound for the size of these models can then be derived from a version of the pumping lemma of formal language theory for classes of trees definable by tree-automata. From this bound on the size, the actual models of φ can easily be computed. $\square$

The previous lemmas together with the Graph Minor Theorem immediatly imply the following corollary which is the basis of the method for computing obstruction sets proposed in [1].

---

[8] The proof presented here follows a suggestion by Bruno Courcelle simplifying the original proof of the result in [1].
[9] Note that the analogous problem for $C \cap D$ is trivial.

Corollary 5.5.19 *There is an algorithm which, given a formula $\varphi \in$ MSO defining a minor ideal C and a $k \in \mathbb{N}$ such that $\mathrm{tw}(H) \leq k$ for all $H \in \mathcal{O}(C)$, computes the set $\mathcal{O}(C)$.*

As an application of the result we show that the obstructions for the union $C \cup \mathcal{D}$ of minor ideals C, $\mathcal{D}$ can be computed from the sets $\mathcal{O}(C)$ and $\mathcal{O}(\mathcal{D})$. For this, we have to show that $C \cup \mathcal{D}$ is MSO-definable and to establish an upper bound on the tree-width of its obstructions.

It is easily seen that for any fixed graph H there is an MSO-formula φ_H which is true in a graph G if, and only if, $H \preccurlyeq G$. This follows immediately from the definition of minors in terms of minor maps and images as presented in Section 5.5.1. To define $C \cup \mathcal{D}$ in MSO note that $G \in C \cup \mathcal{D}$ if, and only if, G either excludes a minor from $\mathcal{O}(C)$ or a minor from $\mathcal{O}(\mathcal{D})$. As we have seen, this is MSO-definable and a corresponding formula can easily be computed. It remains to establish a bound on the tree-width of the obstructions.

Lemma 5.5.20 *Let C and $\mathcal{D}$ be minor ideals and let $\mathcal{U} := C \cup \mathcal{D}$. There is an algorithm which, given $\mathcal{O}(C)$ and $\mathcal{O}(\mathcal{D})$ as input, computes a number $k \in \mathbb{N}$ such that $\mathrm{tw}(H) \leq k$ for all $H \in \mathcal{O}(\mathcal{U})$.*

Proof (sketch). Suppose $G \in \mathcal{O}(\mathcal{U})$. Hence, $G \notin \mathcal{U}$ but $G - v \in \mathcal{U}$ for all $v \in V(G)$. It follows that there are $H \in \mathcal{O}(C)$ and $I \in \mathcal{O}(\mathcal{D})$ such that $H \preccurlyeq G$ and $I \preccurlyeq G$. Let $k := \max\{|H|, |I|\} + 1$ and choose h "large enough", where the meaning of large enough will become clear later.

By the Trinity Lemma 5.5.7, either a) $\mathrm{tw}(G) \leq f(k, h)$ for some computable function f, or b) $K_k \preccurlyeq G$ or c) there is a subset $X \subseteq V(G)$ with $|X| < \binom{k}{2}$, a wall W of height h in $G \setminus X$, and a flat layout of the compass of W in $G \setminus X$.

Suppose c) applies. It follows from a result by Robertson and Seymour in [78] that if h is chosen large enough then there is a vertex v in the wall W (the middle vertex) such that $G - v$ still contains H and I as minors, contradicting the minimality of the obstruction G. Hence, case c) is impossible. The idea to choose the middle vertex is same as in the proof of Theorem 5.5.8 described in Section 5.5.2.

For b), if G contains a K_k minor then there is a strict subgraph $G' \subsetneqq G$ containing a K_{k-1} minor. Hence, by the choice of k, G' contains H and I as minors, contradicting the minimality of G. Thus, case b) is impossible as well.

Finally, in a) the tree width of G is bounded by a computable function in h and k and we have found a uniform upper bound for the tree-width of G which concludes the proof. $\qquad\square$

Corollary 5.5.21 ([1]) *For all minor ideals* $C, \mathcal{D}$ *the set* $\mathcal{O}(C \cup \mathcal{D})$ *is computable from the sets* $\mathcal{O}(C)$ *and* $\mathcal{O}(\mathcal{D})$.

Using a similar approach it was shown in [1] that obstructions can be computed for other natural minor ideals. In particular, if C is a minor ideal whose obstructions are known, then the obstructions can be computed for the class C_{apex} of *apex graphs over* C, defined as

$$C_{\text{apex}} := \{G : \text{there is } v \in V(G) \text{ such that } G - v \in C\}.$$

However, there remain interesting open problems.

Open Problem 5.5.22 *1. Is there an algorithm which, given* $\lambda \geq 0$, *computes the obstructions* $\mathcal{O}(\mathcal{L}(\lambda))$? *See Section 5.5.3 for a definition of* $\mathcal{L}(\lambda)$ *and* $\mathcal{L}(\lambda, \mu)$. *Note that, by using the computability of* $\mathcal{O}(C_{\text{apex}})$ *from* $\mathcal{O}(C)$, *the set* $\mathcal{O}(\mathcal{L}(\lambda, \mu))$ *can be computed from* $\mathcal{O}(\mathcal{L}(\lambda))$, *for all* $\mu \geq 0$.
2. If C *is a minor ideal whose obstructions are given, can we compute the obstructions of the class* $\mathfrak{D}(C)$ *of graphs tree-decomposable over* C?

A solution for both open problems would be particularly interesting as every minor ideal is a subclass of a class $\mathfrak{D}(\mathcal{L}(\lambda, \mu))$ for some $\lambda, \mu \geq 0$.

5.6 Monadic Second-Order Logic Revisited

Recall from Section 5.3.5 that for each k, the MSO_2-theory of the class $\mathcal{T}_k$ of graphs of tree-width at most k is decidable. The aim of this section is to prove a kind of converse, also due to Seese.

Theorem 5.6.1 (*Seese* [81]) *If* C *is a class of graphs with decidable* MSO_2-*theory, then* C *has bounded tree-width.*

The proof of the theorem crucially relies on the excluded grid theorem (Theorem 5.5.4) and the fact that the MSO-theory of grids is undecidable. The latter can easily be established using tiling systems or by a direct encoding of the run of Turing-machines using MSO-formulas (see e.g. [8]).

Suppose C has a decidable MSO_2-theory but unbounded tree-width. Then, by the excluded grid theorem, for all $n \geq 1$, there is a graph $G_n \in C$ containing $G_{n \times n}$ as a minor. The key to the theorem is to show that grid minors can be defined in MSO_2. Hence, the (undecidable) MSO-theory of grids can be reduced to the MSO-theory of C contradicting the assumption that the latter is decidable.

We start by showing how walls can be formalised in MSO_2. The extension to grids follows easily. Let G be a graph and consider an MSO_2-formula formalising the following.

1. There are two sets $\mathcal{H}$ and $\mathcal{V}$ of edges, each of which induces a set of pairwise vertex disjoint paths (which we will think of as horizontal and vertical paths in a wall).

2. For all $P \in \mathcal{H}$ and $Q \in \mathcal{V}$, $P \cap Q$ is a subpath of both, P and Q. Further, $V(P \cap Q) \cap V(H) = \varnothing$ for all $H \in (\mathcal{V} \cup \mathcal{H}) \setminus \{P, Q\}$.

3. There is a path $L \in \mathcal{V}$ such that the intersection of L with each $Q \in \mathcal{H}$ contains an endpoint of Q (L is the left-most vertical path in the wall). Once we have L, we can give the horizontal paths $P \in \mathcal{H}$ a direction, where we say that $p \in V(P)$ is to *the left* of $p' \in V(P)$, if the subpath of P containing p' and a vertex in L also contains p.

4. There is a path $T \in \mathcal{H}$ such that the intersection of T with each $P \in \mathcal{V}$ contains an endpoint of P (T is the top-most horizontal path in the wall). As with horizontal paths, we can now use T to give the vertical paths $P \in \mathcal{V}$ a direction and say that $p \in V(P)$ is *above* $p' \in V(P)$.

5. For each path $P \in \mathcal{V}$ except L there is a path $P' \in \mathcal{V}$ (the path immediately to the left of P) such that for all $Q \in \mathcal{H}$: if $p \in V(P \cap Q)$ and $p' \in V(P' \cap Q)$ are vertices in the intersection of Q and P, P', then p' is to the left of p in Q and there is no $S \in \mathcal{H}$ such that any $s \in V(S \cap Q)$ lies in the subpath of Q between p and p'.

6. The analogue condition for horizontal paths.

Clearly, the various conditions are MSO_2-definable. Now, if $\mathcal{V}$ and $\mathcal{H}$ satisfy the conditions above, then they generate a wall in G and conversely, the disjoint horizontal and vertical paths in a wall satisfy the conditions. Finally, it is easily seen that the class of grids can be defined in the class of walls and hence grid minors are MSO_2-definable in graphs.

Note that here we crucially use the fact the we are working with MSO_2-formulas and hence can quantify over the edge sets of disjoint paths. In MSO_1 we could only try to quantify over the vertex set of disjoint paths. However, if there are sufficiently many edges between these vertices, there is no way we can give the paths an orientation, e.g. define paths being to the left of others. And clearly, we cannot expect clique-minors to be definable in MSO_1 as, by Theorem 5.4.20, the MSO_1-theory of graph classes of bounded clique-width is decidable and hence there are classes with decidable MSO_1-theory but unbounded tree-width.

5.7 First-Order Model-Checking

In Section 5.3.4 and 5.4.3 we showed that the model-checking problem for variants of monadic second-order logic is solvable in linear time for any fixed formula on classes of graphs of bounded tree- or clique-width. As we have argued in Section 5.4.4 and will explore further in Section 5.8 below, there is not much hope for extending these results to other or larger classes of graphs. However, first-order logic is tractable on much larger classes of graphs and in this section we will present tractability results for first-order logic on several special classes of graphs. The important property of first-order logic that makes these results possible is *locality*.

The section is structured as follows. In Section 5.7.1 we introduce the concept of locality and present Gaifman's theorem. In Section 5.7.2 we apply locality to obtain fixed-parameter algorithms for first-order model-checking on graph classes of bounded degree. The algorithms developed in this section can be applied in a much more general context using the concept of localisation of graph invariants. This will be formally defined in Section 5.7.3. In Section 5.7.4 we present fixed-parameter algorithms for first-order model-checking on H-minor free graphs.

5.7.1 Locality of First-Order Logic

Let G be a graph. Recall that the *distance* $d^G(u, v)$ between two vertices $u, v \in V(G)$ is the length of the shortest path from u to v or ∞ if there is no such path. Further, for every $v \in V(G)$ and $r \in \mathbb{N}$ we define the *r-neighbourhood* of v in G as the set

$$N_r^G(v) := \{w \in V(G) : d^G(v, w) \leq r\}$$

of vertices of distance at most r from v. For a set $W \subseteq V(G)$ we set $N_r^G(W) := \bigcup_{v \in W} N_r^G(v)$. We omit the index $\cdot^G$ whenever G is clear from the context.

If σ is a signature and A is a σ-structure, we define the distance $d^A(a, b)$ and the r-neighbourhood $N_r^A(a)$ in terms of the Gaifman-graph $\mathcal{G}(A)$ of A,[10] i.e. $N_r^A(a)$ is the set of elements of distance at most r from a in the Gaifman-graph.

It is easily seen that for any fixed $r \in \mathbb{N}$ "distance at most r" is first-order definable, that is, for every $r \in \mathbb{N}$ there is a formula $\text{dist}_{\leq r}(x, y)$ such that for

---

[10] See Section 5.2 for a definition of Gaifman-graphs.

all structures A and all $u, v \in V(A)$

$$A \models \text{dist}_{\leq r}(u, v) \qquad \text{iff} \qquad d^A(u, v) \leq r.$$

Similarly, there are formulas $\text{dist}_{>r}(x, y)$ and $\text{dist}_{<r}(x, y)$ defining distance $>$ r and $< r$ respectively. To improve readability we will write $\text{dist}(x, y) \leq r$ instead of $\text{dist}_{\leq r}(x, y)$ and likewise for the other formulas.

A first-order formula $\varphi(x)$ is *r-local* if for every structure A and all $a \in V(A)$

$$A \models \varphi(a) \qquad \text{iff} \qquad A\big[N_r^A(a)\big] \models \varphi,$$

where $A\big[N_r^A(a)\big]$ denotes the substructure of A induced by $N_r^A(a)$. Hence, truth of an r-local formula at an element a in a structure only depends on its r-neighbourhood. A formula $\varphi(x)$ is *local* if it is r-local for some $r \in \mathbb{N}$.

A *basic local sentence* is a first-order sentence of the form

$$\exists x_1 \ldots \exists x_k \Big(\bigwedge_{1 \leq i < j \leq k} \text{dist}(x_i, x_j) > 2r \wedge \bigwedge_{i=1}^{k} \vartheta(x_i) \Big)$$

where $\vartheta(x)$ is local. In 1981, Gaifman showed that every first-order sentence is equivalent to a Boolean combination of basic local sentences.

Theorem 5.7.1 (*Gaifman* [48]) *Every first-order sentence is equivalent to a Boolean combination of basic local sentences. Furthermore, there is an algorithm that, given a first-order formula as input, computes an equivalent Boolean combination of basic local sentences.*

A first-order formula is in *Gaifman Normal Form* (GNF), if it is a Boolean combination of basic local sentences. Gaifman's original proof is by an explicit translation of first-order formulas into formulas in GNF. A proof sketch along this lines can also be found in the survey paper [53]. A different, model-theoretical proof can be found in [37, Section 2.5].

The translation of formulas into Gaifman normal form is effective. However, it has recently been shown [23] that this translation may involve a non-elementary blow-up in the size of the sentence.

Theorem 5.7.2 (*Dawar, Grohe, Kreutzer, Schweikardt* [23]) *Let $\sigma := \{E\}$ be the signature of graphs. For every $h \geq 1$ there is an $\text{FO}[\sigma]$-sentence φ_h of size $\mathcal{O}(h^4)$ such that every $\text{FO}[\sigma]$-sentence in Gaifman normal form that is equivalent to φ_h on the class of finite trees has size at least tower(h), where tower(h) denotes a tower of 2s of height h.*

From a practical point of view, this renders algorithms using Gaifman's theorem useless, no matter what their theoretical complexity might be.

Example 5.7.3 *Recall that a dominating set X in a graph G is a set $X \subseteq V(G)$ such that for all $v \in V(G)$, $v \in X$ or there is a $u \in X$ and $\{u, v\} \in E(G)$. For $k \in \mathbb{N}$, the formula*

$$\varphi_k := \exists x_1 \ldots \exists x_k \forall y \Big(\bigvee_{1 \leq i \leq k} \big(x_i = y \vee E y x_i\big) \Big)$$

is true in a graph G if, and only if, G has a dominating set of size at most k.

To convert this into an equivalent sentence in Gaifman normal form, we first observe that no connected graph of diameter at least $3k + 1$ can have a dominating set of size at most k. Here, the diameter of a graph is the maximum of the distance between any two vertices.

Hence, on connected graphs, the formula φ_k above is equivalent to the conjunction of the basic local sentence

$$\psi := \neg \exists x_1 \exists x_2 \mathrm{dist}(x_1, x_2) > 3k + 1,$$

saying that the diameter of G is greater than $3k + 1$, and the basic local sentence $\exists x \chi(x)$, where $\chi(x)$ is the $3k + 1$-local formula

$$\exists y_1 \in N_{3k+1}(x) \ldots \exists y_k \in N_{3k+1}(x) \forall z \in N_{3k+1}(x) \bigvee_{1 \leq i \leq k} \big(y_i = z \vee E z y_i\big).$$

Note that this formula correctly defines the existence of a dominating set of size k only in connected graphs, as in graphs with more than one component there may exist a dominating set of size k even though there are vertices x_1, x_2 of distance greater than $3k + 1$. Adapting the formula to this case requires a little more effort. ⊣

5.7.2 First-Order Logic on Graphs of Bounded Degree

As a first application of the use of Gaifman's locality theorem for algorithmic meta theorems we consider graph classes of bounded degree.

Definition 5.7.4 *A class $\mathcal{C}$ of graphs has* bounded degree *if there is a $d \in \mathbb{N}$ such that $\Delta(G) \leq d$ for all $G \in \mathcal{C}$.*

In 1996, Seese [82] showed that model-checking for a fixed first-order sentence can be done in linear time on graph classes of bounded degree.

Theorem 5.7.5 (*Seese* [82]) *For any class $\mathcal{C}$ of graphs of bounded degree and any fixed first-order sentence it can be decided in linear time whether $G \models \varphi$ for a graph $G \in \mathcal{C}$. In other words, first-order model-checking on $\mathcal{C}$ is fixed-parameter tractable by a linear fpt algorithm.*

1: $L := \varnothing$
2: **while** $Q \neq \varnothing$ **do**
3:     choose $v \in Q$ arbitrarily
4:     $L := L \cup \{v\}$
5:     $Q := Q \cap N_{2r}(v)$
6: **end while**
7: **if** $|L| \geq k$ **then**
8:     accept G
9: **else**
10:     **if** $G[N_{2r}(L)] \models \exists x_1 \ldots x_k (\bigwedge_{i \neq j} \text{dist}(x_i, x_j) > 2r \wedge \bigwedge_i \text{"}x_i \text{ is } red\text{"})$ **then**
11:         accept G
12:     **else**
13:         reject G
14:     **end if**
15: **end if**

Figure 5.17 Algorithm to find k vertices of pairwise distance $> 2r$

Proof. The proof method we use here is essentially the method used by Frick and Grohe to show a similar result for planar graphs.

Let φ and $G \in \mathcal{C}$ be given. We first convert φ into Gaifman normal form, i.e. into a Boolean combination of basic local sentences. As Boolean combinations are easy to deal with, we only need to consider basic local sentences of the form

$$\psi := \exists x_1 \ldots \exists x_k \Big(\bigwedge_{1 \leq i < j \leq k} \text{dist}(x_i, x_j) > 2r \wedge \bigwedge_{i=1}^{k} \vartheta(x_i) \Big)$$

where $\vartheta(x)$ is r-local for some $r \in \mathbb{N}$.

To check whether ψ is true in G we proceed in two steps. First, we test for all $v \in V(G)$ if $G[N_r^G(v)] \models \vartheta$. As G has degree bounded by some constant d, the size of $N_r^G(v)$ is constant and hence this can be decided in constant time. Colour all vertices v *red* for which $G[N_r^G(v)] \models \vartheta$ and let Q be the set of *red* vertices. Now, $G \models \psi$ if Q contains k vertices of pairwise distance $> 2r$.

In the second step we search for k such vertices. For this, we use the greedy algorithm shown in Figure 5.17. The algorithm proceeds as follows. In lines 2–6 of the algorithm, we try to choose k *red* vertices of pairwise distance $> 2r$ greedily. If we succeed, i.e. if the set L contains k elements, then we are done and accept G. Otherwise, we know that L contains fewer than k vertices which are all *red* and of pairwise distance $> 2r$ and also that any other *red* vertex is within distance $\leq 2r$ of an element of L (otherwise we could add the vertex to L). Hence, all *red* vertices of G are contained in the $2r$-neighbourhood $N := N_{2r}[L]$ of L. Again, N is of constant size and hence we can check in constant time whether N contains k *red* vertices of pairwise distance $> 2r$. This is done in line 12 by testing whether the graph induced by the neighbourhood satisfies the first-order formula stating that there are k distinct *red* vertices of pairwise distance $> 2r$. $\quad\square$

The previous theorem gives a simple example how locality can be used to obtain efficient model-checking algorithms for first-order logic. As it turns out, a similar scheme can be employed in many cases.

Theorem 5.7.6 *Let $\mathcal{C}$ be a class of graphs such that the following problem is fixed-parameter tractable:*

Input:	$\varphi \in$ FO, graph $G \in \mathcal{C}$, $v_1, \ldots, v_k \in V(G)$ and $r \in \mathbb{N}$.
Parameter:	$r + k + |\varphi|$.
Problem:	Decide $G\big[N_r^G(v_1, \ldots, v_k)\big] \models \varphi$.

Then model-checking for first-order logic is fixed-parameter tractable on $\mathcal{C}$.

Proof. We proceed as in the proof of Theorem 5.7.5. By Gaifman's theorem, we may assume that φ is a basic local sentence

$$\exists x_1 \ldots \exists x_k \Big(\bigwedge_{i \neq j} \operatorname{dist}(x_i, x_j) > 2r \wedge \bigwedge_i \vartheta(x_i) \Big),$$

where $\vartheta(x)$ is an r-local formula for some $r \in \mathbb{N}$.

In the first step, we compute the set Q of vertices $v \in V(G)$ such that $G\big[N_r(v)\big] \models \vartheta(v)$. By assumption, for each $v \in V(G)$ this can be done in time $f(r + 1 + |\vartheta|) \cdot |G|^{\mathcal{O}(1)}$, for some computable function $f : \mathbb{N} \to \mathbb{N}$, and hence the total running time is $f(r + 1 + |\vartheta|) \cdot |G|^{\mathcal{O}(1)}$.

In the second step we aim to find k vertices in Q whose distance is pairwise $> 2r$. Using the algorithm of Figure 5.17 this can be done in time $f(2r \cdot k + \mathcal{O}(k)) \cdot |G|^{\mathcal{O}(1)}$. Hence, the total running time is $f(2r \cdot k + \mathcal{O}(k)) \cdot |G|^{\mathcal{O}(1)}$. $\quad\square$

While this theorem may appear somewhat artificial, we will see a number of interesting applications of it by considering localisations of graph invariants such as tree-width or rank-width.

5.7.3 Localisation of Graph Invariants

Let GRAPH denote the class of all finite graphs.

Definition 5.7.7 *A* graph invariant *is a function* f : GRAPH $\to$ $\mathbb{N}$. *For every*
$loc_f(G, r)$ graph invariant f *we define its* localisation loc_f : GRAPH $\times$ $\mathbb{N}$ $\to$ $\mathbb{N}$ *as*

$$loc_f(G, r) := \max \Big\{ f\Big(G[N_r(v)]\Big) : v \in V(G) \Big\}.$$

A class C of graphs has bounded local f, *if there is a computable*[11] *function*
$h : \mathbb{N} \to \mathbb{N}$ *such that* $loc_f(G, r) \leq h(r)$ *for all* $G \in C$ *and* $r \in \mathbb{N}$.

That is, to compute $loc_f(G, r)$ we compute the r-neighbourhoods $N :=$
$N_r(v)$ of all vertices $v \in V(G)$ and for each such N the value $f(N)$. $loc_f(G, r)$
is then the maximum of these values. In particular, if the problem: given G and
k, where k is the parameter, to decide whether $f(G) \leq k$ is fixed-parameter
tractable, then so is the problem: given G, r, k, where $r + k$ is the parameter,
to decide if $loc_f(G, r) \leq k$.

Example 5.7.8 *Of particular interest is the localisation of tree-width, called*
local tree-width *(see also the discussion at the end of Section 5.5.3). There
are a number of interesting examples for graph classes with bounded local
tree-width.*

1. *Every graph class of bounded tree-width also has bounded local tree-width
   (bounded by a constant).*
2. *The class of planar graphs has bounded local tree-width. More precisely,
   Robertson and Seymour [77] showed that every planar graph of radius r
   has tree-width $\leq 3r + 1$.*
3. *Any class of graphs of bounded degree. This is easily seen as the r-neigh-
   bourhoods of graphs of degree at most d contain $< d^{r+1}$ vertices.*   ⊣

Similar to local tree-width we can define local rank-width or clique-width,
where we take f : GRAPH $\to$ $\mathbb{N}$ to be the function assigning to each graph its
rank- or clique-width.

Another interesting example is the localisation of the following graph invari-
$mec(G)$ ant. Let mec : GRAPH $\to$ $\mathbb{N}$ *(minimal excluded clique)* be the function assigning
to each graph G the minimal order of a clique that is not a minor of G, i.e.

$$mec(G) := \min\{k : K_k \not\preccurlyeq G\}.$$

Graph classes with locally bounded mec are called graph classes with *locally
excluded minors* and have been studied by Dawar, Grohe and Kreutzer in [21].
Clearly, every graph class C with an excluded minor H also locally excludes

---

[11] As we are asking for h to be computable, we should call this *effectively bounded local f*. But
this would make the notation even more clumsy and we therefore refrain from mentioning
effectiveness in the sequel.

H, i.e. has bounded local *mec*. The converse fails, though, as is witnessed by the following class of graphs. For $k \in \mathbb{N}$ let S_k be the graph obtained from K_k by replacing all edges by internally vertex disjoint paths of length k. Now take $\mathcal{C} := \{S_k : k \in \mathbb{N}\}$. Obviously, the minor closure of $\mathcal{C}$ is the class of all graphs, i.e. $\mathcal{C}$ does not exclude a minor. However, it locally excludes minors, as every k-neighbourhood of graphs $G \in \mathcal{C}$ excludes K_k. Hence, $f :$ GRAPH $\times \mathbb{N} \to \mathbb{N}$ defined as $f(G, r) := r$ dominates the local *mec* of $\mathcal{C}$.

Note, that $\mathcal{C}$ has bounded local tree-width and hence also provides an example separating proper minor ideals and graph classes of bounded local tree-width. It is easily seen that every class of graphs of bounded local tree-width also locally excludes minors. The converse fails again, as not even every minor ideal has bounded local tree-width. This is witnessed by the class of apex graphs defined as

$$\mathcal{C}_{apex} := \{G : \text{ there is } v \in V(G) \text{ such that } G - v \text{ is planar}\}.$$

In particular, this class contains all grids with one additional vertex adjacent to every vertex in the grid. Hence, $\mathcal{C}_{apex}$ has unbounded local tree-width but clearly excludes K_6.

Lemma 5.7.9 *The concept of locally excluded minors strictly generalises both excluded minors and bounded local tree-width. That is, every class of graphs that excludes a minor or has bounded local tree-width, also locally excludes minors. The converse fails in both cases.*

The aim of this section is to prove the following theorem.

Theorem 5.7.10 *Let f be a graph invariant such that the following is fixed-parameter tractable.*

MC(FO, f)	
Input:	Graph G and $\varphi \in$ FO.
Parameter:	$f(G) + |\varphi|$.
Problem:	Decide whether $G \models \varphi$.

Then for every class $\mathcal{C}$ of locally bounded f, the problem MC(FO, $\mathcal{C}$) is fixed-parameter tractable.

Proof. Let $g : \mathbb{N} \to \mathbb{N}$ be a bound for $loc_f(G, \cdot)$ for all $G \in \mathcal{C}$. We first suppose that f is *induced subgraph monotone*, i.e. $f(H) \leq f(G)$ for all H, G such that H is an induced subgraph of G, and further has the property that if G_1, G_2 are vertex disjoint graphs, then $f(G_1 \cup G_2) \leq \max\{f(G_i) : i = 1, 2\}$.

Note that graph invariants such as tree-width, branch-width, clique-width and rank-width all have these properties.

Then the result follows from Theorem 5.7.6 as follows. Given $\varphi \in \mathrm{FO}$, $G \in \mathcal{C}$, $v_1, \ldots, v_k \in V(G)$ and $r \in \mathbb{N}$, we first compute $H := G\left[N_r^G(v_1, \ldots, v_k)\right]$ in polynomial time. Clearly, every component of H has radius at most $k \cdot r$ and hence $f(H) \leq loc_f(G, k \cdot r) \leq g(k \cdot r)$. The assumptions of this lemma then imply that the assumptions of Theorem 5.7.6 are satisfied and thus we can decide $H \models \varphi$ by fpt-algorithms.

If f does not have the properties above, we can no longer apply Theorem 5.7.6 directly. Instead, we have to repeat its proof. We leave the details to the reader. $\square$

Corollary 5.7.11 *First-order model-checking is fixed-parameter tractable on graph classes of*

- *bounded local tree-width*
- *bounded local rank- or clique-width.*

In the next section we will show that first-order model-checking is fixed-parameter tractable on graph classes excluding at least one minor. We will later consider localisation in this context and show an analogous result for graph classes locally excluding a minor.

5.7.4 First-Order Logic on H-Minor Free Graphs

The aim of this section is to show that first-order model-checking is fixed-parameter tractable on every class $\mathcal{C}$ of graphs excluding at least one minor H. If we take $|\varphi|$ as the parameter, this was first shown by Flum and Grohe [45] in 2001. That is, for every fixed H, the problem is tractable under the parametrization $|\varphi|$. However, the exponential of the polynomials occurring in the running time analysis can depend on H. As it turns out, this parametrization is not strong enough to apply our method of localisation to the problem. In [21], therefore, Dawar, Grohe and Kreutzer consider the problem under the parametrization $|\varphi| + |H|$ and show fixed-parameter tractability for this case.

Let us first consider the case where H is fixed and $|\varphi|$ is the parameter. In the light of the previous sections, the proof of the theorem seems rather straightforward: given $G \in \mathcal{C}$, Theorem 5.5.14 tells us that there are $\lambda, \mu \geq 1$ such that G has a tree-decomposition over $\mathcal{L}(\lambda, \mu)$, i.e. a tree-decomposition such that the torsi of its bags have bounded local tree-width after removal of a few elements, and Theorem 5.5.16 tells us how to compute the decomposition in polynomial time. Furthermore, we already know how to deal with graphs in

$\mathcal{L}(\lambda)$ of bounded local tree-width and extending this to graphs in $\mathcal{L}(\lambda, \mu)$ poses no real problem. And indeed, this is the general idea to show that FO model-checking is FPT on H-minor free graphs, although formally implementing the idea requires some care and additional lemmas. To make this precise it is convenient to introduce further notation.

A graph G is the *clique sum* of graphs G_1 and G_2, denoted $G = G_1 \oplus G_2$, *clique sum,* $\oplus$ if $G_1 \cap G_2$ is a complete graph and G is obtained from $G_1 \cup G_2$ by possibly deleting some edges from $E(G_1 \cap G_2)$. Formally, $V(G) = V(G_1) \cup V(G_2)$, $G_1 \cap G_2$ is a clique and there is a (possibly empty) set $X \subseteq E(G_1 \cap G_2)$ such that $E(G) = E(G_1 \cup G_2) - X$. We write $G = G_1 \oplus_{\overline{v}} G_2$ to indicate that G is $\oplus_{\overline{v}}$ the clique-sum of G_1 and G_2 and that $V(G_1 \cap G_2) = \overline{v}$.

Recall that a tree-decomposition of a graph G is *over* a class $\mathcal{C}$ of graphs if the torsi $[B_t]$ of all its bags belong to $\mathcal{C}$, where the torso of a bag B_t is obtained from $G[B_t]$ by turning the intersections of B_t with neighbouring bags B_s into cliques. Hence, the graph G is obtained as the clique-sum of its bags, an observation that we will use in the following proofs.

We begin by proving an extension of Courcelle's theorem, this time not by a reduction to trees but by computing MSO-types directly. Recall the definition of MSO and FO q-types and the Feferman-Vaught theorem from Section 5.2.3.

Lemma 5.7.12 *Let* tp_q *be one of* tp_q^{FO} *and* tp_q^{MSO}. *The following problem is fixed-parameter tractable: given*

- *a labelled graph G of tree-width $\leq k$,*
- *tuples $\overline{v}_i \in V(G)^{r_i}$, $0 \leq i \leq m$ for some m, such that $G[\overline{v}_i]$ is a clique, and*
- *q-types $\Theta_1, \ldots, \Theta_m$,*

compute $\text{tp}_q(G, \overline{v}_0)$ *for all graphs* $G' = G \oplus_{\overline{v}_1} H_1 \oplus_{\overline{v}_2} \cdots \oplus_{\overline{v}_m} H_m$ *such that* $\text{tp}_q(H_i, \overline{v}_i) = \Theta_i$. *The parameter is* $q + k$.

Proof. Given G, we first compute an ordered tree-decomposition $(T, (\overline{b}_t)_{t \in V(T)})$ of G of width at most k (see Definition 5.3.25). Note that, as the $\overline{v}_i$ induce cliques in G, for each i there is at least one t_i such that $\overline{v}_i \subseteq \overline{b}_{t_i}$. Hence, we can assume that for each $0 \leq i \leq m$ there is a leaf $t \in V(T)$ such that $\overline{v}_i = \overline{b}_t$ and that no other leaf contains a vertex from any of the $\overline{v}_i$ for $1 \leq i \leq m$.

For each $t \in V(T)$, let T_t be the subtree of T rooted at t and let $\mathcal{B}_t$ be the set $\mathcal{B}_t := \bigcup_{s \in V(T_t)} \overline{b}_s$. Beginning from the leaves we inductively compute $\text{tp}_q(G[\mathcal{B}_t], \overline{b}_t \overline{v}_0)$ for each node $t \in V(T)$. Here, the notation $\text{tp}_q(G[\mathcal{B}_t], \overline{b}_t \overline{v}_0)$ indicates that in $G[\mathcal{B}_t]$ we compute the type of $\overline{b}_t$ and all vertices of $\overline{v}_0$ contained in $\mathcal{B}_t$. For leaves t with $\overline{b}_t = \overline{v}_i$, for some $1 \leq i \leq m$, we can infer the type $\text{tp}_q(G[\overline{b}_t], \overline{b}_t \overline{v}_0)$ from Θ_i. For other leaves we can compute their types directly, as they only contain at most $k + 1$ elements. For inner nodes t with children t_1, t_2 we apply Lemma 5.2.3. □

As the previous lemma applies to MSO-types, Courcelle's theorem is clearly a special case of it. Hence, the proof here provides an alternative way of establishing Courcelle's theorem. While the two approaches may seem to be somewhat different, the underlying principle is the same. Recall that in our original proof of Courcelle's theorem, we encoded graphs G of tree-width $\leq k$ in labelled trees T and then rewrote the formula φ on G to a new formula φ' on T such that $G \models \varphi$ if, and only, if $T \models \varphi'$. On the tree-encoding, we then applied results from automata theory which establish that MSO model-checking is fixed-parameter tractable on trees. More specifically, the MSO-formula φ' is translated into an automaton $\mathcal{A}_\varphi$ which accepts T if, and only if, $T \models \varphi'$. Although it is not usually proved this way, essentially the automaton has a state for each possible q-type and its transition relation combines types similar to what is done in Lemma 5.2.3.

But back to first-order model-checking on graph classes excluding a minor. Essentially the previous lemma allows us to deal with tree-decompositions over graphs of bounded tree-width, which clearly is not enough for our purposes.

Lemma 5.7.13 *Let* tp_q *denote* $\mathrm{tp}_q^{\mathrm{FO}}$. *The following problem is fixed-parameter tractable for all* λ, μ: *given*

- *a labelled graph* $G \in \mathcal{L}(\lambda, \mu)$,
- *tuples* $\bar{v}_i \in V(G)^{r_i}$, $0 \leq i \leq m$ *for some* m, *such that* $G[\bar{v}_i]$ *is a clique, and*
- *q-types* $\Theta_1, \ldots, \Theta_m$,

compute $\mathrm{tp}_q(G, \bar{v}_0)$ *for all graphs* $G' = G \oplus_{\bar{v}_1} H_1 \oplus_{\bar{v}_2} \cdots \oplus_{\bar{v}_m} H_m$ *such that* $\mathrm{tp}_q(H_i, \bar{v}_i) = \Theta_i$. *The parameter is* q.

Proof. The proof is by induction on μ. For $\mu = 0$, we adapt the proof of Theorem 5.7.6 using Lemma 5.7.12 locally. Now let $\mu > 0$ and let $G \in \mathcal{L}(\lambda, \mu)$, $\bar{v}_i, \Theta_i$ be an instance of the problem. By definition, G contains a vertex $v \in V(G)$ such that $G \setminus v \in \mathcal{L}(\lambda, \mu - 1)$. Note that for all λ', μ', $\mathcal{L}(\lambda', \mu')$ is a minor ideal and hence has a cubic time membership test by Corollary 5.5.3. Thus, in time $\mathcal{O}(|G|^4)$ we can find such a vertex v. Let G_2 be the coloured graph obtained from G by introducing a new colour C by which we label all neighbours of v and then eliminating v from G. By construction, $G_2 \in \mathcal{L}(\lambda, \mu - 1)$. Furthermore, it is an easy exercise to translate first-order formulas φ over G to formulas φ' over G_2 such that $G \models \varphi$ if, and only if, $G_2 \models \varphi'$. Hence, the q-type of $G' = G \oplus_{\bar{v}_1} H_1 \oplus_{\bar{v}_2} \cdots \oplus_{\bar{v}_m} H_m$ can be recovered from the q-type of $G_2' = G_2 \oplus_{\bar{v}_1} H_1 \oplus_{\bar{v}_2} \cdots \oplus_{\bar{v}_m} H_m$, and the latter is computable by the induction hypothesis. $\square$

The previous two lemmas are the main ingredients for the proof of the following theorem.

Theorem 5.7.14 (*Flum, Grohe* [45]) *Let C be a class of graphs excluding at least one minor. Then the following problem is fixed-parameter tractable.*

> MC(FO, C)
>
> *Input:* $G \in C, \varphi \in$ FO.
> *Parameter:* $|\varphi|$.
> *Problem:* Decide $G \models \varphi$.

Proof. Let G and φ be given and let q be the quantifier-rank of φ. Using Theorem 5.5.16, we first compute a tree-decomposition (T, γ) of G over $\mathcal{L}(\lambda, \mu)$, for some λ, μ. We view T as a directed tree with root r.

For each $t \in V(T), t \neq r$, with parent $s \in V(T)$, let $\bar{v}_t := B_t \cap B_s$. Recall that in the torsi of B_t and B_s, $\bar{v}_t$ induces a clique. For the root r we define $\bar{v}_r$ as the empty tuple. Furthermore, for each $t \in V(T)$ let T_t be the subtree of T rooted at t and let $\mathcal{B}_t := \bigcup_{s \in V(T_t)} B_s$. Finally, for $t \in V(T)$ let $G_t := G[\mathcal{B}_t] \cup K[\bar{v}_t]$. Note that for all $t \in V(T), \bar{v}_t \leq k$, where $k := \lambda + \mu$, as $\bar{v}_t$ induces a clique in the torso $[B_t]$ of B_t. As $[B_t] \in \mathcal{L}(\lambda, \mu)$ and graphs in $\mathcal{L}(\lambda, \mu)$ cannot contain a clique of order $> \lambda + \mu$ we obtain $|\bar{v}_t| \leq k$. Hence, as λ, μ only depend on the excluded minor of C and therefore are fixed, we obtain a fixed upper bound for the size of $\bar{v}_t, t \in V(T)$.

To decide $G \models \varphi$, we aim at computing the type $\mathrm{tp}_q(G, \bar{v}_r)$. We can then simply check whether $\varphi \in \mathrm{tp}_q(G, \bar{v}_r)$. Towards this aim, starting at the leaves and proceeding bottom-up, we apply Lemma 5.7.13 at each node to compute the type $\mathrm{tp}_q(G_t, \bar{v}_t)$. $\square$

The previous theorem shows that for every fixed graph H, first-order model-checking is fixed-parameter tractable, with parameter $|\varphi|$, on every class of graphs excluding H. However, the algorithm as described above is not fixed-parameter tractable in the parameter $|H| + |\varphi|$ as we use a non-constructive approach in Lemma 5.7.13 and also the algorithm described in [29] seems to use the minor H in an inappropriate way for parameterized complexity.

We therefore turn to a different parametrization of the problem, where we take the parameter to be $|\varphi| + |H|$. This problem was studied by Dawar, Grohe and Kreutzer in [21]. The approach taken there is similar to the method outlined above. However, instead of using tree-decompositions over $\mathcal{L}(\lambda, \mu)$, [21] uses a slightly weaker form of decompositions, called *weak* decompositions over

$\mathcal{L}(\lambda, \mu)$. The main result in [21] is that for every H, every graph excluding H has a weak decomposition over some $\mathcal{L}(\lambda, \mu)$ (which is relatively straightforward to show) and that these decompositions can be computed by an fpt-algorithm with parameter H (which requires considerably more work). Once this is shown, the proof method outlined above can be adapted to weak decompositions yielding the following result.

Theorem 5.7.15 (*Dawar, Grohe, Kreutzer* [21]) *The following problem is fixed-parameter tractable.*

> p-MC(FO)
> *Input:* G, H such that $H \npreceq G, \varphi \in$ FO.
> *Parameter:* $|\varphi| + |H|$.
> *Problem:*   Decide $G \models \varphi$.

An immediate consequence of the theorem is the following. Recall from Section 5.7.3 the definition of the *minimum excluded clique* number $mec(G)$ of a graph G and of locally excluded minors. For any function $f : \mathbb{N} \to \mathbb{N}$ let $\mathcal{C}_f$ be the class of graphs G such that $mec(G) \leq f(|G|)$.

Corollary 5.7.16 *There is an unbounded function $f : \mathbb{N} \to \mathbb{N}$ such that* MC(FO, $\mathcal{C}_f$) *is fixed-parameter tractable.*

Another consequence of the theorem is that it allows us to apply the framework of localisation as developed in Section 5.7.3 to obtain the following result.

Corollary 5.7.17 *Let $\mathcal{C}$ be a class of graphs locally excluding a minor. Then the problem*

> MC(FO, $\mathcal{C}$)
> *Input:* $G \in \mathcal{C}, \varphi \in$ FO.
> *Parameter:* $|\varphi|$.
> *Problem:*   Decide $G \models \varphi$.

is fixed-parameter tractable.

The previous result has a number of algorithmic applications.

Corollary 5.7.18 *1. The following problem is fixed-parameter tractable.*

---

p-DOMINATING SET
   Input: Given graphs G, H such that $H \not\preceq G$ and $k \in \mathbb{N}$.
  Parameter: $k + |H|$.
   Problem: Decide whether G contains a dominating set of
                 size $\leq k$.

Analogous results hold for all other first-order definable parameterized problems, such as INDEPENDENT SET *and* CLIQUE *and also for problems such as deciding for a fixed graph G' whether G' has a homomorphism into G, or G' is an (induced) subgraph of G, where here the parameter can be taken to be $|H| + |G'|$.*

2. *Let C be a class of graphs locally excluding a minor. Then problem such as* DOMINATING SET, INDEPENDENT SET *etc. are fixed-parameter tractable on C. Furthermore, the problem, given graphs H and G such that $G \in C$, to decide whether H is homomorphic to G or H is an (induced) subgraph of G can be decided by fpt algorithms with parameter $|H|$.*

5.8 Characterising Logical Complexity under Structural Restrictions

The results presented in the previous sections have focussed primarily on methods to establish tractability results of logics on special classes of structures. The aim was to exhibit more and more general classes of structures on which first-order or monadic second-order model-checking becomes tractable. As we have seen in Section 5.2.4, first-order model checking is not fixed-parameter tractable in general (unless FPT = AW[∗]) and hence somewhere there must be a tractability border for the model-checking problem of these logics. Previous research has mostly approached this border from below by establishing tractability results. Quite as important is to establish intractability results, i.e. to approach this tractability border from above. This has so far been studied much less in the literature and the aim of this section is to survey some of the results that have been obtained in this direction.

5.8.1 Classifying Logical Tractability with Respect to Structural Restrictions

In the previous sections we have seen various examples for classes of graphs or structures on which model-checking for first- or monadic second-order logic becomes tractable. The picture described there (and illustrated in Figure 5.18 below) is as yet far from being complete and in particular it is not known whether any of the tractability results are actually strict. Surprisingly, not even for Courcelle's celebrated theorem it is known whether it can be extended to classes of unbounded tree width.

We therefore propose a research program which aims at providing a refined analysis of the complexity of logical formula evaluation with respect to specific classes of structures. More precisely, for the most commonly used logics we aim at identifying a property that precisely captures tractability of the logic in the sense that the logic is tractable on a class of structures if, and only if, the class has this particular property.

Such a classification would give completely new insights into the complexity of the logics and would provide researchers designing new query or specification languages based on these logics with valuable information for designing languages tailored towards their specific application areas.

It may not always be possible to find such a property that excactly characterises tractability of a logic within all classes of structures and possibly we will need to further restrict the admissible classes of structures, such as to classes closed under substructures. For instance, for first-order logic we conjecture that model-checking of FO on a class of structures *closed under substructures* is tractable if, and only if, the class is *nowhere dense* (see below).

There are two different, and somewhat complementary aspects to the results we envisage. The first aspect are tractability results as we have presented them in the previous sections. The other aspect are intractability results where we show that evaluation of formulas is hard whenever a class of structures does not have a particular property. In this context, this aspect has virtually not been studied in the literature before. We will present some recent and new intractability results in the following subsections.

5.8.2 Limits to Monadic Second-Order Model-Checking

Recall Courcelle's theorem (see Theorem 5.3.29 and Corollary 5.3.31) which states that MSO_2-model checking is fixed-parameter tractable on every class of structures of bounded tree-width. We will see in this section that in this generality, Courcelle's theorem can not be extended much beyond bounded tree-width.

The following result by Garey, Johnson and Stockmeyer and the fact that 3-colourability is MSO-definable immediately imply that MSO-model checking is not fixed-parameter tractable on the class of planar graphs.

Theorem 5.8.1 (*Garey, Johnson, Stockmeyer* [49]) *3-colourability is* NP-*complete on the class of planar graphs of degree at most* 4.

However, the class of planar graphs is a very specific class and this result does not rule out that Courcelle's theorem could possibly be extended to classes of unbounded but slowly growing tree-width. To show intractability results for MSO_2-model checking on classes of graphs of unbounded tree-width we first need to classify the degree of "unboundedness".

Definition 5.8.2 *Let* $f : \mathbb{N} \to \mathbb{N}$ *be a non-decreasing function. A class* C *of graphs has* f-*bounded tree width if* $\mathrm{tw}(G) \leq f(|G|)$ *for all* $G \in C$.

Hence, Courcelle's theorem applies to f-bounded classes of graphs for constant functions f. We will particularly be interested in classes of graphs whose tree width grows logarithmically in the size of the graphs and aim at proving that if the tree width C is not bounded logarithmically then MSO_2 model-checking is not tractable on C. A first step towards this direction appeared in [59, 60] where such a result was proved for classes of coloured graphs which we define next.

Let $\Sigma := \{B_1, \ldots, B_k, C_1, \ldots, C_l\}$ be a set of colours, where the B_i are colours of edges and the C_i are colours of vertices. A Σ-coloured graph, or simply Σ-graph, is an undirected graph G where edges may be coloured by $B_1, \ldots, B_k$ and vertices may be coloured by $C_1, \ldots, C_k$. We do not require any additional conditions such as edges having endpoints coloured in different ways, i.e. we do not require the colouring to be *proper* in the graph theoretical sense. To obtain logical structures, we let $\sigma := \{E, B_1, \ldots, B_k, C_1, \ldots, C_l\}$ be the signature containing binary relations $E, B_1, \ldots, B_k$ for edges and their colours and unary relations $C_1, \ldots, C_l$ for vertex colours.

Definition 5.8.3 *A class* C *of* Σ-*graphs is said to be* closed under Σ-*colourings if whenever* $G \in C$ *and* G' *is obtained from* G *by recolouring, i.e. the underlying un-colored graphs are isomorphic, then* $G' \in C$.

A class C *of* σ-*structures is* closed under colourings *if there is a class* C' *of (uncoloured) graphs such that* C *is the class of all* σ-*structures whose Gaifman-graphs are in* C'.

We aim at showing that if C is a class of graphs closed under colourings whose tree width is not bounded by a log-function then MSO_2-model checking is fixed-parameter intractable on C. The proof of this result relies on a reduction from an NP-complete problem to $MC(MSO_2, C)$ and for this to work it is not enough for the tree-width of C not to be bounded by a log-function $f : \mathbb{N} \to \mathbb{N}$, we must also be able to compute witnesses for this large tree-width efficiently. This leads to the following definition of effectively unbounded tree-width.

Definition 5.8.4 *The tree-width of a class C of graphs is effectively unbounded by a function $f : \mathbb{N} \to \mathbb{N}$ if there is a polynomial $p(x)$ such that for all n*

1. *there is a graph $G \in C$ of tree-width between n and $p(n)$ whose tree-width is not bounded by $f(|G|)$ and*
2. *given n, G_n can be constructed in time 2^{n^ε}, for some $\varepsilon < 1$.*

The tree-width of C is effectively unbounded poly-logarithmically if it is effectively unbounded by $\log^c n$, for all c.

We will particularly be interested in classes effectively unbounded by a function $f(n) := \log^c n$ for some small constant c. For such a function the second condition just says that we can compute witnesses for the high tree-width of C in time polynomial in their size, which is what we need for the reduction of an NP-complete problem to work. The first condition says that there are enough witnesses for the large tree-width of C so that there are actually enough graphs to reduce the problem to. The following result was proved in [60] (see also [59]).

Theorem 5.8.5 *Let Σ be a non-empty set of colours including at least one edge and two vertex colours. Let C be any class of Γ-coloured graphs closed under colourings.*

1. *If the tree-width of C is effectively unbounded poly-logarithmically then $\mathrm{MC}(\mathrm{MSO}, C)$ is not in XP, and hence in particular not fixed-parameter tractable, unless all problems in NP (in fact, all problems in the polynomial-time hierarchy) can be solved in sub-exponential time.*
2. *If the tree-width of C is effectively unbounded by $\log^{48} n$ then $\mathrm{MC}(\mathrm{MSO}, C)$ is not in XP unless SAT can be solved in sub-exponential time.*

The theorem together with Courcelle's theorem has the following corollary, as in the classes C_f colours can easily be replaced by suitable gadgets. Note, however, that the corollary also has a much simpler direct proof.

Corollary 5.8.6 *For any non-decreasing function $f : \mathbb{N} \to \mathbb{N}$ let*

$$C_f := \{G : \mathrm{tw}(G) \leq f(|G|)\}.$$

1. *If $f(n) > \log^{48} n$ for all n greater than some $n_0 \in \mathbb{N}$, then $\mathrm{MC}(\mathrm{MSO}_2, C_f) \notin \mathrm{XP}$ unless SAT can be solved in sub-exponential time.*
2. *If f is constant, then $\mathrm{MC}(\mathrm{MSO}_2, C_f) \in \mathrm{FPT}$.*

Theorem 5.8.5 gives a classification of tractability of MSO_2 on classes of coloured graphs. The restriction to coloured graphs is somewhat artificial as coloured graphs do not naturally occur very often. It does show, however, that Courcelle's theorem cannot be extended in full generality beyond logarithmic tree-width.

A much more natural result would be if closure under colours could be replaced by closure under subgraphs. I believe this is possible but it will require much more involved algorithmic techniques.

5.8.3 Limits to First-Order Model-Checking

In this section we will summarise some intractability results for first-order logic. As before, ideally we would like to completely classify the classes C of structures into those where $MC(FO, C)$ is FPT and where it is not. However, with the graph structure properties studied so far, it is unlikely that we can fully explore tractability for first-order model-checking as FO-model-checking is preserved under interpretations whereas properties such as excluding a minor or bounded tree-width are not.

Lemma 5.8.7 *If C is a class of graphs such that $MC(FO, C)$ is fixed-parameter tractable and D is a class of graphs first-order interpretable in C as described in Section 5.2.3, then first-order model-checking is fixed-parameter tractable on D.*

Corollary 5.8.8 *If $MC(FO, C)$ is fixed-parameter tractable then so is $MC(FO, D)$ for the class $D := \{G := (V, V^2 \setminus E) : (V, E) \in C\}$ of graphs whose complements are in C.*

Hence, if there is a graph property that precisely describes when FO model-checking is tractable, it has to be closed under edge-complementation or more generally under first-order interpretations. Note that the analogous result does not hold for MSO_2, as in general MSO_2 formulas on a graph cannot be rewritten to work on the complement graph instead.

In addition to studying further classes of graphs obtained from graph invariants it may therefore be beneficial to consider constructions that allow us to construct new classes C of graphs with tractable model-checking from other, known classes of graphs. For instance, one could try to generalise the constructions using tree-decompositions over classes of graphs. It is easily seen that if C is a class of graphs for which the appropriate version of Lemma 5.7.13 holds, then first-order model-checking is also tractable on the class of graphs that can efficiently be tree-decomposed over C. We refrain from giving a formal definition of this as, so far, its only application seems to be Theorem 5.7.14. Tree-decompositions are a special case where Feferman-Vaught style theorems can be applied. It may be worthwhile to consider further constructions that allow us to define new tractable model-checking intances from the classes we already know.

The previous lemma also has interesting consequences in its negative form, that is, it can be used to show intractability results as demonstrated in the next lemma.

Lemma 5.8.9 *For $k \in \mathbb{N}$ let $\mathcal{AD}_k$ be the class of graphs of maximum average degree at most k, where the* maximum average degree *of a graph G is the maximum of the average degrees of all subgraphs of G. For $k \geq 4$, MC(FO, $\mathcal{AD}_k$) is AW[∗]-hard, i.e. fixed-parameter intractable.*

Proof. Recall from Section 5.2.4 that MC(FO, GRAPH), the model-checking problem for FO on the class of all finite graphs, is AW[∗]-complete. Further, FO model-checking on the class of all graphs G can easily be reduced to FO model-checking on the class of incidence graphs $I(G)$. As incidence graphs have maximum average degree at most 4, the result follows immediately. $\qquad \square$

Hence, graph classes of bounded maximum average degree provide a first non-trivial upper bound for parameterized tractability of FO model-checking.

Towards another graph property that may yield fixed-parameter algorithms for first-order logic, consider again the proof of the previous lemma. Essentially, given a graph G we subdivide every edge once to obtain the incidence graph. For first-order logic, this does not pose much of a problem as we can easily rewrite the formula to deal with the subdivision. Similarly, if we replace every edge by a path of length k, i.e. subdivide a bounded number of times, then again we obtain small maximum average degree but we can easily rewrite first-order formulas to deal with these paths of fixed length.

Note that this essentially means that we replace every vertex by a graph of fixed radius, e.g. in the case of $k = 3$ we replace every vertex by a star. Hence, if we are interested in paramaterized tractability, then we should require our graphs to have bounded maximum average degree even after we contract neighbourhoods of a fixed radius. This idea is formalised in the notion of bounded expansion introduced by Nešetřil and Ossona de Mendez in [65, 66, 67].

An even more general concept of graphs is the concept of graph classes which are *nowhere dense*, introduced by Nešetřil and Ossona de Mendez in [68].

We say that H is a *minor at depth r* of G (and write $H \preccurlyeq_r G$) if H is a minor of G and this is witnessed by a minor map μ of H into G so that every vertex $v \in V(H)$ is mapped to a subgraph $\mu(v) \subseteq G$ which induces a graph of radius at most r. That is, for each $v \in V(H)$, there is a $w \in V(\mu(v))$ such that $\mu(v) \subseteq N_r^{\mu(v)}(w)$.

Definition 5.8.10 ([68]) *A class of graphs C is said to be* nowhere dense *if for every $r \geq 0$ there is a graph H_r such that $H_r \npreccurlyeq_r G$ for all $G \in C$.*

Conversely, if a class C of graphs is not nowhere dense then there is a radius r such that every graph H is a depth r minor of some graph $G_H \in C$. If, furthermore, C is closed under taking subgraphs, then the depth-d image I_H of H in G_H is itself a graph in C. Note that the size of I_H is polynomially bounded in H (for fixed r). Classes which are not nowhere dense are called *somewhere dense* in [68]. Let us call a class *effectively somewhere dense* if, given a graph H, a depth-d image $I_H \in C$ of H in a graph $G_H \in C$ can be computed in polynomial time.

As the following theorem shows, in terms of sparse classes of graphs, nowhere dense classes are the natural border for tractability of first-order logic.

Theorem 5.8.11 *If C is effectively somewhere dense and closed under taking subgraphs, then $MC(FO, C)$ is not fixed-parameter tractable unless $FPT = AW[*]$.*

The proof relies on the fact that we can interpret the class of all graphs in any effectively somewhere dense class of graphs which is closed under subgraphs, as every graph occurs as a depth d minor of a member of C and the depth-d image of this is itself a graph in C. Sub-divisions of a fixed length can be defined in first-order logic and hence model checking for first-order logic on the class of all graphs can be reduced to FO-model-checking on any effectively somewhere dense class of graphs which is closed under subgraphs.

Furthermore, it seems likely that on every nowhere dense class of graphs, first-order model checking is fixed-parameter tractable.

Conjecture 5.8.12 *If C is nowhere dense then $MC(FO, C) \in FPT$.*

If this conjecture could be proved then on subgraph closed classes of graphs, the property of being nowhere dense would exactly characterise the tractable cases.

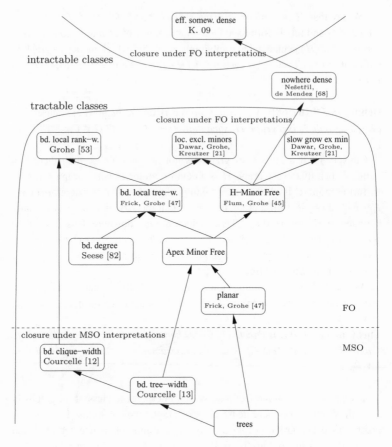

Figure 5.18 Summary of results

5.9 Conclusion

This paper gives an overview of algorithmic meta-theorems developed in recent years. See Figure 5.18 for a diagrammatic summary of the results presented in this paper.

As we have seen, first-order model-checking is fixed-parameter tractable on a wide range of graph classes defined by standard graph invariants such as tree-width or excluded minors. By localising these invariants we obtained even further tractable classes. However, we are still very far from a clear picture of where first-order model-checking is tractable and where it is not. Further research, in particular into intractability results is needed before we can hope for a clean and smooth theory.

Acknowledgements

I want to take this opportunity to express my gratitude towards Martin Grohe for the long discussions we had on graph minors and for his advise, patience and inspiration during the time I worked with him at the Humboldt University in Berlin.

Many thanks to Javier Esparza for proofreading this manuscript and for his patience as an editor. Special thanks for proofreading this manuscript go to Martin Grohe, Paul Hunter, Michael Kreikenbaum, Andras Salamon and Mark Weyer.

References

[1] I. Adler, M. Grohe, and S. Kreutzer. Computing excluded minors. In *Proc. of the 19th ACM-SIAM Symposium on Discrete Algorithms (SODA)*, pages 641–650, 2008.

[2] H. Andréka, J. van Benthem, and I. Németi. Modal languages and bounded fragments of predicate logic. *Journal of Philosophical Logic*, 27:217–274, 1998.

[3] S. Arnborg, D. Corneil, and A. Proskurowski. Complexity of finding embeddings in a *k*-tree. *SIAM Journal on Algebraic Discrete Methods*, 8:277–284, 1987.

[4] S. Arnborg, J. Lagergren, and D. Seese. Easy problems for tree-decomposable graphs. *Journal of Algorithms*, 12(2):308–340, 1991.

[5] Z. Bian and Q.-P. Gu. Computing branch decomposition of large planar graphs. In *Experimental Algorithms*, volume 5038 of *Lecture Notes in Computer Science*, pages 87–100. Springer, 2008.

[6] H. Bodlaender. A linear-time algorithm for finding tree-decompositions of small tree-width. *SIAM Journal on Computing*, 25:1305–1317, 1996.

[7] H. Bodlaender and D. M. Thilikos. Constructive linear time algorithms for branchwidth. In *Proc. of the 24th Int. Col. on Automata, Languages and Programming (ICALP)*, volume 1256 of *Lecture Notes in Computer Science (LNCS)*, pages 627–637. Springer, 1997.

[8] E. Börger, E. Grädel, and Y. Gurevich. *The Classical Decision Problem*. Springer-Verlag, 2001.

[9] J. R. Büchi. Weak second-order arithmetic and finite automata. *Zeitschrift für Mathematische Logik und Grundlagen der Mathematik*, 6:66–92, 1960.

[10] H. Comon, M. Dauchet, R. Gilleron, C. Löding, F. Jacquemard, D. Lugiez, S. Tison, and M. Tommasi. Tree automata techniques and applications. Available on: http://tata.gforge.inria.fr/, 2007. Release 12 October 2007.

[11] T. Cormen, C. Leiserson, R. Rivest, and C. Stein. *Introduction to Algorithms*. MIT Press, 2nd edition, 2001.

[12] B. Courcelle. An axiomatic definition of context-free graph grammars and applications to NLC grammars. *Theoretical Computer Science*, 55:141–181, 1987.

[13] B. Courcelle. Graph rewriting: An algebraic and logic approach. In J. van Leeuwen, editor, *Handbook of Theoretical Computer Science*, volume 2, pages 194 – 242. Elsevier, 1990.

266

[14] B. Courcelle. The monadic second-order theory of graphs VII: Graphs as relational structures. *Theoretical Computer Science*, 101:3–33, 1992.

[15] B. Courcelle, R.G. Downey, and M.R. Fellows. A note on the computability of graph minor obstruction sets for monadic second order ideals. *Journal of Universal Computer Science*, 3:1194–1198, 1997.

[16] B. Courcelle and J. Engelfriet. A logical characterization of the sets of hyper-graphs defined by hyperedge replacement grammars. *Mathematical Systems Theory*, 28:515–552, 1995.

[17] B. Courcelle, J. Engelfriet, and G. Rozenberg. Handle-rewriting hypergraphs grammars. *Journal of Computer and System Sciences*, 46:218–270, 1993.

[18] B. Courcelle, J. Makowski, and U. Rotics. Linear time solvable optimization problems on graphs of bounded clique-width. *Theory of Computing Systems*, 33(2):125–150, 2000.

[19] B. Courcelle and S. Olariu. Upper bounds to the clique width of graphs. *Discrete Applied Mathematics*, 101:77–114, 2000.

[20] B. Courcelle and S.-I. Oum. Vertex-minors, monadic second-order logic, and a conjecture by Seese. *Journal of Combinatorial Theory, Series B*, 97(1):91–126, 2007.

[21] A. Dawar, M. Grohe, and S. Kreutzer. Locally excluding a minor. In *Logic in Computer Science (LICS)*, pages 270–279, 2007.

[22] A. Dawar, M. Grohe, S. Kreutzer, and N. Schweikardt. Approximation schemes for first-order definable optimisation problems. In *Logic in Computer Science (LICS)*, pages 411–420, 2006.

[23] A. Dawar, M. Grohe, S. Kreutzer, and N. Schweikardt. Model theory makes formulas large. In *Proc. of the 34th International Colloquium on Automata, Languages and Programming (ICALP)*, volume 4596 of *Lecture Notes in Computer Science*, pages 913–924, 2007.

[24] E. Demaine and M. Hajiaghayi. Graphs excluding a fixed minor have grids as large as treewidth, with combinatorial and algorithmic applications through bidimensionality. In *Proc. of the 16th ACM-SIAM Symposium on Discrete Algorithms (SODA)*, pages 682–689, 2005.

[25] E. Demaine and M. Hajiaghayi. The bidimensionality theory and its algorithmic applications. *The Computer Journal*, pages 332–337, 2008.

[26] E. Demaine and M. Hajiaghayi. Linearity of grid minors in treewidth with applications through bidimensionality. *Combinatorica*, 28(1):19–36, 2008.

[27] E. Demaine, M. Hajiaghayi, F. Fomin, and D. Thilikos. Bidimensional parameters and local tree-width. *SIAM Journal of Discrete Mathematics*, 2004.

[28] E. Demaine, M. Hajiaghayi, and D. Thilikos. The bidimensional theory of bounded-genus graphs. *SIAM Journal of Discrete Mathematics*, 20(2):357–371, 2006.

[29] E. D. Demaine, M. Hajiaghayi, and K. Kawarabayashi. Algorithmic graph minor theory: Decomposition, approximation, and coloring. In *46th Annual Symposium on Foundations of Computer Science (FOCS)*, pages 637–646, 2005.

[30] R. Diestel. *Graph Theory*. Springer-Verlag, 3rd edition, 2005.

[31] J. Doner. Tree acceptors and some of their applications. *Journal of Computer and System Sciences*, 4:406–451, 1970.

[32] F. Dorn, F. Fomin, and D. Thilikos. Catalan structures and dynamic programming in *H*-minor-free graphs. In *Proc. of the 19th ACM-SIAM Symposium on Discrete Algorithms (SODA)*, pages 631–640, 2008.

[33] F. Dorn, F. Fomin, and D. Thilikos. Subexponential parameterized algorithms. *Computer Science Review*, 2:29–39, 2008.

[34] R. Downey and M. Fellows. Fixed-parameter tractability and completeness I: Basic results. *SIAM Journal on Computing*, 24:873–921, 1995.

[35] R. Downey and M. Fellows. Fixed-parameter tractability and completeness II: On completeness for W[1]. *Theoretical Computer Science*, 141:109–131, 1995.

[36] R. Downey and M. Fellows. *Parameterized Complexity*. Springer, 1998.

[37] H.-D. Ebbinghaus and J. Flum. *Finite Model Theory*. Springer, 2nd edition, 1999.

[38] H.-D. Ebbinghaus, J. Flum, and W. Thomas. *Mathematical Logic*. Springer, 2nd edition, 1994.

[39] J. Engelfriet and V. van Oostrom. Logical description of context-free graph languages. *Journal of Computer and System Sciences*, 3:489–503, 1997.

[40] P. Hliněný and S.-I. Oum. Finding branch-decompositions and rank-decompositions. Available at http://www.math.uwaterloo.ca/~sangil/, 2007.

[41] W. Espelage, F. Gurski, and E. Wanke. Deciding clique-width for graphs of bounded tree-width. In *Proc. of the 7th International Workshop on Algorithms and Data Structures (WADS)*, volume 2125 of *Lecture Notes in Computer Science*, pages 87–98, 2001.

[42] M. R. Fellows, F. A. Rosamond, U. Rotics, and S. Szeider. Clique-width minimization is np-hard. In *38th ACM Symposium on Theory of Computing (STOC)*, pages 354–362, 2006.

[43] M.R. Fellows and M.A. Langston. An analogue of the Myhill-Nerode theorem and its use in computingfinite-basis characterizations. In *Proceedings of the 30th Annual IEEE Symposium on Foundations of Computer Science*, pages 520–525, 1989.

[44] M.R. Fellows and M.A. Langston. On search, decision and the efficiency of polynomial-time algorithms. In *Proc. of the 21st ACM Symposium on Theory of Computing (STOC)*, pages 501–512, 1989.

[45] J. Flum and M. Grohe. Fixed-parameter tractability, definability, and model-checking. *SIAM Journal of Computing*, 31(1):113–145, 2001.

[46] J. Flum and M. Grohe. *Parameterized Complexity Theory*. Springer, 2006. ISBN 3-54-029952-1.

[47] M. Frick and M. Grohe. Deciding first-order properties of locally tree-decomposable structures. *Journal of the ACM*, 48:1148 – 1206, 2001.

[48] H. Gaifman. On local and non-local properties. In J. Stern, editor, *Herbrand Symposium, Logic Colloquium '81*, pages 105–135. North Holland, 1982.

[49] M. R. Garey, D. S. Johnson, and L. Stockmeyer. Some simplified NP-complete problems. In *Proc. of the 6th ACM Symposium on Theory of Computing (STOC)*, pages 47–63, 1974.

[50] J.F. Geelen, A.M.H. Gerards, N. Robertson, and G. Whittle. On the excluded minors for the matroids of branch-width *k*. *Journal of Combinatorial Theory, Series B*, 88:261–265, 2003.

[51] E. Grädel, C. Hirsch, and M. Otto. Back and forth between guarded and modal logics. *ACM Transactions on Computational Logics*, 3(3):418–463, 2002.
[52] M. Grohe. Local tree-width, excluded minors, and approximation algorithms. *Combinatorica*, 23(4):613–632, 2003.
[53] M. Grohe. Logic, graphs, and algorithms. In T.Wilke J.Flum, E.Grädel, editor, *Logic and Automata – History and Perspectives*. Amsterdam University Press, 2007.
[54] Q-P. Gu and H. Tamaki. Optimal branch-decomposition of planar graphs in $O(n^3)$ time. In *Proc. of the 32nd International Colloquium on Automata, Languages and Programming (ICALP)*, pages 373–384, 2005.
[55] F. Gurski and E. Wanke. The tree-width of clique-width bounded graphs without k_n, n. In *Proc. of the 27th International Workshop on Graph-Theoretic Concepts in Computer Science (WG)*, volume 1928 of *Lecture Notes in Computer Science*, pages 196–205, 2001.
[56] F. Gurski and E. Wanke. Minimizing NLC-width is NP-complete. In *Proc. of the 31st International Workshop on Graph-Theoretic Concepts in Computer Science (WG)*, pages 69–80, 2005.
[57] W. Hodges. *A shorter model theory*. Cambridge University Press, 1997.
[58] V. Jelínek. The rank-width of the square grid. In *Proc. of the 34th International Workshop on Graph-Theoretic Concepts in Computer Science (WG)*, 2008.
[59] S. Kreutzer. On the parameterised intractability of monadic second-order logic. In *Proc. of Computer Science Logic (CSL)*, pages 348–363, 2009.
[60] S. Kreutzer and S. Tazari. On brambles, grid-like minors, and parameterized intractability of monadic second-order logic. In *Symposium on Discrete Algorithms (SODA)*, 2010. to appear.
[61] K. Kuratowski. Sur le probléme des courbes gauches en topologie. *Fundamentae Mathematicae*, 15:271 – 283, 1930.
[62] J. Lagergren. Upper bounds on the size of obstructions and intertwines. *Journal of Combinatorial Theory, Series B*, 73:7–40, 1998.
[63] L. Libkin. *Elements of Finite Model Theory*. Springer, 2004.
[64] J. Makowski. Algorithmic uses of the Feferman-Vaught theorem. *Annals of Pure and Applied Logic*, 126:159–213, 2004.
[65] J. Nešetřil and P. Ossona de Mendez. Grad and classes with bounded expansion I. Decompositions. *European Journal of Combinatorics*, 29(3):760–776, 2008.
[66] J. Nešetřil and P. Ossona de Mendez. Grad and classes with bounded expansion II. Algorithmic aspects. *European Journal of Combinatorics*, 29(3):777–791, 2008.
[67] J. Nešetřil and P. Ossona de Mendez. Grad and classes with bounded expansion III. Restricted graph homomorphism dualities. *European Journal of Combinatorics*, 29(4):1012–1024, 2008.
[68] J. Nešetřil and P. Ossona de Mendez. On nowhere dense graphs. to appear in European Journal on Combinatorics, 2008.
[69] R. Niedermeier. *Invitation to Fixed-Parameter Algorithms*. Oxford University Press, 2006. ISBN 0-19-856607-7.
[70] S.-I. Oum. Rank-width is less than or equal to branch-width. *Journal of Graph Theory*, 57(3):239–244, 2008.
[71] S.-I. Oum and P.D. Seymour. Approximating clique-width and branch-width. *Journal of Combinatorial Theory, Series B*, 96:514 – 528, 2006.

[72] C. Papadimitriou. *Computational Complexity*. Addison-Wesley, 1994.

[73] N. Robertson, P. Seymour, and R. Thomas. Quickly excluding a planar graph. *Journal of Combinatorial Theory, Series B*, 1994.

[74] N. Robertson and P. D. Seymour. Graph minors V. Excluding a planar graph. *Journal of Combinatorial Theory, Series B*, 41(1):92–114, 1986.

[75] N. Robertson and P. D. Seymour. Graph minors X. Obstructions to tree-decompositions. *Journal of Combinatorial Theory B*, 52:153–190, 1991.

[76] N. Robertson and P.D. Seymour. Graph minors I–XXIII. Appearing in Journal of Combinatorial Theory, Series B since 1982.

[77] N. Robertson and P.D. Seymour. Graph minors III. Planar tree-width. *Journal of Combinatorial Theory, Series B*, 36:49–64, 1984.

[78] N. Robertson and P.D. Seymour. Graph minors XIII. The disjoint paths problem. *Journal of Combinatorial Theory, Series B*, 63:65–110, 1995.

[79] N. Robertson and P.D. Seymour. Graph minors XVI. Excluding a non-planar graph. *Journal of Combinatorial Theory, Series B*, 77:1–27, 1999.

[80] N. Robertson and P.D. Seymour. Graph minors XX. Wagner's conjecture. *Journal of Combinatorial Theory, Series B*, 92:325–357, 2004.

[81] D. Seese. The structure of models of decidable monadic theories of graphs. *Annals of Pure Applied Logic*, 53(2):169–195, 1991.

[82] D. Seese. Linear time computable problems and first-order descriptions. *Mathematical Structures in Computer Science*, 5:505–526, 1996.

[83] P. Seymour. A bound on the excluded minors for a surface, 1995. Unpublished manuscript.

[84] P. Seymour and R. Thomas. Call routing and the ratcatcher. *Combinatorica*, 14(2):217–241, 1994.

[85] J. W. Thatcher and J. B. Wright. Generalised finite automata theory with an application to a decision problem of second-order logic. *Mathematical Systems Theory*, 2:57–81, 1968.

[86] W. Thomas. Languages, automata, and logic. In G. Rozenberg and A. Salomaa, editors, *Handbook of Formal Languages*, volume III, pages 389–455. Springer, 1997.

[87] C. Thomassen. A simpler proof of the excluded minor theorem for higher surfaces. *Journal of Combinatorial Theory, Series B*, 70:306–311, 1997.

[88] M. Vardi. On the complexity of relational query languages. In *Proc. of the 14th Symposium on Theory of Computing (STOC)*, pages 137–146, 1982.

[89] M. Vardi. On the complexity of bounded variable queries. In *Proc. of the 14th ACM Symposium on Principles of Database Systems (PODS)*, pages 266–276, 1995.

[90] K. Wagner. Über eine Eigenschaft der ebenen Komplexe. *Mathematische Annalen*, 114:570–590, 1937.

[91] E. Wanke. k-NLC graphs and polynomial algorithms. *Discrete Applied Mathematics*, 54:251–266, 1994.

6

Model theoretic methods for fragments of FO and special classes of (finite) structures

Abstract

Some prominent fragments of first-order logic are discussed from a game-oriented and modal point of view, with an emphasis on model theoretic techniques for the non-classical context. This includes the context of finite

[a] Technische Universität Darmstadt, otto@mathematik.tu-darmstadt.de

model theory as well as the model theory of other natural non-elementary classes of structures. We stress the modularity and compositionality of the games as a key ingredient in the exploration of the expressive power of logics over specific classes of structures. The leading model theoretic theme is expressive completeness – or the characterisation of fragments of first-order logic as expressively complete over some class of (finite) structures for first-order properties with some prescribed semantic preservation behaviour. In contrast with classical expressive completeness arguments, the emphasis here is on explicit model constructions and transformations, which are guided by the game analysis of both first-order logic and of the imposed semantic constraints.

keywords: finite model theory, model theoretic games, bisimulation, modal and guarded logic, expressive completeness, preservation and characterisation theorems

6.1 Introduction

6.1.1 Expressiveness over restricted classes of structures

The purpose of this survey is to highlight game-oriented methods and explicit model constructions for the analysis of fragments of first-order logic, in particular in restriction to non-elementary classes of structures. The following is meant to highlight and preview some key points in terms of both the material to be covered and the perspective that we want to adopt in its presentation. All these points will be addressed in a more self-contained manner in the technical sections; an outline of the structure of the technical sections concludes this preview.

Varying the class of structures   The class of all finite structures is one prominent non-elementary class of interest, but recent developments in finite model theory have broadened the perspective. While the first tier of results in finite model theory, which set the stage and clarified much of the specifics of finite model theory, brought predominantly negative results ('failures' in comparison to classical model theory, the first and foremost being the 'failure of compactness' in finite model theory), a much more positive picture has emerged with a focus on specific classes of well-behaved finite structures rather than the class of all finite structures (cf. Weinstein's *tame fragments and tame classes* [51]). What good behaviour means for classes of structures, may of course depend on the model theoretic issue at hand. Nevertheless, there are some interesting recurring themes, revolving around tree-likeness on the one side and locality criteria on the other side, in delineating *well-behaved classes* of (finite) structures.

Expressiveness and expressive completeness   Our leading model theoretic theme in terms of results is that of expressive completeness. We regard expressive completeness results as classical hallmarks in the study of expressiveness of fragments of first-order logic. Think of a classical example like the Łos–Tarski existential preservation theorem (cf. Theorem 6.5.2) that a first-order formula is preserved under extensions if, and only if, it is logically equivalent to an existential formula. The preservation claim in this statement – that existential formulae are preserved under extensions – is a trivial exercise in syntactic induction, and its truth carries over to any restricted class of structures. The expressive completeness statement – that within first-order, the existential fragment is *expressively complete* for properties preserved under extensions – requires real model theoretic proof. The classical proof in [11] uses elementary extensions, whose availability hinges on the use of the compactness theorem for first-order logic. So that proof does not relativise to arbitrary restricted classes, and in fact the relativisation to the class of all finite structures is a typical example of a 'failure in finite model theory' (due to Tait and Gurevich, see for instance [15]). Preservation of a first-order property under extensions among finite structures does not imply expressibility in existential first-order logic over finite structures. Some instances of classical preservation theorems, like Łos–Tarski, fail in restriction to the class of all finite structures, but are true – with totally new proofs – in interesting restricted classes of finite structures (cf. Theorem 6.5.9 for results pertaining to extension preservation, from [3]). Other instances, like van Benthem's theorem concerning preservation under bisimulation (cf. Corollary 6.3.5), or, more classically, the Lyndon–Tarski theorem (cf. Theorem 6.5.3), which associates preservation under homomorphisms with the existential positive fragment, do have literal analogues in restriction to the class of all finite structures as well as to some other restricted classes of structures of interest (cf. sections 6.3.2 and 6.5.2) – with new proofs that do not draw on the classical proofs but shed interesting new light on the classical results as well. And in some few instances we know of expressive completeness results over restricted classes of (finite) structures that require more expressive fragments than the classical analogue; a recent example concerning bisimulation preservation is discussed in section 6.3.2.

Explicit model constructions and transformations   Compactness, and with it many of the typical model constructions prevalent in classical expressive completeness results, are typically not available over the restricted classes of structures under consideration. Where expressive completeness results can be obtained over non-elementary classes, the methods are very different from the classical ones. The technical crux of many expressive completeness results, classical or otherwise, consists in an *upgrading* of transfer or equivalence relations between structures. For instance, in the case of preservation under some equivalence relation $\leftrightharpoons$ like bisimulation associated with expressibility in the fragment L: here preservation under $\leftrightharpoons$ must be linked to preservation under finitary approximations $\leftrightharpoons^\ell$ to L-equivalence, finitary in the sense of finite index and in the sense that its classes are L-definable (think of approximations parameterised, e.g., by quantifier rank ℓ). As these finitary approximations $\leftrightharpoons^\ell$ are rougher than full $\leftrightharpoons$, the task of showing that every first-order property φ preserved under $\leftrightharpoons$ is even preserved under some $\leftrightharpoons^\ell$, involves model theoretic transformations that allow us to boost $\leftrightharpoons^\ell$ either to $\leftrightharpoons$ or to some other equivalence under which φ is preserved (e.g., on account of being first-order of a certain quantifier rank). The classical treatment of the Łos–Tarski theorem, for instance, can similarly be viewed as an upgrading of a transfer relationship $\mathfrak{A} \Rightarrow_\exists \mathfrak{B}$ (existential sentences true in $\mathfrak{A}$ are also true in $\mathfrak{B}$), or of its finitary approximations, to a substructure relationship between elementarily equivalent companion structures of $\mathfrak{A}$ and $\mathfrak{B}$. (In this case, $\mathfrak{B}$ admits an elementary extension that embeds $\mathfrak{A}$ as a substructure, by compactness.) It follows that any first-order φ preserved under extensions is preserved under $\Rightarrow_\exists$, and – by another compactness argument – therefore also under some finite quantifier rank approximation $\Rightarrow_\exists^\ell$ to $\Rightarrow_\exists$.

As will be discussed in section 6.3.2, such upgrading arguments tend to proceed in orthogonal directions of entirely different character, depending on whether they are based on classical compactness arguments (often involving elementary chains and saturation) or on explicit and finitary model transformations, which may also be carried out within some restricted, non-elementary class of structures like the class of just all finite structures. Explicit model constructions and transformations can thus sometimes replace the sweeping classical compactness arguments that guarantee the existence of nice and smooth (but typically infinite) representatives of the structures at hand, in which crucial technicalities (e.g., back-and-forth arguments) can be dealt with more elegantly. But there is also something to be gained, even from the classical point of view, from the more explicit, more controlled and more constructive nature of the alternative model transformations: in key examples of expressive completeness results to be discussed below, for instance, bounds on the quantifier rank of the target formulae are an integral part of the proofs based on explicit model constructions and transformations. In this sense, the alternative approach, which is necessitated by the loss of compactness in finite model theory, can offer a new perspective and sometimes extra information on classical results.

Model theoretic games The equivalences and transfer relations between structures underlying semantic preservation properties on the one hand, and logical equivalences or transfer relations induced by fragments of first-order logic on the other hand, are closely linked to model theoretic games or back-and-forth systems. As pointed out above, upgrading arguments between these equivalences and suitable finitary approximations, which are themselves naturally cast as game equivalences, play a crucial role in expressive completeness proofs. The methodological importance of model theoretic games, both to understand the semantics and expressive power of logics and to guide the desired explicit model constructions or transformations (over restricted classes of structures), is being put at the centre of this presentation. We shall here especially discuss variants of the classical Ehrenfeucht–Fraïssé game and the first-order model checking game for several fragments of first-order logic. A prominent place among these variants is given to the modal Ehrenfeucht–Fraïssé game, or bisimulation game. In section 6.3, bisimulation games and model transformations that respect bisimulation feature prominently in the discussion of expressive completeness results for modal logics over various classes of Kripke structures. Also locality of first-order logic in the sense of Gaifman's theorem (cf. Theorem 6.2.13) is presented in terms of the modularity of the first-order Ehrenfeucht–Fraïssé game w.r.t. locality in the Gaifman graph. Locality-based approximations to first-order equivalence also play a role in some of the expressive completeness results for modal logics, or in the upgrading between approximate levels of bisimulation and first-order equivalence. Structurally, the concept of locality will also be important in connection with classes of structures defined in terms of wideness criteria in section 6.5.

Bisimulation as the game of games Putting games – model checking games that define the semantics of a logic and Ehrenfeucht–Fraïssé model comparison games – at the centre of the analysis of fragments of first-order logic, it becomes very natural to adopt a modal perspective [9, 10] and to relate other fragments and their games to the bisimulation game. We thus draw on bisimulation games and bisimulation equivalence not just in the study of modal fragments but also on its role as an equivalence between game graphs that encapsulate the semantics of other fragments. The connection is made by looking at the natural game graphs associated with model checking games or Ehrenfeucht–Fraïssé games as Kripke structures. The elements of these Kripke structures are formed by the *observable configurations* in the underlying structures, their accessibility relations reflect the transitions between game positions, which in turn reflect the available quantification patterns of the fragment at hand. For the modal fragment itself, the structure (Kripke structure, transition system) *is* its own game graph, in which the elements can be navigated along the edges (of the given accessibility or transition relation). Richer fragments have access to more complex types of configurations within structures and possibly more complex rules for navigation between configurations. For instance, in the k-variable fragment $\mathsf{FO}^k \subseteq \mathsf{FO}$ we deal with arbitrary configurations consisting of up to k elements, while in the guarded fragment $\mathsf{GF} \subseteq \mathsf{FO}$ the configurations need to be covered by some relational ground atom. This view may not directly offer new technical insights, but has the advantage of making explicit a unifying and, I think, intuitive framework whose specialisations to individual fragments are of course very well understood.

Structure of the paper   The overall structure of the paper is as follows. In section 6.2 we review model checking and model comparison games for FO and some of its fragments from a modal perspective; we also discuss Gaifman locality in relation to the FO Ehrenfeucht–Fraïssé game. Section 6.3 deals with expressive completeness issues for modal logics over specific classes of transition systems. The extension of the concept of bisimulation from graphs to hypergraphs, its relationship with the guarded fragment and a connection with extension properties for partial automorphisms is discussed in section 6.4. In section 6.5 we turn to locality based techniques for special classes of relational structures, and to expressive completeness for preservation under extensions and homomorphisms.

Sections 6.2 and 6.3 are meant to be fairly expository, and may serve either as a brief introduction to the fragments and methods discussed, or as an invitation to re-discover some rather familiar concepts in a slightly different light. Sections 6.4 and 6.5 are more technical and also less self-contained. To a large extent they may, on the other hand, also be considered independently of the first part. The intention is to give at least some high-level account of some more recent results and developments in the framework of this survey.

6.1.2 Basic terminology and notational conventions

Structures and assignments   Throughout we only consider relational structures. Typically τ will be a finite relational signature, and we refer to the maximal arity of relations in τ as its *width*. A τ-structure with universe A will usually be denoted as $\mathfrak{A} = (A, (R^{\mathfrak{A}})_{R \in \tau})$, but we often omit superscripts where the structure is clear from context.

Within a τ-structure $\mathfrak{A}$, we look at (partial) assignments (to an official set of first-order variables $x_1, x_2, \ldots$), described by partial functions $\beta: (x_i) \to A$. Assignments to finite tuples of variables are often regarded as momentarily fixed parameter tuples, like $\mathbf{a} = (a_1, \ldots, a_k) \in A^k$ as an assignment $\beta: (x_i \mapsto a_i)_{i=1,\ldots,k}$. Such (finite) assignments will also play a role in games as *configurations* (tuples of marked elements) within a structure, often directly associated also with the substructure induced on the subset $[\mathbf{a}] := \{a_1, \ldots, a_k\} \subseteq A$. Because we do not want to clutter terminology with a fine distinction between tuples and assignments, we also think of assignments (which officially are assignments to variables x_i) as partial functions $\beta: i \mapsto \beta(i)$ over a domain of positive natural numbers. Notation for modifications of assignments is as in $\beta \frac{a}{i}$, for the assignment obtained by changing (or extending) β at i (at x_i) to take the value a. For the semantics of formulae $\varphi(\mathbf{x})$ with free variables among those listed in the tuple $\mathbf{x}$, notations $\mathfrak{A}, \mathbf{a} \models \varphi$, $\mathfrak{A} \models \varphi[\mathbf{a}]$, and $\mathfrak{A}, \beta \models \varphi$ are used interchangeably, if β is an assignment to (at least) the free variables of φ and assigns $\mathbf{a}$ to $\mathbf{x}$.

Among important specific types of structures we mention the following to clarify terminology. Other more specific classes of structures will be introduced at appropriate places.

Directed and undirected graphs are structures over relational vocabularies of width 2, i.e., we admit several binary relations (edge-labelled directed graphs) and unary predicates (vertex colours). More traditional plain directed graphs are a special case, with just a single binary edge relation. We also view directed edge-labelled and vertex-coloured graphs as *transition systems*, with several transition relations and atomic state predicates. Such transition systems are just a terminological variant of *Kripke structures*, as the structures for modal logics. Undirected graphs are graphs with a single edge relation that is symmetric and irreflexive, viewed as a special case of directed graphs.

A *(directed) tree* is a directed graph that has a *root* w.r.t. the union of its binary relations such that every other element is reachable on a unique edge-labelled directed path from this node. Note that this implies irreflexivity (no loops), antisymmetry (no edges in opposite directions, not even with different labels) and that there are no multiple edges (with different labels). More generally, a directed graph or transition system is called *simple* if it has no loops and no multiple edges (not even in opposite directions).[1]

---

[1] In section 6.3.2 we also discuss transitive tree structures, which are trees in the partial order sense, not in the graph sense, but that will be highlighted there.

Hypergraphs, which are at the centre of section 6.4, are not regarded as relational structures but as second-order structures of the format $H = (A, S)$ with a universe A and a subset of the power set $S \subseteq \mathcal{P}(A)$ as the set of hyperedges. We shall encounter hypergraphs as auxiliary combinatorial structure, induced by relational structures, but will not look at logics over hypergraphs.

Gaifman graph and distance   With any structure in a finite relational vocabulary τ we associate an undirected graph, its Gaifman graph.

Definition 6.1.1   The *Gaifman graph* of the τ-structure $\mathfrak{A}$ is the undirected graph $G(\mathfrak{A}) = (A, E^{G(\mathfrak{A})})$ with the same universe A and an edge $(a, b) \in E^{G(\mathfrak{A})}$ for $a \neq b$ if a and b occur together in some tuple within some relation $R^{\mathfrak{A}}$, $R \in \tau$.

The associated notion of *Gaifman distance* is just ordinary graph distance (minimal length of a connecting path, or infinity) between elements in $G(\mathfrak{A})$. We denote this distance as $d(\cdot, \cdot)$. Finite distance relations like $d(x, y) \leqslant k$ are clearly FO-definable in $\mathfrak{A}$. In graphs (τ finite and of width 2), $d(x, y) \leqslant 1$ is quantifier free definable, while in general the required quantifier rank is the width of τ minus 2. An easy induction shows that $d(x, y) \leqslant 2^q$ is definable by a first-order formula $\varphi(x, y)$ for any finite τ.

Definition 6.1.2   The *Gaifman neighbourhood of radius ℓ*, or *ℓ-neighbourhood* for short, of an element a in $\mathfrak{A}$ is the subset $N^\ell(a) = \{b \in A \colon d(a, b) \leqslant \ell\} \subseteq A$. By extension, the ℓ-neighbourhood of a tuple $\mathbf{a} = (a_1, \ldots, a_k)$ in $\mathfrak{A}$ is the union of the $N^\ell(a_i)$.

A subset (or tuple) in $\mathfrak{A}$ is *ℓ-scattered* if its elements (or components) have pairwise distance greater than 2ℓ (i.e., if their ℓ-neighbourhoods are disjoint).

By the above considerations, ℓ-neighbourhoods of tuples, or the property of a tuple to be ℓ-scattered, are all first-order definable, for every $\ell \in \mathbb{N}$ and for any fixed finite τ.

A relational structure is called *acyclic* if its Gaifman graph is acyclic; for directed graphs as relational structures, this is different from the usual notion which only forbids directed cycles.

A directed graph or transition system is *ℓ-acyclic* if its Gaifman graph is acyclic in every ℓ-neighbourhood (this rules out *undirected* cycles of lengths up to $2\ell + 1$).

Logics   We write FO for first-order logic, or more specifically FO[τ] for the set of first-order formulae over vocabulary τ. The set of free variables of a first-order formula φ is denoted free(φ). Notation as in $\varphi = \varphi(\mathbf{x})$ indicates that free(φ) $\subseteq$ [$\mathbf{x}$] (the set of variables listed as components of the tuple $\mathbf{x}$).

Quantifier-rank is defined as usual for first-order formulae, and denoted qr(φ). Atomic and quantifier-free types of tuples $\mathbf{a}$ in a τ-structure $\mathfrak{A}$ provide full descriptions of $\mathbf{a}$ at the quantifier-free level. Formally we may define the atomic type of $\mathbf{a}$ (in a matching tuple of variables, so that $\beta \colon \mathbf{x} \mapsto \mathbf{a}$ is appropriate as an assignment) as the set of all atomic and negated atomic formulae $\alpha(\mathbf{x})$ in variables $\mathbf{x}$ for which $\mathfrak{A} \models \alpha[\mathbf{a}]$. It is clear that the correspondingly defined quantifier-free type is fully determined by the atomic type, and that both can be summarised by a single quantifier-free formula in case τ is finite. The atomic or quantifier-free type of $\mathbf{a}$ in $\mathfrak{A}$ fully determines the isomorphism type of $\mathfrak{A} \!\restriction\! [\mathbf{a}]$ (of configuration $\mathbf{a}$ in $\mathfrak{A}$).

FO^k stands for the k-variable fragment of FO, which uses only the variable symbols $x_1, \ldots, x_k$. The finite variable fragments have played a very prominent role in the development of finite model theory as witnessed for instance in [15, 33]; we shall not focus on these fragments very much here, but treat the associated k-pebble games as a typical and natural example in the exposition of section 6.2.

Apart from fragments of FO, we occasionally look at its infinitary extension FO_∞ (classically denoted $L_{\infty\omega}$), which extends the syntactic framework of FO by allowing disjunctions and conjunctions over arbitrary sets of formulae. Connectedness of graphs, for instance, becomes definable in FO_∞ with the use of an infinite disjunction to express "$d(x, y) < \infty$" as "$\bigvee \{d(x, y) \leqslant n \colon n \in \omega\}$". Formulae in FO_∞ have ordinal quantifier rank, defined by the usual inductive clauses extended by taking suprema for infinite disjunctions or conjunctions. The quantifier-rank of the formula "$d(x, y) < \infty$" would thus be ω, that of the natural sentence defining connectivity $\omega + 2$. Similar infinitary extensions naturally arise, e.g., for the modal fragment to be discussed next.

Basic modal logic is denoted ML, or $\mathsf{ML}[\tau]$ for a given vocabulary of width 2 appropriate for transition systems (Kripke structures). We typically use a τ with binary transition relations E_α (regarding the indices α as edge labels) and unary predicates P_j (associated to atomic state properties or atomic propositions p_j). The formulae of $\mathsf{ML}[\tau]$ are generated from the atomic propositions p_j by means of boolean connectives and modal quantifications with $\Diamond_\alpha$ or $\Box_\alpha$. The defining clause for the semantics of $\varphi = \Diamond_\alpha \psi$, say at a state a in a τ-structure $\mathfrak{A}$, is

$$\mathfrak{A}, a \models \varphi \qquad \text{iff} \qquad \mathfrak{A}, b \models \psi \text{ for some } b \text{ such that } (a, b) \in E_\alpha^{\mathfrak{A}},$$

and dually for $\Box_\alpha \psi$, which is equivalent to $\neg \Diamond_\alpha \neg \psi$. We also view $\mathsf{ML}[\tau]$ as a fragment of $\mathrm{FO}[\tau]$, having only formulae in one free variable, via the standard translation that associates p_j with $P_j x$ and $\Diamond_\alpha \psi$ with $\exists y(R_\alpha xy \wedge \psi(y))$ so that, dually, $\Box_\alpha \psi$ is associated with $\forall y(R_\alpha xy \rightarrow \psi(y))$. This is briefly reviewed in connection with the model checking game for modal logic in section 6.2.3.

The extension of basic modal logic with modal quantification backward along E_α (*inverse modalities*) is denoted ML^-; the extension by a *global modality*, corresponding to the introduction of modal quantification associated with the full binary relation, is denoted $\mathsf{ML}^\vee$; the combined extension with both these additions is $\mathsf{ML}^{-\vee}$. For background in connection with our treatment of modal logics and much more material on the model theory of modal logics see in particular [17].

The *guarded fragment* GF is defined to be a syntactic fragment of FO consisting of formulae in which all quantifications are relativised as in

$$\varphi(\mathbf{x}) = \exists \mathbf{y}(\alpha(\mathbf{x}') \wedge \psi(\mathbf{x}')), \text{ or}$$
$$\varphi(\mathbf{x}) = \forall \mathbf{y}(\alpha(\mathbf{x}') \rightarrow \psi(\mathbf{x}')),$$

where $\alpha(\mathbf{x}')$ is an atomic τ-formula (a relational atom, or an equality: the *guard atom*) such that $\mathrm{free}(\psi) \subseteq \mathrm{var}(\alpha)$ (and $\mathbf{y}$ is a sub-tuple of $\mathbf{x}'$ such that $[\mathbf{x}'] \setminus [\mathbf{y}] \subseteq [\mathbf{x}]$).

The quantification pattern of guarded logic extends that of modal logic. For a modal vocabulary τ, $\mathsf{GF}[\tau]$ properly contains (the standard first-order translations of) $\mathsf{ML}[\tau]$ and even $\mathsf{ML}^{-\vee}[\tau]$. One motivation for the study of the guarded fragment stems from the analogy with modal logic, and the extension of modal quantification patterns from Kripke structures to more general relational structures. Guarded fragments were proposed in [2] with a view to explaining the good algorithmic and model theoretic properties of modal logics in a richer fragment of first-order logic and other than the 2-variable fragment [23]; see [21]. In many ways the guarded fragment has been shown to be a rather well-behaved intermediary between first-order and modal logic, in terms of its model theoretic and algorithmic properties. For instance (like modal logic and unlike FO^k for $k \geqslant 3$), GF has the finite model property and is decidable: the satisfiability problem for $\mathsf{GF}[\tau]$ is complete for deterministic exponential time if τ is fixed (more precisely, for any fixed bound on the width of τ), and complete for doubly exponential time without this constraint [21]. Similarly to the tree model property of modal logic (which is a consequence of bisimulation invariance and the model transformation of tree unfolding, see in particular section 6.3.1), GF has a generalised tree model property, which similarly stems from invariance under guarded bisimulation and the availability of guarded tree unfoldings. For these considerations we refer to the discussion in section 6.4.2, where we interpret these phenomena in the light of a generalisation of bisimulations from graphs to hypergraphs. For further results concerning the model theory of GF and some of its generalisations see [21, 31, 25, 24, 8, 32] among many others.

The semantics of the above-mentioned fragments, though assumed familiar, will be reviewed again in section 6.2.3 when we discuss the associated model checking games. There we shall proceed in the order of increasing specialisation, from FO to FO^k to GF to (variants of) ML.

6.2 Model theoretic games and bisimulation

As mentioned above, we adopt a non-standard perspective of looking at first-order logic (and some of its fragments) through modal eyes. Connections are made through games, at two levels: at the level of *model checking games*, which capture the semantics, and at the level of *model comparison games*, which capture degrees of logical indistinguishability between structures.

No technical knowledge of model checking games and Ehrenfeucht-Fraïssé games is assumed. The reader who has some familiarity with model checking games and the Ehrenfeucht-Fraïssé technique for various fragments and extensions of FO on the other hand, will recognise the familiar notions in a slightly different perspective.

6.2.1 The semantic game: verifier vs. falsifier

We take a look at the first-order model checking game from a modal point of view. We shall then want to present some fragments of first-order logic in terms of restricted game boards; the same view will uniformly be applied to the Ehrenfeucht–Fraïssé model comparison games in the next section.

A transition system of observable configurations With the relational vocabulary τ associate the vocabulary τ^* consisting of binary transition relations E_i for $i \geqslant 1$ and unary predicates P_θ for atomic τ-types $\theta = \theta(\mathbf{x})$ in finite tuples of variables from $(x_i)_{i\geqslant1}$. With a τ-structure $\mathfrak{A}$ associate the following τ^* transition system $\mathcal{O}(\mathfrak{A})$ of *observable configurations* over $\mathfrak{A}$:

– the universe of $\mathcal{O}(\mathfrak{A})$ is the set of partial assignments to variables $(x_i)_{i\geqslant1}$;

– E_i is interpreted as $\{(\beta, \beta\frac{a}{i}): a \in A\}$ (modifications of assignments at x_i);

– P_θ as the set of assignments β satisfying θ (in particular $\text{var}(\theta) \subseteq \text{dom}(\beta)$).

In a straightforward manner one obtains a uniform translation from $FO[\tau]$ over $\mathfrak{A}$ to $ML[\tau^*]$ over the associated $\mathcal{O}(\mathfrak{A})$. This translation,

$$FO[\tau] \longrightarrow ML[\tau^*]$$
$$\varphi(\mathbf{x}) \longmapsto \varphi^*,$$

is such that for all β with $free(\varphi) \subseteq dom(\beta)$:

$$\mathfrak{A}, \beta \models \varphi \quad \Leftrightarrow \quad \mathcal{O}(\mathfrak{A}), \beta \models \varphi^*.$$

At the quantifier-free level, $\varphi = \varphi(\mathbf{x})$ translates into

$$\varphi^* := \bigvee \{ P_\theta : \varphi \in \theta, var(\theta) = var(\varphi) \};$$

the translation is compatible with boolean connectives; and existential quantification translates into a modal diamond in a natural manner, as in

$$\varphi = \exists x_i \psi(\mathbf{x}) \quad \longmapsto \quad \varphi^* = \Diamond_i \psi^*.$$

Note that the modal vocabularies involved are a priori infinite; this can be avoided if we restrict attention to the k-variable fragment $FO^k[\tau]$ for fixed k and fixed finite relational vocabulary τ. In this case, there are only finitely many P_θ corresponding to atomic τ-types in variables $\mathbf{x} = (x_1, \ldots, x_k)$; we may restrict attention to full assignments to all the variables $\{x_1, \ldots, x_k\}$, which can be identified with A^n; and we just retain k transition relations E_i for $1 \leqslant i \leqslant k$. Further natural restrictions to be discussed in section 6.2.3 lead to modal and guarded logics.

The model checking game   The idea to associate a two-person game with the semantics of first-order logic goes back at least to Lorenz' and Lorenzen's dialogue games [40, 41] between a proponent and an opponent of some assertion. The current interest in these games stems not from foundational issues but from their algorithmic content, or more precisely from their conceptual strengths towards the design of efficient model checking algorithms, see, e.g., [22, 50].

With formulae φ and τ-structures $\mathfrak{A}$ with partial assignments β we associate a game played by two players, V (verifier) and F (falsifier) such that the winning positions in the game determine whether or not $\mathfrak{A}, \beta \models \varphi$.

We present this basic and simple idea in a modular fashion that uses the transition system of observable configurations as one constituent of the game (representing the structure input to the model checking problem). The other constituent is essentially the syntax tree of the formulae to be checked (representing the formula input to the model checking problem). For a transparent account of the algorithmic content of this game, and its complexity analysis, compare [22].

Let $\Phi \subseteq \mathrm{FO}[\tau]$ be a set of negation normal form formulae that is closed under subformulae (negation normal form restricts the occurrence of negations to negated atoms). Let $S(\Phi)$ be the transition system whose universe is Φ, with transition relations $E_\vee$, $E_\wedge$, $E_{\exists x_i}$ and $E_{\forall x_i}$ ($i \geqslant 1$) interpreted as follows.

$E_\vee$ contains the pairs (φ, φ_1) and (φ, φ_2) for $\varphi = \varphi_1 \vee \varphi_2 \in \Phi$; similarly for $E_\wedge$;

$E_{\exists x_i}$ consists of all pairs (φ, ψ) for $\varphi = \exists x_i \psi \in \Phi$; similarly for $E_{\forall x_i}$.

The game graph $\mathbb{G} := \mathbb{G}(\mathfrak{A}, \Phi)$ for the Φ model checking game over $\mathfrak{A}$ may then be interpreted in a subsystem of the product system

$$\mathcal{O}(\mathfrak{A}) \times S(\Phi).$$

More specifically, the universe of $\mathbb{G}(\mathfrak{A}, \Phi)$ is the set of all syntactically appropriate assignment/formula pairs, $\{(\beta, \varphi) : \mathrm{free}(\varphi) \subseteq \mathrm{dom}(\beta)\}$. The relevant transition relations of $\mathbb{G}(\mathfrak{A}, \Phi)$ are

in $\mathbb{G}(\mathfrak{A}, \Phi)$		in $\mathcal{O}(\mathfrak{A})$		in $S(\Phi)$	
$E_\vee$	$:=$	id	$\times$	$E_\vee$	(disjunctive moves)
$E_\wedge$	$:=$	id	$\times$	$E_\wedge$	(conjunctive moves)
$E_{i,\exists}$	$:=$	E_i	$\times$	$E_{\exists x_i}$	(existential moves)
$E_{i,\forall}$	$:=$	E_i	$\times$	$E_{\forall x_i}$	(universal moves)

As atomic predicates we use P_V and P_F, which partition the universe of $\mathbb{G}(\mathfrak{A}, \Phi)$ according to:

$$P_F^{\mathbb{G}} = \big\{(\beta, \varphi) : \varphi = \varphi_1 \wedge \varphi_2 \text{ or } \varphi = \forall x_i \psi\big\}$$
$$\cup \big\{(\beta, \varphi) : \varphi \text{ atomic or negated atomic}, \mathfrak{A}, \beta \models \varphi\big\},$$
$$P_V^{\mathbb{G}} = \big\{(\beta, \varphi) : \varphi = \varphi_1 \vee \varphi_2 \text{ or } \varphi = \exists x_i \psi\big\}$$
$$\cup \big\{(\beta, \varphi) : \varphi \text{ atomic or negated atomic}, \mathfrak{A}, \beta \not\models \varphi\big\}.$$

The rules of the game are then simply the following, according to which the players move a pebble in the game graph $\mathbb{G}$:
- Positions in P_V require a move by V:
  V moves along any $E_\vee$- or $E_{i,\exists}$-edge (as available in current position);
  V loses when stuck for a move.
- Positions in P_F require a move by F:
  F moves along any $E_\wedge$- or $E_{i,\forall}$-edge (as available in current position);
  F loses when stuck for a move.

As formula complexity is strictly reduced in each move, all plays are finite. Positions in which neither player can move are terminal positions for the game and the player who ought to move has lost. This happens exactly in positions associated with atomic or negated atomic formulae, and here the attribution of these nodes to V and F is such that V wins (because F ought to move) if $\mathfrak{A}, \beta \models \varphi$, and vice versa. Clearly the game is positionally determined, and the following is proved by an easy induction on the structure of the formula (or on the length of the remaining game).

Lemma 6.2.1 *The verifier V has a winning strategy in the model checking game on $\mathfrak{A}$ precisely in those positions (β, φ) for which $\mathfrak{A}, \beta \models \varphi$.*

Let us sketch part of the game graph in one tiny example. For a binary relation R consider the formula $\varphi(x) = \exists y\big(Rxy \wedge \forall x(Rxx \vee Rxy)\big)$ over the R-structure $\mathfrak{A}$ with two elements a and b and with R-edges as indicated by arrows:

$$a \longrightarrow b \, \circlearrowright$$

The model checking game to determine whether $\mathfrak{A} \models \varphi[a]$ has positions (β, ψ) where ψ is one of the subfurmulae of φ and β a (partial) assignment to variables x, y. We may represent β by an $\{a, b, \cdot\}$-word of length 2 and enumerate the subformulae ψ as $\varphi_0 := Rxx$, $\varphi_1 := Rxy$, $\varphi_2 = \varphi_0 \vee \varphi_1$, $\varphi_3 := \forall x \varphi_2$, $\varphi_4 := \varphi_1 \wedge \varphi_3$ such that $\varphi = \exists y \varphi_4$.

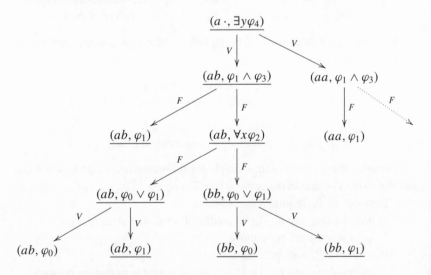

In this partial sketch of the game tree, winning positions for V are underlined.

There is a natural variant of the model checking game that does not restrict formulae to negation normal form. The transition corresponding to the elimination of a negation, say from $\neg\varphi$ to φ, corresponds to a swap of players' roles. Let us therefore call the players neutrally player 1 and player 0. Positions in the game graph are extended by an extra component $\wp \in \{0, 1\}$ to indicate which of the two players acts as verifier; the opponent, $\bar{\wp}$, correspondingly acts as falsifier. The two component games, $\mathbb{G}(\mathfrak{A}, \Phi) \times \{0\}$ and $\mathbb{G}(\mathfrak{A}, \Phi) \times \{1\}$ are each as before (but not insisting on negation normal form formulae, and with player $\wp$ in the role of V), and linked by $E_\neg$-edges from $(\beta, \neg\varphi, \wp)$ to $(\beta, \varphi, \bar{\wp})$. $E_\neg$-edges prescribe forced moves (for player $\wp$ say, but it does not matter) from configurations in which the leading connective of φ is a negation. Then the winning positions of player 1 are those $(\beta, \varphi, \wp)$ in which either $\wp = 1$ and $\mathfrak{A}, \beta \models \varphi$ or $\wp = 0$ and $\mathfrak{A}, \beta \not\models \varphi$.

It is also straightforward to adapt the model checking game to deal with FO_∞ rather than FO. $E_\vee$ and $E_\wedge$ can have infinite out-degree reflecting the syntax of infinitary disjunctions and conjunctions; everything else remains just the same; in particular plays are still finite, albeit not necessarily with a uniform finite bound.

6.2.2 The comparison game: back and forth

The familiar Ehrenfeucht–Fraïssé style model comparison games are two player games played over two structures. A game configuration in these games may be seen as a pairing between two observable configurations, one from each structure. The game is such that the winning positions determine whether or not (or to which degree) these two observable configurations are logically indistinguishable. We present the basic idea in the slightly non-standard terminology of (pairings between) observable configurations in order to highlight the connection between the comparison games and the model checking games. This point of view will contribute to a rather uniform presentation of fragments via restrictions imposed at the level of observable configurations.

The first-order Ehrenfeucht–Fraïssé game Consider two τ-structures $\mathfrak{A}$ and $\mathfrak{A}'$ over the same finite relational vocabulary τ. For partial assignments β, β' to the same (finite) subset of variables $(x_i)_{i \geqslant 1}$ in $\mathfrak{A}$ and $\mathfrak{A}'$, respectively, we write

$$\mathfrak{A}, \beta \equiv_q \mathfrak{A}', \beta'$$

for FO-equivalence up to quantifier-rank q, i.e., $\mathfrak{A}, \beta \models \varphi \Leftrightarrow \mathfrak{A}', \beta' \models \varphi$ for all $\varphi \in FO[\tau]$ such that free$(\varphi) \subseteq \text{dom}(\beta) = \text{dom}(\beta')$ and qr$(\varphi) \leqslant q$. If $\mathfrak{A}$ and

$\mathfrak{A}'$ are clear from the context, we also write just

$$\beta \equiv_q \beta'.$$

The coarsest of these equivalences, $\mathfrak{A}, \beta \equiv_0 \mathfrak{A}', \beta'$ corresponds to a local isomorphism: $\pi: \beta(i) \mapsto \beta'(i)$ for $i \in \text{dom}(\beta) = \text{dom}(\beta')$ being an isomorphism between the induced substructures $\mathfrak{A}\restriction\text{image}(\beta)$ and $\mathfrak{A}'\restriction\text{image}(\beta')$, which is the same as equality of quantifier-free types.

Elementary equivalence, $\mathfrak{A}, \beta \equiv \mathfrak{A}', \beta'$, without the restriction on quantifier-rank, is similarly defined. Note that $\equiv$ is the limit (coarsest common refinement) of the approximations $(\equiv_q)_{q \in \omega}$.

Further, $\mathfrak{A}, \beta \equiv_\infty \mathfrak{A}', \beta'$ stands for equivalence w.r.t. infinitary logic FO_∞.[2]

The first-order Ehrenfeucht–Fraïssé game over $\mathfrak{A}$ and $\mathfrak{A}'$ is played by two players, whom we call player **I** and player **II**. We describe the game protocol in terms of rounds, each round consisting of an exchange of moves: challenge by **I**/response by **II**.

The game board: positions. Positions between rounds are pairs (β, β') of assignments to the same finite subset of variables $(x_i)_{i \geq 1}$. Only locally isomorphic assignments will be admissible for player **II**; we speak of *sound positions*:
Sound positions. Position (β, β') is sound if $\mathfrak{A}, \beta \equiv_0 \mathfrak{A}', \beta'$, i.e., if the correspondence $\beta(i) \mapsto \beta'(i)$ describes a local isomorphism. In terms of $\mathcal{O}(\mathfrak{A})$ and $\mathcal{O}(\mathfrak{A}')$: $\beta \in P_\theta \Leftrightarrow \beta' \in P_\theta$ for all atomic θ.

Single round and overall protocol. A single round consists of a challenge/response exchange of moves as follows. In position (β, β'),

- **I** chooses $i \geq 1$ and makes a move $\begin{cases} \text{either along an } E_i\text{-edge in } \mathcal{O}(\mathfrak{A}) \text{ from } \beta, \\ \text{or along an } E_i\text{-edge in } \mathcal{O}(\mathfrak{A}') \text{ from } \beta'. \end{cases}$

- **II** must make a move along an E_i-edge in the opposite structure.

This exchange of moves results in an overall transition from position (β, β') to some successor position (γ, γ'), where $\gamma = \beta\frac{a}{i}$ for some $a \in A$ and $\gamma' = \beta'\frac{a'}{i}$ for some $a' \in A'$.

We distinguish different levels of the game according to how many rounds are played.

The q-round game $\mathbb{G}_q(\mathfrak{A}; \mathfrak{A}')$ *(for fixed $q \in \omega$):* play continues from an initial position through q rounds (or until a position is reached that is not sound).

---

[2] Equivalence up to quantifier-rank α in FO_∞ can be defined, for every ordinal α. For finite relational vocabularies, $\equiv$ coincides with $\equiv_\omega$, equivalence up to quantifier-rank ω in FO_∞. Note, however, that finitary and infinitary first-order equivalences do not coincide even at quantifier-rank 1 for infinite relational vocabularies.

The finite-round game $\mathbb{G}_\omega(\mathfrak{A}; \mathfrak{A}')$: in the initial position, player **I** first selects some $q \in \omega$, then play continues in $\mathbb{G}_q(\mathfrak{A}; \mathfrak{A}')$ from the initial position.

The infinite game $\mathbb{G}_\infty(\mathfrak{A}; \mathfrak{A}')$: play continues through an infinite number of rounds (or until a position is reached that is not sound).

In each variant, **II** loses as soon as the position is not sound. Maintaining soundness of the evolving position is in fact the only commitment for **II**: **II** wins the q-round game $\mathbb{G}_q$ after completion of round q if this final position is sound; similarly **II** wins the finite-round game $\mathbb{G}_\omega$ if she wins $\mathbb{G}_q$ for the q initially selected by **I**; and she wins the infinite game $\mathbb{G}_\infty$ if play continues indefinitely without violation of soundness.[3]

In all of these games we typically also specify the initial position as in $\mathbb{G}_q(\mathfrak{A}, \beta; \mathfrak{A}', \beta')$. For instance, we say that **II** has a winning strategy in $\mathbb{G}_q(\mathfrak{A}, \beta; \mathfrak{A}', \beta')$ if (β, β') is a winning position for player **II** in $\mathbb{G}_q(\mathfrak{A}; \mathfrak{A}')$ (or in $\mathbb{G}_q(\mathfrak{A}, \beta; \mathfrak{A}', \beta')$).

It is obvious that plays of $\mathbb{G}_q$ and $\mathbb{G}_\omega$ are finite and end in a position in which one of the players has won; hence $\mathbb{G}_q$ and $\mathbb{G}_\omega$ are positionally determined. But also $\mathbb{G}_\infty$ is rather easily shown to be positionally determined, without recourse to deeper results from game theory, as part of the model theoretic analysis underpinning the following theorem. The core of this well-known analysis can be summarised as follows.

Theorem 6.2.2 (Ehrenfeucht–Fraïssé and Karp) *For all structures of the same finite relational vocabulary, $\mathfrak{A}$ and $\mathfrak{A}'$, winning positions in games characterise levels of first-order equivalence in the sense of the following equivalences.*

*(a) (β, β') is a winning position for **II** in $\mathbb{G}_q(\mathfrak{A}; \mathfrak{A}')$ if, and only if, $\mathfrak{A}, \beta \equiv_q \mathfrak{A}', \beta'$.*

*(b) (β, β') is a winning position for **II** in $\mathbb{G}_\omega(\mathfrak{A}; \mathfrak{A}')$ if, and only if, $\mathfrak{A}, \beta \equiv \mathfrak{A}', \beta'$.*

*(c) (β, β') is a winning position for **II** in $\mathbb{G}_\infty(\mathfrak{A}; \mathfrak{A}')$ if, and only if, $\mathfrak{A}, \beta \equiv_\infty \mathfrak{A}', \beta'$.*

We sketch the game-oriented skeleton of the underlying arguments in their most rudimentary form to highlight this aspect (and deliberately ignoring some of the logical niceties, like characteristic formulae, which the more thorough analysis presented in textbooks typically yields).

(i) For the direction from left to right, one shows that logical *in*equivalence yields a winning strategy for player **I**. This follows from the observation that **I**

---

[3] Clearly a variant formulation to essentially the same effect would restrict the game board to sound positions right away, making **II** lose when she is stuck for a response. This formulation, however, has the slight disadvantage of restricting us to sound initial positions, too.

can choose his challenge in a single round from a sound position such that, no matter what response **II** chooses, the resulting position is logically inequivalent at a lower quantifier-rank.

Why is that? A glance at the model checking game helps to illustrate the point. For instance, if $\beta \not\equiv_{m+1} \beta'$ (but $\beta \equiv_0 \beta'$), then this inequivalence manifests itself in some formula $\exists x_i \psi$ with ψ of quantifier-rank at most m. Suppose w.l.o.g. that $\mathfrak{A}, \beta \models \exists x_i \psi$ while $\mathfrak{A}', \beta' \models \forall x_i \neg \psi$. Then a good move for the verifier in position $(\beta, \exists x_i \psi)$ in the model checking game over $\mathfrak{A}$ obviously makes a good move for **I** in this game.[4]

(ii) In the opposite direction, player **II** always has a strategy, for her response to **I**'s challenge in a single round, to maintain the required level of logical equivalence. For instance towards (a) or (b), for a challenge $\gamma = \beta \frac{a}{i}$ in a position (β, β') such that $\beta \equiv_{m+1} \beta'$, **II** can find $a' \in A$ such that $\beta \frac{a}{i} \equiv_m \beta' \frac{a'}{i}$. Otherwise, there would have to be a distinguishing formula $\psi_{a'}$ of quantifier-rank m for every choice of $a' \in A'$, such that $\mathfrak{A}, \beta \frac{a}{i} \models \psi_{a'}$ while $\mathfrak{A}', \beta' \frac{a'}{i} \not\models \psi_{a'}$. But then the formula $\exists x_i \bigwedge_{a'} \psi_{a'}$ would distinguish β and β' at quantifier-rank $m + 1$.

If the underlying structures (and hence the branching degree of the transition systems of observable configurations) are infinite, this argument crucially uses the fact that, for a fixed tuple of free variables there are only finitely many formulae of quantifier-rank m over a fixed finite relational vocabulary, up to logical equivalence – this is what brings $\exists x_i \bigwedge_{a'} \psi_{a'}$ into first-order, even if A' is infinite.[5] We note that the corresponding claims in (a) and (b) of the theorem actually fail for infinite relational vocabularies, even over finite structures. For (c) on the other hand, to which the above argument is readily adapted, finiteness (of the conjunction or of the vocabulary) is not essential.

The equally familiar description in terms of *back-and-forth systems* corresponds to a delineation of a winning region for **II** with the appropriate closure conditions (the *back-and-forth conditions*) that guarantee that player **II** has responses to keep the game within the prescribed region, against all challenges by **I**. The essential difference between the finite and the infinite game is that, in the finite games, winning regions are stratified according to how many rounds are still to be survived. The winning region for the infinite game, on the other

---

[4] Entirely analogous reasoning applies towards (c) and for inequivalence in FO_∞, w.r.t. its ordinal-valued quantifier-rank.

[5] While this is easily proved by induction on quantifier-rank, these preparatory considerations are clearly not even required for the argument if we deal just with finite models.

hand, is static, corresponding to an invariant that needs to be maintained indefinitely (this is the classical notion of back-and-forth equivalence or *partial isomorphism* in model theory, see for instance [30]).

Example 1: finite linear orderings   The first example illustrating the usefulness of the first-order Ehrenfeucht–Fraïssé game in almost any textbook presentation concerns the limitations of FO in expressing properties of finite linear orderings (or discrete linear orderings more generally). We just state the following well-known result in order to stress its technical affinity with simple locality based arguments to be considered later.

Lemma 6.2.3   *Consider two finite linear orderings $\mathfrak{A} = (\mathbb{N}, <)\restriction[0, m]$ and $\mathfrak{A}' = (\mathbb{N}, <)\restriction[0, m']$ with assignments to tuples*

$$\beta = \mathbf{n} = (n_0, \ldots, n_k) \text{ where } 0 = n_0 < n_1 < \cdots < n_{k-1} < n_k = m \text{ and}$$
$$\beta' = \mathbf{n}' = (n'_0, \ldots, n'_k) \text{ where } 0 = n'_0 < n'_1 < \cdots < n'_{k-1} < n'_k = m'.$$

We write $d_i := n_{i+1} - n_i$ and $d'_i := n'_{i+1} - n'_i$ for distances between consecutive points in these assignments. Then the following are equivalent for any $q \geqslant 1$:

   (i) $\mathfrak{A}, \beta \equiv_q \mathfrak{A}', \beta'$
   (ii) for $0 \leqslant i < k$: $d_i = d'_i$ or both $d_i, d'_i \geqslant 2^q$.

For the naked finite linear orderings one obtains that

$$\mathfrak{A} \equiv_q \mathfrak{A}' \quad \Longleftrightarrow \quad |A| = |A'| \text{ or } |A|, |A'| \geqslant 2^q - 1.$$

For (ii) $\Rightarrow$ (i) in the lemma, consider the first round in a game played from a position satisfying the distance constraints (ii) with critical distance 2^q. It suffices to exhibit a strategy for player **II** to respond to any challenge by player **I** in such a manner that the resulting position satisfies the analogous distance constraints (ii), but now with critical distance 2^{q-1} instead of 2^q. W.l.o.g. we may assume that **I** extends the configuration β by some new element $n \in I_i = (n_i, n_{i+1})$. The case that **I** plays in $\mathfrak{A}'$ instead is symmetric. In case $d_i = d'_i$ (the pair of intervals concerned have exactly the same length), **II** may select an element $n' \in I'_i = (n'_i, n'_{i+1})$ at precisely the same distances from the end points in I'_i as n has in I_i; the resulting position even satisfies the distance constraints with critical distance 2^q again.

In the more interesting case, we have $d_i \neq d'_i$ but $d_i, d'_i \geqslant 2^q$. We consider cases, as to the sub-division of the interval $I_i = (n_i, n_{i+1})$ by n:

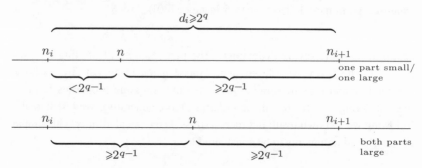

As the distances of n from the end points of I_i add up to d_i, at most one of these distances can be less than 2^{q-1}; if one distance is 'small' in this sense, **II** may copy this distance exactly to find a matching $n' \in I'_i$ (the other distance will automatically be 'large', i.e., $\geqslant 2^{q-1}$ just as on the side of I_i); if both distances are at least 2^{q-1}, then **II** similarly finds $n' \in I'_i$ which is at least that far from both end points of I'_i.

It is a nice exercise to formalise sentences in quantifier rank q that, over finite linear orderings, require at least $2^q - 1$ elements, thus showing that the given bounds are tight.

It is also useful to draw on the compositionality of strategies for **II** w.r.t. concatenation of linearly ordered intervals (slightly more generally, strategies for player **II** are compatible with ordered sums of linearly ordered structures in an otherwise monadic vocabulary; or with concatenation of word structures). The implicit decomposition of the game into subgames on intervals in the above strategy considerations reflects this.

Remark The above game argument illustrates the well-known fact that, for instance, no FO sentence can distinguish even length from odd length finite linear orderings. Any sentence φ proposed for the purpose is defeated by the example of linear orderings of lengths 2^q and $2^q - 1$ for $q := \mathrm{qr}(\varphi)$.

Maybe somewhat unexpectedly (and disturbing only from a didactic point of view), this particular finite model theory assertion can also be shown by classical means. Suppose there were a sentence $\varphi \in \mathrm{FO}[<]$ such that a *finite* linear ordering satisfies φ if, and only if, it is of even length. Let $[\varphi]^{\leqslant x}$ be the relativisation of φ to the initial segment formed by x. Let $\psi_0 \in \mathrm{FO}[<]$ be the usual characterisation of discrete linear orderings with first and without last element; $\psi_1 \in \mathrm{FO}[<]$ the assertion that precisely every other element x

satisfies $[\varphi]^{\leqslant x}$. Then $\psi_0 \wedge \psi_1$ would characterise the order type of $(\omega, <)$, which is impossible by compactness. To see that $\psi_0 \wedge \psi_1$ forces the standard model, consider any non-standard model $(A, <)$ of ψ_0 as in the sketch. Since the non-standard part of $(A, <)$ consists of an ordered sum of parts ordered like $(\mathbb{Z}, <)$, the successor operation induces an automorphism of the non-standard part. Therefore $[\varphi]^{\leqslant x}$ cannot distinguish next neighbours within the non-standard part, and $(A, <) \not\models \psi_0 \wedge \psi_1$.

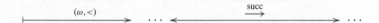

This argument immediately also shows that the unary predicate P consisting of every other element of a finite linear ordering and starting with the minimal element cannot be explicitly definable in FO[$<$] over the class of all finite linear orderings. Since the given specification of P translates into an obvious implicit definition in FO[$<$, P], the example directly refutes the finite model theory analogue of Beth's theorem. In this light, the given automorphism argument just shows that the implicit definition of P over *finite* linear orderings does not extend to any implicit definition that would be good over the class of all discrete linear orderings with first element.

Example 2: a simple locality argument   (also compare section 6.2.5) Let τ be a finite relational vocabulary. A formula $\varphi(\mathbf{x}) \in$ FO[τ] is called ℓ-*local* if, in any τ-structure $\mathfrak{A}$, whether $\mathfrak{A} \models \varphi[\mathbf{a}]$ is fully determined by $\mathfrak{A} {\upharpoonright} N^\ell[\mathbf{a}]$ (the ℓ-neighbourhood of $\mathbf{a}$):

$$\mathfrak{A} \models \varphi[\mathbf{a}] \quad \Leftrightarrow \quad \mathfrak{A} {\upharpoonright} N^\ell[\mathbf{a}] \models \varphi[\mathbf{a}].$$

Similarly $\varphi(\mathbf{x})$ is *invariant under disjoint unions* if for all $\mathfrak{A}$, $\mathbf{a}$ and $\mathfrak{B}$,

$$\mathfrak{A} \models \varphi[\mathbf{a}] \quad \Leftrightarrow \quad \mathfrak{A} \oplus \mathfrak{B} \models \varphi[\mathbf{a}],$$

where $\mathfrak{A} \oplus \mathfrak{B}$ is the disjoint union of $\mathfrak{A}$ and $\mathfrak{B}$.

Lemma 6.2.4 *If $\varphi \in$ FO[τ] is invariant under disjoint unions, then φ is ℓ-local for $\ell = 2^{\mathrm{qr}(\varphi)} - 1$.*

Remark: the bound on ℓ is optimal, since there is, for every q, a quantifier-rank q formula $\varphi_q(x) \in$ FO[E, P] asserting that $N^{2^q-1}(x) \cap P \neq \emptyset$.[6]

Proof. Let φ be invariant under disjoint unions, $q := \mathrm{qr}(\varphi)$ and $\ell := 2^q - 1$. For $\mathbf{a} \in \mathfrak{A}$ and $\mathfrak{A}_0 := \mathfrak{A} {\upharpoonright} N^\ell(\mathbf{a})$ it suffices to show that $\mathfrak{A} \models \varphi[\mathbf{a}]$ iff

[6] One obtains $\varphi_q(x)$ inductively, based on $\varphi_{q+1}(x) := \exists y(d(x, y) \leqslant 2^q \wedge \varphi_q(y))$.

$\mathfrak{A}_0 \models \varphi[\mathbf{a}]$. By invariance under disjoint unions, moreover, it suffices to establish an equivalence of the form $\mathfrak{A}, \mathbf{a} \oplus \mathfrak{C} \equiv_q \mathfrak{A}_0, \mathbf{a} \oplus \mathfrak{C}$ for a suitable structure $\mathfrak{C}$. Taking $\mathfrak{C}$ to be the disjoint union of q further disjoint isomorphic copies each of $\mathfrak{A}$ and of $\mathfrak{A}_0$, we argue this equivalence:

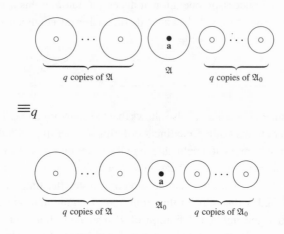

In the game on these structures, **II** wins the q-round game as follows. We use $d_m := 2^{q-m}$ as a critical distance to be observed in round m. **II** is to play such that the configurations resulting from round m are linked by a component-wise trivial isomorphism between their $(d_m - 1)$-neighbourhoods. This condition is satisfied at the start, for $m = 0$; for $m = q$ it still guarantees a local isomorphism between the final configurations, hence a win for **II**.

Here is how to maintain the condition through round m of the game for $m \geqslant 1$:

(i) if **I**'s challenge goes to some element at distance greater than d_m from the current configuration, then **II** responds with the same element in a new isomorphic component on the opposite side (new in the sense of not yet involved in the current configuration; such are always left).

(ii) if **I**'s challenge goes to an element within distance d_m of the current configuration, then **II** finds a response via the trivial local isomorphism between the $(d_{m-1} - 1)$-neighbourhoods of the current configurations. We note that $d(x, y) \leqslant d_m$ implies $N^{d_m-1}(y) \subseteq N^{d_{m-1}-1}(x)$, as $d(x, z) \leqslant d(x, y) + d(y, z) \leqslant d_m + d_m - 1 = d_{m-1} - 1$. $\qquad \square$

6.2.3 Natural restrictions/variations

Several of the most natural fragments of FO can be presented in terms of restrictions or modifications of the system $\mathcal{O}(\mathfrak{A})$ of observable configurations associated with structure $\mathfrak{A}$. The k-variable fragment FO^k of FO, for instance, exactly corresponds to the restriction that only up to k elements of $\mathfrak{A}$ are "simultaneously observable" – we just need to restrict the assignments to size k. While this is a uniform, purely quantitative restriction, the modal and guarded fragments of first-order logic are based on structural, qualitative restrictions. In the guarded fragment GF, access to observable configurations is restricted by the requirement that the target configuration be guarded, i.e., covered by some relational ground atom (which is explicitly reflected in the syntax of guarded quantification). In the more basic modal fragment ML of FO, $\mathfrak{A}$ itself *is* the system of observable configurations – in this case the transition relations between the (trivial, one-point) configurations are the key to the restrictions imposed in modal quantification. Below we treat k-variable logic, guarded logic and modal logic in this order of increasing specialisation, with particular emphasis on modal logic and its comparison game, the bisimulation game.

The k-variable fragment and the k-pebble game

The k-variable fragment $\mathsf{FO}^k \subseteq \mathsf{FO}$ in a relational vocabulary τ consists of all first-order formulae in which only the variable symbols $x_1, \ldots, x_k$ occur, bound or free. Assignments over τ-structures $\mathfrak{A}$ can thus be restricted and normalised to be full assignments to these k variables. We therefore identify assignments with k-tuples.

Correspondingly, let $\mathcal{O}^k(\mathfrak{A})$ be the restriction of $\mathcal{O}(\mathfrak{A})$ to $\{\beta : |\beta| = k\} = A^k$.

It is easy to see that the restriction of both the model checking game and of the comparison game that ensues if $\mathcal{O}(\mathfrak{A})$ is consistently replaced with $\mathcal{O}^k(\mathfrak{A})$ are adequate for the semantics of FO^k and for the induced notions of k-variable equivalence. The k-variable Ehrenfeucht–Fraïssé game is just the k-pebble game: moves along E_i-edges in $\mathcal{O}^k(\mathfrak{A})$ correspond to the re-positioning of the i-th pebble on $\mathfrak{A}$. The correspondence between the different levels of the game and of k-variable equivalence are the following, for finite relational vocabularies τ:

$\equiv_q^k$: FO^k-equivalence up to quantifier-rank q;
  captured by the q-round k-pebble game $\mathbb{G}_q^k$.

$\equiv^k$: FO^k-equivalence;
  captured by the finite-round k-pebble game $\mathbb{G}_\omega^k$.

$\equiv_\infty^k$: FO_∞^k-equivalence;
  captured by the infinite k-pebble game $\mathbb{G}_\infty^k$.

It is important to note that $\mathcal{O}^k(\mathfrak{A})$ is of finite type for each finite τ, and of polynomial size in the size of $\mathfrak{A}$, for finite $\mathfrak{A}$. For the model checking implications see [23, 22] and also some related remarks in section 6.5.1. In particular, the *combined model checking complexity* for FO^k is complete for Ptime, while for FO it is complete for Pspace.

The guarded fragment and the guarded bisimulation game

The characteristic feature of the guarded fragment GF of first-order logic [2] is the relativisation of first-order quantification to guarded tuples – similar to the restriction along accessibility edges in modal logic. Also compare the remarks in section 6.1.2 where GF was introduced as a fragment of FO.

We start out with a discussion of a very liberal setting for the guarded fragment that most naturally reflects the syntactic freedom allowed in the standard formalisation of GF as given in section 6.1.2. Afterwards we also indicate some more succinct alternative formulations that correspond to certain syntactic normalisations (e.g., regarding the number of variables used) to which GF can be subjected without impairing its expressive power; such less liberal formalisations can be of technical advantage in the model theoretic analysis of GF and its relatives.

Recall that a subset $s \subseteq A$ is *guarded* in the τ-structure $\mathfrak{A}$ if it is a singleton set or, if for one of the relations $R \in \tau$ there is some tuple $\mathbf{a} = (a_1, \ldots, a_r) \in R^{\mathfrak{A}}$ for which $s = [\mathbf{a}]$.[7] In particular, the cardinality of guarded subsets is bounded by the width of the vocabulary τ. A *tuple* $\mathbf{b}$ in $\mathfrak{A}$ is called guarded if $[\mathbf{b}] \subseteq s$ for some guarded subset s. The same terminology applies to assignments β in $\mathfrak{A}$.

Call a tuple $\mathbf{b}$ or an assignment β in $\mathfrak{A}$ *strictly guarded* if $[\beta]$ is itself a guarded subset. More specifically, for an atomic τ-formula α, we say the assignment β is *strictly guarded by* α if $\text{var}(\alpha) = \text{dom}(\beta)$ and $\mathfrak{A}, \beta \models \alpha$, which implies that $[\beta]$ is indeed a guarded subset. (In order to capture also guarded singleton sets, we allow α to be an equality atom.)

A system of observable configurations for GF   We work with the following system of observable configurations $\mathcal{O}^{G}(\mathfrak{A})$ over the set of all finite (partial) assignments over $\mathfrak{A}$ with new binary transitions relations $E_{\alpha, \rho}$ (see below) and unary predicates P_{θ} (as before). The universe of $\mathcal{O}^{G}(\mathfrak{A})$ is the same as in $\mathcal{O}(\mathfrak{A})$ for FO (this is for the liberal, redundant formalisation).

The transition relations of $\mathcal{O}^{G}(\mathfrak{A})$ describe passages from some assignment β to a new assignment β' where the target assignment β' is required to be strictly guarded by some atomic formula α. Each transition relation specifies

---

[7] Recall that we denote as $[\mathbf{b}]$ the set of components of a tuple $\mathbf{b}$, and similarly write $[\beta]$ for the image set of an assignment β.

both the atomic formula α and a set of identities between components of the old and the new assignment. As both β and β' are finite partial functions on the positive integers, a set of identities between components can be specified as a finite set ρ of pairs of positive integers. We write $\beta \overset{\rho}{=} \beta'$ if $\beta(i) = \beta'(j)$ for all $(i, j) \in \rho$. Then for every ρ and α, let $E_{\alpha,\rho}$ be interpreted as the following transition relation on $\mathcal{O}^{\mathsf{G}}(\mathfrak{A})$:

$$E_{\alpha,\rho} = \left\{ (\beta, \beta') \colon \beta \overset{\rho}{=} \beta', \beta' \text{ strictly guarded by } \alpha \right\}.$$

Unary predicates P_θ for atomic types $\theta(\mathbf{x})$ are as in the basic system $\mathcal{O}(\mathfrak{A})$.

Guarded model checking The game graph for the model checking of formulae in GF is obtained from $\mathcal{O}^{\mathsf{G}}(\mathfrak{A})$ and a suitable formalisation of the syntax of guarded quantification in close analogy to the basic case. With the formation rule of existential guarded quantification, for instance,

$$\varphi(\mathbf{x}) = \exists \mathbf{y}(\alpha(\mathbf{x}') \wedge \psi(\mathbf{x}')),$$

where $\mathbf{y}$ is a subtuple of $\mathbf{x}'$, associate an $E_{\alpha,\rho,\exists}$-edge in the syntax tree from $\varphi(\mathbf{x})$ to $\psi(\mathbf{x}')$, where $\rho = \{(i, j) \colon x_i = x'_j\}$. In the game graph $\mathbb{G}^{\mathsf{G}}(\mathfrak{A}, \Phi)$, correspondingly, there are $E_{\alpha,\rho,\exists}$-edges from positions $(\beta, \varphi(\mathbf{x}))$ to positions $(\beta', \psi(\mathbf{x}'))$ such that $\mathbf{x} \subseteq \mathrm{dom}(\beta)$, $\mathbf{x}' \subseteq \mathrm{dom}(\beta')$, $(\beta, \beta') \in E_{\alpha,\rho}$ in $\mathcal{O}^{\mathsf{G}}(\mathfrak{A})$. Similarly, universal guarded quantifications $\varphi(\mathbf{x}) = \forall \mathbf{y}(\alpha(\mathbf{x}') \rightarrow \psi(\mathbf{x}'))$ give rise to edges in $E_{\alpha,\rho,\forall}$ in the syntax tree, and induce transition relations $E_{\alpha,\rho,\forall}^{\mathsf{G}}$ in $\mathbb{G}^{\mathsf{G}}(\mathfrak{A}, \Phi)$.

Note that existential and universal quantification of variables in GF proceeds in batches (so as to cover a guarded successor set fully in one step) rather than element-wise. Correspondingly, first-order quantifier-rank is replaced by the nesting depth of guarded quantification steps for an appropriate analysis of quantifier complexity. This is important for the induced levels of GF equivalence, which are considered in connection with the comparison game of guarded bisimulation below.

Guarded bisimulation In line with the general idea, positions between rounds in the guarded Ehrenfeucht-Fraïssé game $\mathbb{G}^{\mathsf{G}}(\mathfrak{A}, \mathfrak{A}')$ are matching pairs of assignments (β, β') in $\mathfrak{A}$ and $\mathfrak{A}'$. With the possible exception of the initial position of the game, which we choose to ignore in the following, we may restrict attention to positions in which both β and β' are strictly guarded (this is guaranteed for successor positions after the first round, by the rules below).

Soundness means that the induced correspondence $\beta(i) \mapsto \beta'(i)$ for $i \in \mathrm{dom}(\beta) = \mathrm{dom}(\beta')$ is a local isomorphism; insofar as the assignments are strictly guarded in their structures, the correspondence is a bijection between guarded subsets and thus a local isomorphism between induced substructures on guarded subsets $s = [\beta]$ and $s' = [\beta']$. Challenge/response pairs of moves responsible for taking the game through a single round are governed by **I**'s selection of an $E_{\alpha,\rho}$ and an $E_{\alpha,\rho}$ successor γ of β in $\mathcal{O}^{\mathsf{G}}(\mathfrak{A})$ or an $E_{\alpha,\rho}$ successor γ' of β' in $\mathcal{O}^{\mathsf{G}}(\mathfrak{A}')$, and thus, together with **II**'s response, to a new local isomorphism between substructures induced on a new pair of guarded subsets $t = [\gamma]$ and $t' = [\gamma']$ (insofar as the successor position is sound again, i.e., unless **II** has lost).

A conceptually smoother, equivalent formulation therefore is the following, which we take as the preferred description of the *guarded bisimulation game*. Positions in the game are local bijections $\sigma : s \to s'$ between guarded subsets $s \subseteq A$ and $s' \subseteq A'$. In a single round played from position $\sigma : s \to s'$, **I** proposes either a guarded subset $t \subseteq A$ or a guarded subset $t' \subseteq A'$; **II** has to respond with a guarded subset in the opposite structure (call this other subset $t' \subseteq A'$ or $t \subseteq A$, as the case may be) and a bijection $\rho : t \to t'$ that is compatible with σ. Compatibility of ρ with σ means that ρ needs to agree with σ on $s \cap t$ if **I** chose t; and on $s' \cap t'$ if **I** chose t'. **II** loses if there is no such ρ or if ρ is not a local isomorphism.

Either formulation of the game supports the usual analysis, which, as expected, establishes correspondences between winning positions for **II** in the different levels of the game and equivalence in GF. For finite relational vocabularies τ these are:

$\equiv_q^{\mathsf{G}}$: GF-equivalence up to guarded nesting depth q;
    captured by the q-round guarded bisimulation game $\mathbb{G}_q^{\mathsf{G}}$.

$\equiv^{\mathsf{G}}$: GF-equivalence;
    captured by the finite-round guarded bisimulation game $\mathbb{G}_\omega^{\mathsf{G}}$.

$\equiv_\infty^{\mathsf{G}}$: GF$_\infty$-equivalence;
    captured by the infinite guarded bisimulation game $\mathbb{G}_\infty^{\mathsf{G}}$.

More succinct representations   Another, much more succinct view on the observable configurations can be based on more restricted classes of assignments: it essentially suffices to admit strictly guarded assignments with domain $\{1, \ldots, k\}$ where k is the width of τ. This second aspect corresponds to the normalisation of variables to $x_1, \ldots, x_k$ as in FO^k.[8] Here we use strictly guarded assignments to variables $x_1, \ldots, x_k$, or surjective partial maps from $\{1, \ldots, k\}$ onto guarded subsets of $\mathfrak{A}$.

---

[8] Even more restrictively, [25] for technical convenience uses a format with only injective assignments, there called *guarded lists*.

The type of the resulting system of guarded observable configurations is finite for finite τ. The model checking game obtained in analogy with the above, by making the obvious changes and restrictions regarding the syntax of formulae, then really is for (a specific syntactic variant of) $\mathsf{GF}^k := \mathsf{GF} \cap \mathsf{FO}^k$.

A closer analysis of the Ehrenfeucht–Fraïssé games and notions of guarded equivalence resulting from the two different formalisations would show that there is no loss of expressiveness as far as properties of (strictly) guarded tuples are concerned. The only real restriction concerns expressiveness at the quantifier-free level and in boolean combinations, and this is inessential for many purposes. The difference arises, trivially, because GF does not impose any restrictions on boolean combinations. Analysis of the game shows, however, that any formula of GF (in the liberal format) is logically equivalent to a boolean combination of quantifier-free formulae and strictly guarded formulae (each of which can, up to a necessary renaming of variables, be formalised in the above fragment GF^k).

Corollary 6.2.5 *Any formula in* $\mathsf{GF}[\tau]$ *with explicitly guarded free variables is equivalent to a formula in* $\mathsf{GF} \cap \mathsf{FO}^k$ *where k is the width of τ.*

The modal fragment and the bisimulation game

Modal logic is naturally interpreted over transition systems (Kripke structures in traditional terminology). Having chosen a modal perspective for our analysis of fragments, we may now choose the transition system $\mathfrak{A}$ itself – as a relational structure in a given vocabulary τ with binary relations E_α and unary predicates P_j – as the system of modally observable configurations, putting $\mathcal{O}^{\mathsf{M}}(\mathfrak{A}) = \mathfrak{A}$. To keep in line with the general framework we may want to replace the individual P_j in $\mathfrak{A}$ by P_θ that are complete propositional types in the p_j / P_j (in first-order terms: atomic P_j-types in single variables x, containing for each P_j either the atomic formula $P_j x$ or its negation $\neg P_j x$).

Modal model checking The modal model checking game over structure $\mathfrak{A}$ is played in a game graph based on $\mathfrak{A}$ and the syntax tree of the modal formulae under consideration. With the formation rule of existential modal quantification

$$\varphi = \Diamond_\alpha \psi$$

we associate an $E_{\Diamond_\alpha}$ edge in the syntax tree from φ to ψ. In the game graph $\mathbb{G}^{\mathsf{M}}(\mathfrak{A}, \Phi)$, this induces edges from positions (a, φ) to positions (b, ψ) for $(a, b) \in E_\alpha^{\mathfrak{A}}$. Analogously for $\Box_\alpha$ quantification: edges in $E_{\Box_\alpha}$ from $\varphi = \Box_\alpha \psi$ to ψ in the syntax tree give rise to transitions in $\mathbb{G}^{\mathsf{M}}(\mathfrak{A}, \Phi)$ from (a, φ) to (b, ψ) for every $(a, b) \in E_\alpha^{\mathfrak{A}}$.

It is clear that the model checking game for FO^2 emulates the modal model checking game, via the standard translation of ML into FO^2:

$$(\Diamond_\alpha \psi)_x \; = \; \exists y (E_\alpha x y \wedge \psi_y),$$
$$(\Box_\alpha \psi)_x = \forall y (E_\alpha x y \rightarrow \psi_y),$$

where $\{x, y\} = \{x_1, x_2\}$. In terms of this translation, a move along an E_α edge (a, b) in the $\mathfrak{A}$ component of $\mathbb{G}^M(\mathfrak{A}, \Phi)$ is simulated by an E_2 move from any position of the form $(a, *)$ to (a, b) or by an E_1 move from any $(*, a)$ to (b, a) in the $\mathcal{O}^2(\mathfrak{A})$ component of $\mathbb{G}^2(\mathfrak{A}, FO(\Phi))$. At the same time this emulation can be interpreted in $\mathbb{G}^G(\mathfrak{A}, FO(\Phi))$, since $\{a, b\}$ is a strictly guarded assignment and (a, a) is linked to (a, b), for instance, by an $E_{\alpha, \rho}$ edge in $\mathcal{O}^2(\mathfrak{A})$ for $\rho = \{(1, 1)\}$.

Bisimulation   The bisimulation game is the Ehrenfeucht–Fraïssé game for modal logic. It also has a special status because of its fundamental nature as the quintessential back-and-forth game – game equivalence of game graphs – to be discussed in the following section.

In line with the general approach, the positions (between rounds) in $\mathbb{G}^M(\mathfrak{A}, \mathfrak{A}')$ are pairs of observable configurations in $\mathcal{O}(\mathfrak{A}) = \mathfrak{A}$ and $\mathcal{O}(\mathfrak{A}') = \mathfrak{A}'$, i.e., pairs $(a, a') \in A \times A'$. The challenge/response exchange that constitutes a single round is as follows:

- **I** selects a transition relation E_α, and $\begin{cases} \text{either some } E_\alpha \text{ successor } b \text{ of } a \text{ in } \mathfrak{A}, \\ \text{or some } E_\alpha \text{ successor } b' \text{ of } a' \text{ in } \mathfrak{A}'. \end{cases}$
- **II** has to respond by selecting an E_α successor in the opposite structure.

Overall this results in a successor position (b, b') for which $(a, b) \in E_\alpha^{\mathfrak{A}}$ and $(a', b') \in E_\alpha^{\mathfrak{A}'}$. A position (a, a') is sound if a and a' satisfy exactly the same predicates P_j (atomic propositions p_j in modal terminology), which clearly corresponds to quantifier-free indistinguishability in $ML[\tau]$.

Because of their immediate importance we introduce the usual dedicated notation for the levels of equivalence that are defined in terms of winning positions for player **II** in the different levels of this bisimulation game. As above, the q-round, finite-round, and infinite bisimulation game on $\mathfrak{A}$ and $\mathfrak{A}'$ are denoted $\mathbb{G}_q^M(\mathfrak{A}, \mathfrak{A}')$, $\mathbb{G}_\omega^M(\mathfrak{A}, \mathfrak{A}')$, and $\mathbb{G}_\infty^M(\mathfrak{A}, \mathfrak{A}')$. We then define

$\mathfrak{A}, a \sim_q \mathfrak{A}', a'$	iff	(a, a') is a winning position for **II** in $\mathbb{G}_q^M(\mathfrak{A}, \mathfrak{A}')$;
$\mathfrak{A}, a \sim_\omega \mathfrak{A}', a'$	iff	(a, a') is a winning position for **II** in $\mathbb{G}_\omega^M(\mathfrak{A}, \mathfrak{A}')$;
$\mathfrak{A}, a \sim \mathfrak{A}', a'$	iff	(a, a') is a winning position for **II** in $\mathbb{G}_\infty^M(\mathfrak{A}, \mathfrak{A}')$.

Note that $\sim$ is the classical notion of bisimulation equivalence – equivalence w.r.t. the infinite bisimulation game, and as such the modal counterpart of partial isomorphism.

We denote the relevant levels of equivalence in modal logic as $\equiv_q^M$ (up to modal nesting depth q), $\equiv^M$ (full equivalence in finitary ML), and $\equiv_\infty^M$ (equivalence in the infinitary extension ML_∞). The associated Ehrenfeucht–Fraïssé and Karp theorems then state, for finite modal vocabularies τ, the following equivalences:

$$\mathfrak{A}, a \sim_q \mathfrak{A}', a' \quad \Leftrightarrow \quad \mathfrak{A}, a \equiv_q^M \mathfrak{A}', a'.$$
$$\mathfrak{A}, a \sim_\omega \mathfrak{A}', a' \quad \Leftrightarrow \quad \mathfrak{A}, a \equiv_\omega^M \mathfrak{A}', a'.$$
$$\mathfrak{A}, a \sim \mathfrak{A}', a' \quad \Leftrightarrow \quad \mathfrak{A}, a \equiv_\infty^M \mathfrak{A}', a'.$$

Modal variations The simple extensions of basic modal logic by inverse modalities and/or global modality, ML^-, $\mathsf{ML}^\vee$ and $\mathsf{ML}^{-\vee}$, are matched by corresponding variations in $\mathcal{O}(\mathfrak{A})$ and $\mathbb{G}(\mathfrak{A}, \mathfrak{A}')$. To deal with inverse modalities, $\mathcal{O}(\mathfrak{A})$ is enriched with the converse relations to the E_α, $(E_\alpha^-)^{\mathfrak{A}} = \{(b, a) \colon (a, b) \in E_\alpha^{\mathfrak{A}}\}$; to deal with the global modality, $\mathcal{O}(\mathfrak{A})$ is expanded by the full binary relation $U^{\mathfrak{A}} = A \times A$. Everything else, including associated Ehrenfeucht–Fraïssé and Karp theorems, is then set up by straightforward analogy and we leave the details as an exercise. For later use, we denote the levels of *two-way global bisimulation equivalence* corresponding to the combined extension by inverse modalities and the global modality by $\approx_q$, $\approx_\omega$ and $\approx$.

Bisimulations as relations and back-and-forth systems We also want to use the notational variants corresponding to back-and-forth systems for bisimulation games. Infinitary bisimulation equivalence (the modal counterpart of partial isomorphism) between the nodes of two structures $\mathfrak{A}$ and $\mathfrak{A}'$, in particular, is captured by the relation $Z \subseteq A \times A'$ comprising exactly the winning positions for $\mathbf{II}$ in $\mathbb{G}_\infty^M(\mathfrak{A}, \mathfrak{A}')$ (known as the *largest bisimulation relation* between $\mathfrak{A}$ and $\mathfrak{A}'$, cf. [9, 17]). Any other relation $Z \subseteq A \times A'$ that delineates an appropriately closed winning region for $\mathbf{II}$ is also a bisimulation relation, and necessarily a subset of the largest such. Corresponding finite bisimulation levels are described by stratified back-and-forth systems in the usual manner. Again, natural and straightforward adaptations for, e.g., two-way global bisimulations are obtained. The difference lies in the closure conditions (back-and-forth conditions), which reflect the nature of the challenges that $\mathbf{I}$ is allowed, since $\mathbf{II}$ must have responses to all of them within the prescribed collection of positions.

A particular variant of bisimulation relationships is realised by homomorphisms whose graphs are bisimulation relations (*bounded morphisms* in classical modal terminology, cf. [9, 17]). For instance, in the case of the two-way global bisimulation relation $\approx$, we write

$$\pi : \mathfrak{A}, a \xrightarrow{\approx} \mathfrak{A}', a'$$

to indicate that $\pi : A \to A'$ is a map sending a to a' and such that its graph is a bisimulation relation with the back-and-forth closure conditions appropriate for global two-way bisimulation game (in particular π needs to be a surjective homomorphism).

Saturation and Hennessy–Milner properties   We shall later look at the relationship between equivalence w.r.t. the infinite game $\mathbb{G}_\infty$ and the finite approximations to the finite-round game $\mathbb{G}_\omega$ induced by the q-round games $(\mathbb{G}_q)_{q \in \omega}$ also for games other than bisimulation. It is therefore interesting to understand under which conditions there is no gap between the limit of the finite approximations and full infinitary equivalence. In the modal situation, or for the bisimulation game, this situation is particularly transparent, and at the same time holds the key to the general situation for other fragments in the game-oriented analysis.

Definition 6.2.6   Let $\mathfrak{A}$ be a τ transition system with transition relations E_α.
(i) $\Phi \subseteq \mathsf{ML}[\tau]$ is called a $\Diamond_\alpha$-*type* at $a \in \mathfrak{A}$ if $\mathfrak{A}, a \models \Diamond_\alpha \bigwedge \Phi_0$ for every finite $\Phi_0 \subseteq \Phi$; it is *realised* at $a \in \mathfrak{A}$ if there is some b such that $(a, b) \in E_\alpha^{\mathfrak{A}}$ and $\mathfrak{A}, b \models \Phi$.
(ii) $\mathfrak{A}$ is called *modally saturated* if, for all α and all $a \in \mathfrak{A}$, every $\Diamond_\alpha$-type at a is realised at a.

It is not hard to see that ω-saturated transition systems, and in particular finite transition systems are modally saturated. But a very simple argument also shows that even all finitely branching transition systems are modally saturated. In the case of a structure $\mathfrak{A}$ that is finitely branching (w.r.t. E_α) at a, consider some $\Diamond_\alpha$-type Φ at a. Suppose Φ were not realised at a. This means that, for every E_α successor b of a there must be some $\varphi_b \in \Phi$ not satisfied at b. But then the finite subset Φ_0 of these φ_b would violate the defining condition for a $\Diamond_\alpha$-type at a: $\mathfrak{A}, a \models \Box_\alpha \bigvee_b \neg\varphi_b$, whence $\mathfrak{A}, a \not\models \Diamond_\alpha \bigwedge \Phi_0$.

For this and also for the reasoning behind the lemma below, compare part (ii) of the argument indicated in connection with Theorem 6.2.2.

Definition 6.2.7   A class of transition systems has the *Hennessy–Milner property* if over this class, modal equivalence $\equiv^M$ coincides with full bisimulation $\sim$.

Note that, since even for not necessarily finite vocabularies τ, $\sim_\omega$ implies $\equiv^M$, the Hennessy–Milner property implies that in particular also finite bisimulation equivalence coincides with full bisimulation equivalence. The following lemma also implies that for modally saturated transition systems, modal equivalence, finite and full bisimulation equivalence all fall into one, even for infinite vocabularies.

Lemma 6.2.8 *The class of modally saturated transition systems has the Hennessy–Milner property.*

The straightforward game argument for this is again suggested by the reasoning underlying Theorem 6.2.2, part (ii), but finiteness of τ is not required. Playing over modally saturated structures, **II** can maintain modal equivalence between configurations. Consider a position (a, a') in the game $\mathbb{G}^M_\infty(\mathfrak{A}, \mathfrak{A}')$ for which $\mathfrak{A}, a \equiv^M \mathfrak{A}', a'$, and think of a challenge played by **I**, with a move along $(a, b) \in E^{\mathfrak{A}}_\alpha$ say. In general (and even for finite vocabulary) modal equivalence $\mathfrak{A}, a \equiv^M \mathfrak{A}', a'$ (or even $\mathfrak{A}, a \sim_\omega \mathfrak{A}', a'$) would only provide **II** with responses b' that are good for surviving q further rounds, where this could be a separate response for each individual q. Now, however, the full modal theory of b in $\mathfrak{A}$ constitutes a $\Diamond_\alpha$-type at a in $\mathfrak{A}$, and modal equivalence $\mathfrak{A}, a \equiv^M \mathfrak{A}', a'$ is good enough to ensure that it therefore also is a $\Diamond_\alpha$-type at a' in $\mathfrak{A}'$. By modal saturation, therefore, this $\Diamond_\alpha$-type is realised at a' in $\mathfrak{A}'$, and any such realisation gives **II** a valid response in the game which maintains $\equiv^M$. But maintaining $\equiv^M$ equivalence throughout the game, **II** cannot lose; so this gives her a strategy in $\mathbb{G}^M_\infty$.

6.2.4 Bisimulation as the master game

An analysis of whole families of fragments of FO w.r.t. their notions of finite and infinitary equivalence can very nicely be based on the analysis of the bisimulation game over the transition systems of observable configurations associated with the particular fragment.

The possible advantage of this perspective lies in the conceptual separation of the game theoretic commonality, which is here uniformly described in terms of bisimulation, and the particular constraints of the fragment under consideration, which enters the picture through the right formalisation of the observable configurations. The natural criterion for the *right* formalisation lies in the adequacy of the induced model checking game for the semantics of the given fragment.

The treatment of FO and fragments like FO^k, GF and ML (and some of its simple variants) can be put in a uniform format as follows. Let $L \subseteq FO$ be a fragment associated with systems of observable configurations $\mathcal{O}^{\mathsf{L}}(\mathfrak{A})$ over relational structures $\mathfrak{A}$ in a finite relational vocabulary τ. Together with the overhead that links syntax of L with moves in the model checking game with structure inputs $\mathcal{O}^{\mathsf{L}}(\mathfrak{A})$, this model checking game can be taken as a specification of the semantics of L. The bisimulation game between $\mathcal{O}^{\mathsf{L}}(\mathfrak{A})$ and $\mathcal{O}^{\mathsf{L}}(\mathfrak{A}')$ then *is* a representation of the Ehrenfeucht-Fraïssé or model comparison game for L. This representation is adequate at a round-by-round level in terms of a syntactic notion of depth in L that corresponds to the number of quantification rounds required in model checking a formula in L. The specification of the model checking game is in turn reflected in the format of $\mathcal{O}^{\mathsf{L}}(\mathfrak{A})$. As an example for the latter point, consider GF as presented above: we deliberately chose transitions in $\mathcal{O}^{\mathsf{G}}(\mathfrak{A})$ to link any two strictly guarded patches in one transition rather than a sequence of transitions corresponding to one-new-element-at-a-time moves as in $\mathcal{O}(\mathfrak{A})$. The latter option would have turned FO quantifier-rank into our measure of semantic complexity in GF whereas the chosen stipulation relates to the coarser but more intuitive notion of guarded nesting depth. With the appropriate notion of depth that is implicit in the granularity of the model checking game based on $\mathcal{O}^{\mathsf{L}}(\mathfrak{A})$ come the notions of $\equiv_q^{\mathsf{L}}$ as finite approximations to $\equiv^{\mathsf{L}}$, and (for finite vocabulary) an Ehrenfeucht–Fraïssé theorem of the format

$$\mathfrak{A}, \beta \equiv_q^{\mathsf{L}} \mathfrak{A}', \beta' \quad \Leftrightarrow \quad \mathcal{O}^{\mathsf{L}}(\mathfrak{A}), \beta \sim_q \mathcal{O}^{\mathsf{L}}(\mathfrak{A}'), \beta', \text{ for } q \in \omega, \text{ and}$$

$$\mathfrak{A}, \beta \equiv^{\mathsf{L}} \mathfrak{A}', \beta' \quad \Leftrightarrow \quad \mathcal{O}^{\mathsf{L}}(\mathfrak{A}), \beta \sim_\omega \mathcal{O}^{\mathsf{L}}(\mathfrak{A}'), \beta'.$$

At the same time, a notion of infinitary L-equivalence is induced by the full bisimulation relation, $\mathcal{O}(\mathfrak{A}), \beta \sim \mathcal{O}^{\mathsf{L}}(\mathfrak{A}'), \beta'$, supporting a Karp theorem of the format

$$\mathfrak{A}, \beta \equiv_\infty^{\mathsf{L}} \mathfrak{A}', \beta' \quad \Leftrightarrow \quad \mathcal{O}^{\mathsf{L}}(\mathfrak{A}), \beta \sim \mathcal{O}^{\mathsf{L}}(\mathfrak{A}'), \beta',$$

which can now also be seen as a specification of what L_∞ (in terms of its model checking game) needs to be.

Beyond a uniform perspective on the games and equivalences themselves, the modal perspective on fragments of FO can also indicate what the right transfer of other game-related notions to fragments should be. As one example we state the following observation concerning ω-saturation (in the usual first-order context).

Observation 6.2.9 $\mathfrak{A}$ *is ω-saturated if, and only if, $\mathcal{O}(\mathfrak{A})$ is modally saturated.*

Similar correspondences can then be taken to define the appropriate notion of ω-saturation in the context of fragments $\mathsf{L} \subseteq \mathsf{FO}$ (e.g., for FO^k or GF), in terms of modal saturation of the corresponding $\mathcal{O}^{\mathsf{L}}(\mathfrak{A})$. This allows us to extrapolate to a range of in-between fragments from the Hennessy–Milner property of modal logic to other fragments with the appropriate notion of ω-saturation. In particular, the right types to be considered for this notion of saturation are derived from the modal $\Diamond$-types in the $\mathcal{O}^{\mathsf{L}}(\mathfrak{A})$.

On the other hand, for many natural fragments including FO^k, GF^k and all the modal fragments, classical first-order ω-saturation implies ω-saturation (and the Hennessy–Milner property) in the sense of L. This is due to the following.

Observation 6.2.10 *For any fragment $\mathsf{L} \subseteq \text{FO}$ for which the system of observable configurations $\mathcal{O}^{\mathsf{L}}(\mathfrak{A})$ is uniformly first-order interpretable in $\mathfrak{A}$ itself, ω-saturation of $\mathfrak{A}$ implies ω-saturation of $\mathcal{O}^{\mathsf{L}}(\mathfrak{A})$, which (by the previous observation) implies modal saturation of $\mathcal{O}^{\mathsf{L}}(\mathfrak{A})$, and hence the analogue of the Hennessy–Milner property for L over the class of ω-saturated structures.*

Note that this modal view is based on imposing the modal picture and the bisimulation game on richer fragments of first-order logic, uniformly via the appropriate system of observable configurations and games. Alternatively, one may think of a specialisation of the classically well understood situation for first-order and its infinitary counterpart, their links with classical Ehrenfeucht–Fraïssé games and Karp's theorem (cf. Theorem 6.2.2). In connection with the last observation for instance, ω-saturation (in the classical sense, w.r.t. FO-types) implies ω-saturation in the sense of L for a fragment $\mathsf{L} \subseteq \text{FO}$, since L-types are (partial) FO-types; a Hennessy–Milner property for ω-saturated structures then follows because player **II** has a strategy to maintain L-equivalence in the infinite L-game starting from L-equivalent configurations. But this, and how L-types are to be defined so that they can be transferred between L-equivalent configurations as required for this argument, may be best understood systematically in terms of the game and its observable configurations as discussed above.

6.2.5 Locality and modularity of the first-order game

Games and the Ehrenfeucht–Fraïssé method are well suited to the exploration of the expressive power of FO not just classically but equally well over restricted classes of structures, and also to understanding the nature of fragments within FO. Such explorations typically depend on the availability of suitable structures over which the game can be usefully analysed. In order to facilitate the analysis, and equally importantly also as an indication of where to look for the right candidate structures, one can often use the modularity of the game w.r.t. Gaifman locality. We saw a glimpse of that aspect in Lemma 6.2.4 above.

For Gaifman's theorem, we want to establish that position $(\mathbf{a}, \mathbf{a}')$ in $\mathbb{G}_q(\mathfrak{A}; \mathfrak{A}')$ is a winning position for **II**, i.e., that $\mathfrak{A}, \mathbf{a} \equiv_q \mathfrak{A}', \mathbf{a}'$, on the basis of

– suitable global conditions on $\mathfrak{A}$ and $\mathfrak{A}'$ (without reference to $\mathbf{a}$ and $\mathbf{a}'$), and

– purely local conditions on these parameters within their structures of the form

$$\mathfrak{A}{\restriction}N^\ell(\mathbf{a}), \mathbf{a} \equiv_r \mathfrak{A}'{\restriction}N^\ell(\mathbf{a}'), \mathbf{a}'$$

for values of ℓ and r that are recursively determined as functions of q.

Towards an understanding of the nature of the global requirement, and for a gradation of both the local and global equivalences involved, we need the following definition.

Definition 6.2.11   (i) For any $\varphi(\mathbf{x})$ we write $\varphi^\ell(\mathbf{x})$ for its relativisation to the (FO-definable) ℓ-neighbourhood of its free variables, $\varphi^\ell(\mathbf{x}) := [\varphi]^{N^\ell(\mathbf{x})}$. If $q = \mathrm{qr}(\varphi)$, we refer to φ^ℓ as a *local formula* of Gaifman rank (ℓ, q).

(ii) A *basic ℓ-local sentence* is a sentence of the form

$$\exists x_1 \ldots \exists x_m \bigwedge_{i<j} d(x_i, x_j) > 2\ell \,\wedge\, \bigwedge_i \psi^\ell(x_i),$$

asserting the existing of an ℓ-scattered m-tuple whose components satisfy the ℓ-local formula $\psi^\ell(x)$. If $q = \mathrm{qr}(\psi)$, we regard the above basic local sentence as one of Gaifman rank (ℓ, q, m).

Definition 6.2.12   The configurations $\mathfrak{A}, \mathbf{a}$ and $\mathfrak{A}', \mathbf{a}'$ are (ℓ, q, m)-Gaifman-equivalent, denoted as $\mathfrak{A}, \mathbf{a} \equiv_{q,m}^\ell \mathfrak{A}', \mathbf{a}'$, if:

(i) $\mathfrak{A}{\restriction}N^\ell(\mathbf{a}), \mathbf{a} \equiv_q \mathfrak{A}'{\restriction}N^\ell(\mathbf{a}'), \mathbf{a}'$, i.e., $\mathbf{a}$ and $\mathbf{a}'$ satisfy the same ℓ-local formulae φ^ℓ for $\mathrm{qr}(\varphi) \leqslant q$ (*local condition*).

(ii) $\mathfrak{A}$ and $\mathfrak{A}'$ satisfy the same basic local sentences of ranks (ℓ', q', m') for all $\ell' \leqslant \ell, q' \leqslant q$ and $m' \leqslant m$ (*global condition*).

For fixed finite relational vocabulary and fixed arity of the tuples $\mathbf{a}$, each $\equiv_{q,m}^\ell$ has finite index, and respects $\equiv$. Clearly also $\equiv_{q,m}^\ell$ is monotone w.r.t. the ranks (ℓ, q, m). Gaifman's theorem says that $\equiv_{q,m}^\ell$ approximates full first-order equivalence $\equiv$ well, in the sense that $\equiv$ is the common refinement or limit of all levels $\equiv_{q,m}^\ell$.

Theorem 6.2.13 (Gaifman)   *Any* FO-*formula is preserved under* $\equiv_{q,m}^\ell$ *for suitable* (ℓ, q, m). *Equivalently: any formula of* FO *is logically equivalent to a boolean combination of local formulae and basic local sentences.*

Gaifman's original proof establishes the second statement by induction on the FO formula under consideration. The link with the modularity of the Ehrenfeucht–Fraïssé game, however, is brought out more clearly in an argument given in [15], which we adapt to give a brief sketch. To prove the first of the statements in the theorem, it inductively suffices to establish the following assertion about good responses for **II**.

Claim 6.2.14 *If $\mathfrak{A}$ and $\mathfrak{A}'$ are (L, Q, m)-Gaifman-equivalent* [9] *for values of L and Q that are sufficiently large in relation to ℓ and q, and if $\mathbf{a}$ and $\mathbf{a}'$ of arity less than m are such that*

$$\mathfrak{A} \restriction N^L(\mathbf{a}), \mathbf{a} \equiv_Q \mathfrak{A}' \restriction N^L(\mathbf{a}'), \mathbf{a}' \qquad \textit{local pre-condition}$$

then for any $b \in A$ there is some $b' \in A'$ such that

$$\mathfrak{A} \restriction N^\ell(\mathbf{a}b), \mathbf{a}b \equiv_q \mathfrak{A}' \restriction N^\ell(\mathbf{a}'b'), \mathbf{a}'b', \qquad \textit{local post-condition}$$

and, symmetrically, with the roles of b and b' exchanged.

The claim is established on the basis of a case distinction w.r.t. the distance of b from $\mathbf{a}$. Suitable conditions on the choices of L and Q are extracted along the way. Choosing $L \geqslant 3\ell + 1$ and $Q \geqslant q + 1$ at least, any $b \in N^{2\ell+1}(\mathbf{a})$ can be dealt with according to the local pre-condition. For b that are further away from $\mathbf{a}$, $\mathfrak{A} \restriction N^\ell(\mathbf{a}b)$ is the disjoint union of $\mathfrak{A} \restriction N^\ell(\mathbf{a})$ and $\mathfrak{A} \restriction N^\ell(b)$. Due to modularity of the game w.r.t. disjoint unions, it suffices to find $b' \in A'$ that is also far from $\mathbf{a}'$ and such that $\mathfrak{A}' \restriction N^\ell(b'), b' \equiv_q \mathfrak{A} \restriction N^\ell(b), b$. In this case we rely on the global condition on $\mathfrak{A}$ and $\mathfrak{A}'$ for a further case distinction. We use the global condition for scattered tuples w.r.t. a quantifier-rank q formula $\psi(x)$ that characterises $\mathfrak{A} \restriction N^\ell(b), b$ up to $\equiv_q$. We need to guarantee that $\mathfrak{A}'$ has a matching b', i.e., we seek some $b' \notin N^{2\ell+1}(\mathbf{a}')$ satisfying ψ^ℓ.

Firstly, if $\mathfrak{A}$ and hence also $\mathfrak{A}'$ have $(2\ell + 1)$-scattered m-tuples of elements satisfying ψ^ℓ, then one of the components of any such tuple in $\mathfrak{A}'$ will serve as b'.

If, on the other hand, there are no such m-tuples, then the maximal size $n < m$ of $(2\ell + 1)$-scattered tuples for ψ^ℓ is the same in $\mathfrak{A}$ and $\mathfrak{A}'$. Now a comparison with n_0, the maximal size of $(2\ell + 1)$-scattered tuples for ψ^ℓ within $N^{2\ell+1}(\mathbf{a})$ can help to locate b', provided $L \geqslant 3\ell + 1$ and provided Q is large enough to force the same n_0 to work in $\mathfrak{A}' \restriction N^{2\ell+1}(\mathbf{a}')$ (via the local pre-condition).

---

[9] Due to the absence of parameters this involves only the global condition (ii) of Definition 6.2.12.

If $n_0 < n$, then there must be realisations of ψ^ℓ outside $N^{2\ell+1}(\mathbf{a}')$ and any such is a good choice for b'.

The remaining subcase that $n_0 = n$ (no surplus of realisations of ψ^ℓ beyond $N^{2\ell+1}(\mathbf{a}')$), implies in particular that $d(\mathbf{a}, b) \leqslant 6\ell + 3$ and the existence of such an element satisfying ψ^ℓ at distance greater than $2\ell + 1$ but at most $6\ell + 3$ is covered by the local pre-condition, provided $L \geqslant 7\ell + 3$ and Q is large enough to cover this (under the local pre-condition), too.

6.3 Special classes of transition systems

Up to bisimulation, every transition system is equivalent to a tree via a bisimilar tree unfolding, just as every game graph can be replaced by the associated game tree, typically making the representation structurally simpler though less succinct. Correspondingly, any *bisimulation invariant* logic (logic whose formulae are preserved under bisimulation equivalence) has the tree model property. Because cycles are unfolded into infinite paths, bisimulation equivalent tree models may necessarily be infinite even though the original model was finite. So bisimilar unfoldings into tree models are typically not available within classes of finite models. In the investigation of the model theoretic relationship between bisimulation invariant fragments of FO with FO itself, however, Gaifman locality can be used to replace acyclicity by local acyclicity in key arguments. We briefly review the classical construction of bisimilar unfoldings into tree models and then review a construction of locally acyclic bisimilar companion structures from [42]. These are used to establish variants of the classical model theoretic characterisations of modal fragments of FO in terms of bisimulation preservation (van Benthem's theorem, cf. Corollary 6.3.5 below) over natural, restricted classes of transition systems in section 6.3.2.

6.3.1 Tree unfoldings and locally tree-like systems

Bisimulation invariance and the tree model property

Let $\mathfrak{A}$ be a transition system in a finite vocabulary τ consisting of binary relations E_α and unary predicates P_j. With $a \in \mathfrak{A}$ we associate the following *bisimilar unfolding of $\mathfrak{A}$ at a, $\mathfrak{A}_a^*$*. The universe of $\mathfrak{A}_a^*$ is the set of all finite, edge-labelled paths from a in $\mathfrak{A}$, $\sigma = (a_0, \alpha_1, a_1, \ldots, \alpha_n, a_n)$, where $a_0 = a$ and $(a_{i-1}, a_i) \in E_{\alpha_i}^{\mathfrak{A}}$. The transition relation E_α of $\mathfrak{A}_a^*$ corresponds to path extensions by single $E_\alpha^{\mathfrak{A}}$ edges; the unary predicate P_j in $\mathfrak{A}_a^*$ consists of those paths that end in $P_j^{\mathfrak{A}}$. Then the map that associates to every path its last element, viewed as a map $\pi : \mathfrak{A}_a^* \to \mathfrak{A}$, induces a bisimulation:

$$\pi : \mathfrak{A}_a^* \xrightarrow{\sim} \mathfrak{A} \qquad \mathfrak{A}_a^*, \sigma \sim \mathfrak{A}, \pi(\sigma).$$

It follows that every bisimulation invariant logic has the *tree model property*: satisfiability implies satisfiability in a tree model. The tree model property has important algorithmic consequences. Since it reduces satisfiability issues to problems over trees, strong classical results like Rabin's decidability result for the MSO theory of trees [45] and in particular automata theoretic methods can be brought to bear, see also [49]. The example below illustrates the usefulness of this simple insight for the (classical) model theory of modal logic, in giving an alternative proof for van Benthem's classical characterisation theorem for modal logic (a *preservation theorem* in classical model theoretic terminology). We first discuss the classical argument, though, with emphasis on the more interesting aspect of expressive completeness.

Theorem 6.3.1 (van Benthem)   *Any bisimulation invariant first-order formula $\varphi(x) \in \mathrm{FO}[\tau]$ is equivalent to a formula of $\mathrm{ML}[\tau]$ (and, conversely, this condition is sufficient to guarantee bisimulation invariance).*

A simple compactness argument shows that, if φ is not expressible in ML, then there are $\mathfrak{A}, a \equiv^M \mathfrak{A}', a'$ such that $\mathfrak{A} \models \varphi[a]$ while $\mathfrak{A}' \not\models \varphi[a']$. In ω-saturated elementary extensions $\hat{\mathfrak{A}} \succeq \mathfrak{A}$ and $\hat{\mathfrak{A}}' \succeq \mathfrak{A}'$, which are modally saturated, one automatically upgrades $\mathfrak{A}, a \sim_\omega \mathfrak{A}', a'$ and $\mathfrak{A}, a \equiv^M \mathfrak{A}', a'$ to $\hat{\mathfrak{A}}, a \sim \hat{\mathfrak{A}}', a'$ (cf. the Hennessy–Milner property in Lemma 6.2.8), whence $\hat{\mathfrak{A}} \models \varphi[a]$ and $\hat{\mathfrak{A}}' \not\models \varphi[a']$ refutes preservation under $\sim$.

We turn to alternative approaches that work with explicit model constructions and transformations. We shall later see how this alternative approach relativises to many restricted classes (in particular also of finite models) where compactness is not available. But even in the classical context, and working over the class of all frames, such an explicit and game-based approach yields extra benefits.

Example: van Benthem's theorem via explicit constructions   The following auxiliary observation is straightforward from the bisimulation game: any common upper bound on the lengths of directed paths from the elements in a bisimulation game position is also a bound on the number of rounds that can be played by **I**.

Observation 6.3.2   *For directed, rooted trees $\mathfrak{A}, a$ and $\mathfrak{A}', a'$ of depths $\leqslant \ell$:*

$$\mathfrak{A}, a \sim_\ell \mathfrak{A}', a' \quad \Rightarrow \quad \mathfrak{A}, a \sim \mathfrak{A}', a'.$$

Combining this with the tree model property, we find the following.

Claim 6.3.3   *Any ℓ-local $\varphi(x) \in \mathrm{FO}[\tau]$ that is invariant under $\sim$ is invariant under $\sim_\ell$.*

Proof. We need to show for $\mathfrak{A}, a \sim_\ell \mathfrak{A}', a'$ that $\mathfrak{A} \models \varphi[a] \Leftrightarrow \mathfrak{A}' \models \varphi[a']$. Replacing both structures by their bisimilar unfoldings from the distinguished nodes (and appealing to $\sim$ invariance of φ), then truncating both tree structures at depth ℓ (and appealing to ℓ-locality of φ), we have transformed the given situation into

$$\mathfrak{A}, a \simeq^{(\ell)} \hat{\mathfrak{A}}, \hat{a} \sim \hat{\mathfrak{A}}', \hat{a}' \simeq^{(\ell)} \mathfrak{A}', a',$$

where $\simeq^{(\ell)}$ stands for isomorphism up to depth ℓ from the distinguished node. The central bisimulation equivalence is based on Observation 6.3.2. But now $\mathfrak{A} \models \varphi[a] \Leftrightarrow \mathfrak{A}' \models \varphi[a']$ follows by $\sim$ invariance and ℓ-locality. $\square$

Claim 6.3.4   *If $\varphi(x) \in \mathrm{FO}[\tau]$ is preserved under $\sim$, then it is preserved under $\sim_\ell$ for $\ell = 2^{\mathrm{qr}(\varphi)} - 1$.*

Proof. As $\sim$ invariance implies invariance under disjoint unions, Lemma 6.2.4 shows that φ is ℓ-local, thus $\sim_\ell$ invariant by Claim 6.3.3. $\square$

As $\sim_\ell$ is of finite index, and each $\sim_\ell$ class definable in ML at nesting depth ℓ, we directly have the following version of van Benthem's theorem, which even gives a tight bound on the modal nesting depth which is not implicit in the classical proof.

Corollary 6.3.5   *Any quantifier-rank q formula $\varphi(x) \in \mathrm{FO}[\tau]$ that is preserved under bisimulation is equivalent to a formula of $\mathrm{ML}[\tau]$ of nesting depth $\leqslant 2^q - 1$.*

It may be worth representing the overall strategy of *upgrading a concrete level of $\sim^\ell$* to preservation of φ in this approach. The transformations, from top to bottom in the diagram, involve firstly a tree unfolding and secondly truncation at depth ℓ. The first step preserves $\sim$, the second simultaneously preserves $\sim_\ell$ and φ (by Lemma 6.2.4). Consequently φ is preserved all along the vertical, but also along the bottom horizontal (as here $\sim_\ell$ guarantees full $\sim$ equivalence, by Observation 6.3.2). Thus φ is shown to be preserved along the top horizontal, too.

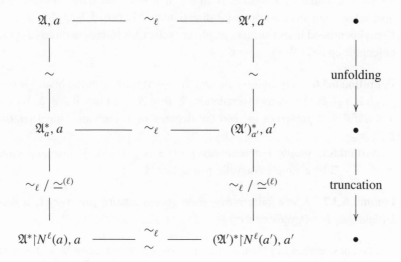

The construction of unfoldings shows that every τ transition system is bisimilar to a τ-tree, and (by taking disjoint unions of unfoldings at different elements as appropriate) globally bisimilar to a τ-forest. Obvious variations of these constructions provide acyclic companion structures that are (globally) two-way bisimilar.

As pointed out above, not every finite transition system is bisimilar to a finite acyclic system. Note that, for instance, the above proof of van Benthem's theorem fails to yield the finite model theory version (due to Rosen [46]): the argument crucially uses bisimulation invariance of φ in the transition from $\mathfrak{A}, a$ to $\mathfrak{A}_a^*$, where the target structure may be infinite.

In the case of Corollary 6.3.5 there is in fact an easy way out: the full (and potentially infinite) tree unfoldings of the given finite structures in the proof of Claim 6.3.3 can in that context be replaced by truncations to depth ℓ with isomorphic copies of the finite original structures attached at the cut-off points to yield fully bisimilar companions that are both finite and tree-like up to depth ℓ. This simple modification yields a proof of Rosen's finite model theory analogue of van Benthem's theorem [46], including the tight bound on nesting depth in our version [42].

In connection with stronger and, in particular, global notions of bisimulation equivalence, however, better approximations to acyclicity in finite models are required. The upgrading will lead from suitable levels of finitary game equivalence to appropriate levels of local FO equivalence (Gaifman equivalence).

Locally acyclic bisimilar covers

Recall that a transition system is *simple* if it does not have loops or multiple edges (not even in opposite directions); it is called ℓ-*acyclic* if every ℓ-neighbourhood in its Gaifman graph is acyclic (this forbids undirected cycles of lengths up to $2\ell + 1$).

Definition 6.3.6   A *bisimilar cover* $\pi : \hat{\mathfrak{A}} \xrightarrow{\approx} \mathfrak{A}$ is a homomorphism π whose graph is a global two-way bisimulation: $\hat{\mathfrak{A}}, \hat{a} \approx \mathfrak{A}, \pi(\hat{a})$ for all $\hat{a} \in \hat{\mathfrak{A}}$. We call π *faithful* if it preserves in- and out-degrees w.r.t. each individual relation $E_\alpha \in \tau$.

A (faithful) *simple ℓ-acyclic cover* of $\mathfrak{A}$ is a (faithful) bisimilar cover $\pi : \hat{\mathfrak{A}} \xrightarrow{\approx} \mathfrak{A}$ by a simple ℓ-acyclic τ-structure $\hat{\mathfrak{A}}$.

Lemma 6.3.7   *Every finite τ transition system admits, for every ℓ, a finite faithful simple ℓ-acyclic cover.*

The construction in [42] uses for $\hat{\mathfrak{A}}$ a product of the given $\mathfrak{A}$ with a finite group G which has a generator g_e for every edge $e \in \bigcup_\alpha E_\alpha^{\mathfrak{A}}$ and such that the Cayley graph of G w.r.t. this set of generators has girth greater than $2\ell + 1$ (compare [1] for such groups) – much as the tree unfolding could be described in terms of a product with the infinite free group of this set of generators. Over the cartesian product $A \times G$ one puts an E_α-edge precisely from (a, h) to (b, k) if $e = (a, b) \in E_\alpha^{\mathfrak{A}}$ and $k = h \circ g_e$. In this fashion, any cycle in the product projects to a cycle in the Cayley graph of G, and hence its length is bounded from below by the girth of that graph.

The following is a simple auxiliary observation towards an ℓ-local upgrading of ℓ-bisimulation equivalence to $\equiv_q$. A natural strategy for **II** can be based on maintaining full isomorphism of the substructures generated by the paths connecting the elements of the current configurations to the roots [14].

Observation 6.3.8   *Let $\mathfrak{A}, a \sim_\ell \mathfrak{A}', a'$ be two directed τ-trees of depths $\leqslant \ell$ with roots a and a', such that every node apart from the root is one of at least q bisimilar siblings. Then $\mathfrak{A}, a \equiv_q \mathfrak{A}', a'$. The same holds w.r.t. two-way ℓ-bisimulation equivalence in acyclic ℓ-neighbourhoods $\mathfrak{A} \upharpoonright N^\ell(a)$ and $\mathfrak{A}' \upharpoonright N^\ell(a')$ with at least q equivalent siblings to choose from in every node.*

Structures that have at least q equivalent successors/predecessors in every node are easily obtained by taking products with $\{1, \ldots, q\}$ in the natural manner. We write $\mathfrak{A} \mapsto \mathfrak{A} \otimes q$ for this transformation, and identify a distinguished element a with $(a, 1)$ in the new structure were appropriate.

Faithful bisimilar covers preserve this property, and can be used to achieve local acyclicity and therefore local $\equiv_q$-equivalence, viz. $\equiv_{q,0}^{\ell}$, by the above observation.

Example: van Benthem–Rosen once more Combining the passage to $\mathfrak{A} \otimes q$ (boosting multiplicities) with a bisimilar unfolding, one obtains a variant proof of Claim 6.3.4 (and through it Corollary 6.3.5 and its finite model theory analogue, too). Let $\mathrm{qr}(\varphi) = q$ and $\ell := 2^q - 1$. Let $\hat{\mathfrak{A}}$ be the tree unfolding from $(a, 1)$ in $\mathfrak{A} \otimes q$ (or the truncation of this unfolding glued with copies of $\mathfrak{A}$ if we want to deal with finite structures exclusively), similarly for $\mathfrak{A}'$, a'.

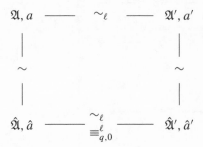

Now $\equiv_{q,0}^{\ell}$ equivalence in the bottom horizontal follows from Observation 6.3.8; preservation of φ along the bottom horizontal additionally uses Lemma 6.2.4 again.

Acyclic bisimilar covers really come into their own in upgradings to some target level $\equiv_{q,m}^{\ell}$ of Gaifman equivalence with $m > 0$, i.e., if the first-order property at hand really does express non-trivial global conditions on the existence or non-existence of certain local types – global in the sense of not only involving the ℓ-neighbourhood of the distinguished element.[10]

We look, as a typical example, at the characterisation of $\mathsf{ML}^{-\vee} \subseteq \mathsf{FO}$ in terms of invariance under $\approx$ (global two-way bisimulation) [42]. Again, we stress the expressive completeness phenomenon, as preservation of $\mathsf{ML}^{-\vee}$ under $\approx$ is obvious.

Theorem 6.3.9 *Both classically and in the sense of finite model theory: any first-order formula $\varphi(x) \in \mathsf{FO}[\tau]$ that is preserved under $\approx$ is equivalent to a formula of $\mathsf{ML}^{-\vee}[\tau]$.*

---

[10] See [42] for a discussion that for any $\varphi(x) \in \mathsf{FO}$ that is invariant under disjoint sums (over finite structures, or indeed over some other class which itself is closed under disjoint sums), only $\equiv_{q,m}^{\ell}$ for $m = 0, 1$ can matter.

This follows from the following claim, based on an upgrading of $\approx_\ell$ to $\equiv_{q,m}^\ell$ in an explicit $\approx$ preserving model transformation, under which in particular the class of finite structures is closed.

Claim 6.3.10   *If $\varphi(x) \in \mathrm{FO}[\tau]$ is preserved under $\approx$ (over finite structures), then it is preserved under $\approx_\ell$ and hence expressible in $\mathrm{ML}^{-\vee}[\tau]$ at nesting depth ℓ, for some ℓ. Any ℓ such that φ is preserved under $\equiv_{q,m}^\ell$ for some q, m will do, i.e., the Gaifman locality radius of φ gives a bound on the necessary modal nesting depth.*[11]

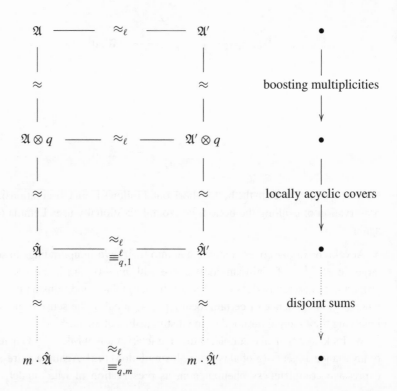

Proof. We just mention the upgrading steps towards the proof of the claim, also indicated in the diagram below.

---

[11] For simplicity, the modal nesting depth in $\mathrm{ML}^{-\vee}$ discounts $\forall/\exists$ quantifiers, which w.l.o.g. can be eliminated from within the scope of modal quantifications so that they only occur 'on the outside.'

The first step, passage to a product with $\{1, \ldots, q\}$, serves to boost all multiplicities to at least q: every E_α successor or predecessor of any node belongs to a group of at least q siblings related by automorphisms of the entire structure.

The second step yields an ℓ-acyclic bisimilar cover of the resulting structures so that the ℓ-neighbourhood of any node will be acyclic, and maintains the at-least-q-similar-siblings property due to the preservation of in- and out-degrees in faithful covers. In these circumstances, the $\approx_\ell$ relationship between the two structures guarantees $\equiv_{q,1}^\ell$ equivalence, by Observation 6.3.8.

Finally we can, if we wish, upgrade $\equiv_{q,1}^\ell$ further to $\equiv_{q,m}^\ell$, for any desired level m, by just passing to m disjoint copies of the structures obtained so far. This step guarantees that any local isomorphism type that is realised at all is a member of a scattered set of at least m many nodes of the same local isomorphism type, so that $\equiv_{q,1}^\ell$ implies $\equiv_{q,m}^\ell$. As pointed out above, however, this last upgrading can be made redundant by showing right away that φ must be preserved under some $\equiv_{q,1}^\ell$ (i.e., $m = 1$ suffices). $\qquad\square$

It is clear that arguments of the kind explored here may have entirely different relativisations from the classical arguments. While classical model theoretic arguments based on compactness go through in restriction to any elementary class of structures, the above argument goes through, for instance, in restriction to any class of (finite) transition systems that is closed under $\approx$. But while this upgrading argument, and hence the expressive completeness result, does relativise to the class of all finite transition systems, it does for instance not immediately relativise to the class of connected or rooted (finite) transitions systems: clearly the last step does not preserve connectivity (and there is no immediate reason why a first-order formula $\varphi(x)$ that is invariant under $\approx$ over connected structures should be preserved by some $\equiv_{q,1}^\ell$), and even the first step does not preserve rootedness.

6.3.2 Non-classical modal characterisation theorems

The general format

Analogues of the van Benthem theorem in classical and finite model theory for stronger and in particular global forms of bisimulation in the style of Theorem 6.3.9 are pursued in [42]. Many further natural variations of the underlying class of (finite) structures are explored in [14], with an emphasis also on methodological distinctions. In all these cases, concrete and explicit model transformations adapted to the classes at hand are used, which in many cases also provide alternative routes to characterisations over some interesting elementary classes of not necessarily finite structures.

We highlight the general format of a characterisation theorem for a fragment L of FO of this kind. Let $L \subseteq FO$ be a fragment of FO with

(1) equivalences $\leftrightharpoons_q$ for the relation of L-equivalence up to rank q, which we assume to have finite index; it follows that $\leftrightharpoons_q$ classes are L-definable at rank q.

($\leftrightharpoons_q$ is induced by the q-round game $\mathbb{G}_q^L$.)

(2) the common refinement of the $(\leftrightharpoons_q)_{q \in \omega}$, $\leftrightharpoons_\omega$, capturing $\equiv^L$.

($\leftrightharpoons_\omega$ is induced by $\mathbb{G}_\omega^L$.)

(3) the full infinitary equivalence $\leftrightharpoons$ associated with $\mathbb{G}_\infty^L$.

The assumptions that each $\leftrightharpoons_q$ has finite index and that $\equiv^L$ is the limit of these finitary game equivalences reflect the 'finitary nature' of L. In this context we want to show, over a given class $\mathcal{C}$ of τ-structures, that the following are equivalent for $\varphi(x) \in FO[\tau]$:

(i) φ is preserved under $\leftrightharpoons$ over $\mathcal{C}$, i.e.,

for all $\mathfrak{A}, a$ and $\mathfrak{A}', a'$ in $\mathcal{C}$: $\mathfrak{A}, a \leftrightharpoons \mathfrak{A}', a' \Rightarrow (\mathfrak{A} \models \varphi[a] \Leftrightarrow \mathfrak{A}' \models \varphi[a'])$.

(ii) φ is equivalent over $\mathcal{C}$ to a formula $\tilde{\varphi} \in L[\tau]$, i.e.,

there is some $\tilde{\varphi} \in L[\tau]$ s.t. for all $\mathfrak{A}, a$ in $\mathcal{C}$: $\mathfrak{A} \models \varphi[a] \Leftrightarrow \mathfrak{A} \models \tilde{\varphi}[a]$.

It is worth looking at the two implications separately:

Preservation, (ii) $\Rightarrow$ (i), is a trivial consequence of the game analysis of L-equivalence (our assumptions above). Moreover, the validity of this implication over the class of all structures trivially implies its validity in restriction to any subclass $\mathcal{C}$. In particular a *preservation* statement trivially implies its finite model theory analogue.

Expressive completeness, (i) $\Rightarrow$ (ii), is the crucial and non-trivial part of the equivalence, which is sensitive to the class $\mathcal{C}$. In particular, expressive completeness does not generally relativise to subclasses, and a classical result cannot generally be expected to persist in the sense of finite model theory.[12]

If $\leftrightharpoons_\omega$ coincides with $\leftrightharpoons_\infty$ in ω-saturated structures, as is typically the case,[13] then expressive completeness of L for first-order properties that are preserved

---

[12] An easy example of a known failure of the finite model theory version of a classically valid expressive completeness result close to our concerns is provided by $\equiv_\infty^2$ and FO^2: the class of all finite linear orderings is closed under $\equiv_\infty^2$ within the class of finite structures, but not definable in FO^2.

[13] Our discussion of saturation and the Hennessy–Milner property in section 6.2 and especially section 6.2.4 indicates that this is true whenever the corresponding $\mathcal{O}^L(\mathfrak{A})$ is uniformly FO-interpretable over $\mathfrak{A}$.

under $\leftrightharpoons$ over the class of all τ-structures follows from the assumptions (along the lines of the classical proof outlined for van Benthem's theorem above, for instance).

Under the assumptions made, expressibility of φ in L (over C) is equivalent to preservation of φ under some level of $\leftrightharpoons_\ell$ (over C). Therefore, the expressive completeness of L for $\leftrightharpoons$ invariance over C is equivalent (for any C) to the implication

$$\varphi(x) \text{ preserved under } \leftrightharpoons \text{ (over } C\text{)}$$
$$\Rightarrow \varphi(x) \text{ preserved under } \leftrightharpoons_\ell \text{ (over } C\text{) for some } \ell \in \omega,$$

which is a particular 'compactness property' that may or may not be valid, depending on the nature of $\leftrightharpoons$ and C. The classical manner of establishing this compactness property, as well as the alternative explicit and game-oriented constructions indicated above may both be cast as upgradings of equivalences, albeit in orthogonal directions. The juxtaposition of the generic diagrams below may serve to make this distinction apparent. While the classical upgrading involves a transformation of structures up to full FO equivalence (passage to ω-saturated elementary extensions say) to boost $\leftrightharpoons_\omega$ to $\leftrightharpoons$, the alternative upgrading consists of a transformation of structures up to full (infinitary) $\leftrightharpoons$ to boost a concrete finitary level of $\leftrightharpoons_\ell$ to an approximate level $\equiv$ of first-order equivalence that is good enough to preserve φ. In the examples encountered here, $\equiv$ is either some level $\equiv_q$ or $\equiv_{q,m}^\ell$. The following two sections will review and summarise some of the results obtained along these lines in [14].

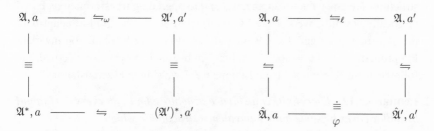

Explicit upgrading through local control

By approximating FO equivalence by a concrete level of Gaifman equivalence we shift the emphasis to local control over FO equivalence. This allows us to make use of explicit model constructions that lead to locally acyclic structures, as in Lemma 6.3.7, which means that *locally* $\approx^\ell$ can be upgraded to $\equiv_q$ (if multiplicities have been boosted in preparation) via Observation 6.3.8. For characterisations of $\sim_\vee$ invariance rather than $\approx$ (global but only forward bisimulation, related to $\mathsf{ML}^\vee$), a correspondingly higher level of global ℓ_0-bisimulation equivalence can first be upgraded (in a transformation up to full global forward bisimulation $\sim_\vee$) to $\approx^{\ell_1}$, which can then be further upgraded to some $\equiv^\ell_{q,m}$ as above. In this manner, for example the expressive completeness results below are proved in [14].

A *rooted* structure is a τ-structure $\mathfrak{A}, a$ with distinguished element a as a *root* from which all elements of $\mathfrak{A}$ are reachable on directed paths. For tree structures compare section 6.1.2. Note that even the class of not necessarily finite rooted structures is not elementary. Also note that for rooted structures, the full infinitary equivalences $\sim_\vee$ and $\sim$ coincide at the roots, while the finite levels clearly do not.

Theorem 6.3.11 $\mathsf{ML}^\vee$ *is expressively complete for first-order properties that are preserved under $\sim$ over the following classes C of structures:*

   (i)   *the class of rooted structures.*
   (ii) *the class of finite rooted structures.*
   (iii) *the class of tree structures.*
   (iv) *the class of finite tree structures.*

Another natural and classically important class of transition systems (as Kripke structures in the context of knowledge representation) is the class of *equivalence structures*: τ-structures in which all transition relations E_α are interpreted as equivalence relations. And even though transitivity requirements tend to trivialise locality analysis (also compare the next section), equivalence structures are amenable to an analysis and to upgrading transformations based on locally acyclic covers. Here FO interpretations can be used to adapt both the construction of suitable covers and the analysis of bisimulation invariant FO properties. As far as local acyclicity in bisimilar covers is concerned, the following can be obtained from Lemma 6.3.7 via simple FO translations.

Lemma 6.3.12 *Every finite equivalence structure admits, for every ℓ, a faithful bisimilar cover by some finite equivalence structure in which*

   (i) *any two equivalence classes (w.r.t. to distinct E_α) intersect in at most one element,*
   (ii) *all cycles of lengths up to $2\ell + 1$ stay within a single E_α class for some α.*

Over such essentially ℓ-acyclic structures, an analogue of Observation 6.3.8 is available to show that global ℓ-bisimulation can be upgraded to $\equiv^\ell_{q,m}$ for any required level of q and m. Therefore, $\sim$ invariance implies $\sim_\ell$ invariance also over the class of finite equivalence structures.

Corollary 6.3.13 $ML^\vee$ *is expressively complete for first-order properties that are preserved under global bisimulation* $\sim_\vee$ *over the class of finite equivalence structures.*

Explicit upgrading through decomposition

Locality arguments cannot be used to great effect over structures that trivialise Gaifman locality. For instance, the Gaifman graph of directed transitive trees (trees with a partial order) has diameter 2, and $\equiv^\ell_{q,m}$ is essentially just $\equiv_q$, for $\ell \geqslant 1$. On some related and particularly interesting classes of transition systems with one transitive transition relation, however, one may instead base expressive completeness proofs for modal fragments on another classical constructive approach to the analysis of games: composition arguments w.r.t. order. We saw a glimpse of this in the Ehrenfeucht–Fraïssé analysis of finite linear orderings in section 6.2.2 (Lemma 6.2.3).

We consider the example of rooted, irreflexive transitive tree structures with a single transition relation E: $\mathfrak{A} = (A, E^{\mathfrak{A}}, (P_i^{\mathfrak{A}}))$ with distinguished root a, with a transitive and irreflexive partial order relation $E^{\mathfrak{A}}$ such that the set of E-predecessors of any element $b \in A$ is well-ordered by $E^{\mathfrak{A}}$ with minimal element a. For succinctness we refer to such structures as $\prec$-trees. The class of all $\prec$-trees (finite and infinite ones) is non-elementary (due to the well-foundedness condition); and so is the class of all finite $\prec$-trees (due to the finiteness condition).

We review the key decomposition idea from [14] that allows us to upgrade ℓ-bisimulation equivalence between (finite) $\prec$-trees $\mathfrak{A}, a \sim_\ell \mathfrak{A}', a'$ to quantifier-rank q first-order equivalence $\equiv_q$ through a transformation that preserves full bisimulation equivalence.

In a preparatory step, we boost multiplicities and unravel in order to achieve some homogeneity w.r.t. paths in $\prec$-trees.

For a given q let the $\prec$-trees $\mathfrak{A}_0^q$ and $\mathfrak{A}_{q-1}^q$ (an expansion of $\mathfrak{A}_0^q$ by colours for certain $\equiv_{q-1}$ types) be obtained from $\mathfrak{A}, a$ as follows.

The universe and the interpretation of the unary predicates of $\mathfrak{A}_0^q$ are those of the bisimilar unfolding of $\mathfrak{A} \otimes \{1, \ldots, q\}$ from one of the representatives of the root a (say we identify a with $(a, 1)$); for its transition relation we pass to the transitive closure of the transition relation in the unfolding. It is easily checked that this transformation leads to a bisimilar $\prec$-tree $\mathfrak{A}_0^q$, which is finite if $\mathfrak{A}$ is. Even for infinite $\mathfrak{A}$ the $\prec$-tree $\mathfrak{A}_0^q$ has predecessor sets that are finite linear orderings rather than arbitrary well-orderings. In addition, due to the unfolding step in its construction, $\mathfrak{A}_0^q$ has the following useful representation property for its paths. Any path $a_0 = a, a_1, \ldots, a_n$ from the root in $\mathfrak{A}_0^q$, can be matched with some *full path* $\hat{a}_0 = a, \hat{a}_1, \ldots, \hat{a}_n$ consisting of the full predecessor set of the target node $\hat{a}_n$ in $\mathfrak{A}_0^q$, such that a_i and $\hat{a}_i$ are not only bisimilar but even are the roots of isomorphic subtrees.

Towards an inductive analysis of $\equiv_q$, we use $\mathfrak{A}_{q-1}^q$, which is the expansion of $\mathfrak{A}_0^q$ with new unary predicates that colour every node with the $\equiv_{q-1}$-class of the subtree rooted at this node in $\mathfrak{A}_0^q$.

In order to show how suitable levels of ℓ-bisimulation between $\prec$-trees $\mathfrak{A}, a$ and $\mathfrak{A}', a'$ can be upgraded to $\equiv_q$ equivalence in bisimilar $\prec$-trees, we firstly replace $\mathfrak{A}$ and $\mathfrak{A}'$ by the $\prec$-trees $\mathfrak{A}_0^q, a \sim \mathfrak{A}, a$ and $(\mathfrak{A}')_0^q, a' \sim \mathfrak{A}', a'$. It then suffices to show, in the context of an induction on q, that for some sufficiently large ℓ (depending on q):

$$(*) \qquad \mathfrak{A}_{q-1}^q, a \sim_\ell (\mathfrak{A}')_{q-1}^q, a' \quad \Rightarrow \quad \mathfrak{A}_0^q, a \equiv_q (\mathfrak{A}')_0^q, a'.$$

For this, a composition argument can be used towards a reduction to the analysis of Ehrenfeucht–Fraïssé games over finite coloured linear orderings. We associate with an element b in $\mathfrak{A}^q, a$ the coloured finite linear ordering $\mathfrak{I}_b$ induced on the interval $[a, b]$ in $\mathfrak{A}^q$; similarly $\mathfrak{I}'_{b'}$ with any b' in $(\mathfrak{A}')^q, a'$. Then

$$\mathfrak{I}_b, a, b \equiv_{q-1} \mathfrak{I}'_{b'}, a', b' \quad \Rightarrow \quad \mathfrak{A}_{q-1}^q, a, b \equiv_{q-1} (\mathfrak{A}')_{q-1}^q, a', b',$$

due to compositionality of strategies in the games. A winning strategy for **II** in the remaining $(q - 1)$-round game on the $\prec$-trees can be based on

(a) a strategy in the $(q - 1)$-round game on the induced linear orderings: this provides a match between subtrees rooted along the coloured paths $[a, b]$ and $[a', b']$.

(b) strategies to play within colour-matched subtrees based on their $\equiv_{q-1}$ equivalence.

Therefore, it suffices to guarantee that for every b there is some b' (and vice versa, for every b' a b) such that $\mathfrak{I}_b, a, b \equiv_{q-1} \mathfrak{I}'_{b'}, a', b'$, provided only that $\mathfrak{A}_{q-1}^q, a \sim_\ell (\mathfrak{A}')_{q-1}^q, a'$. A bound on such an ℓ can now be extracted from the Ehrenfeucht–Fraïssé game on finite coloured linear orderings. The following is a consequence of the compatibility of the game with ordered sums or concatenation (we leave it as a nice exercise; see [15] and also [14] for details).

Observation 6.3.14 *There is a bound N (depending on q and the number of colours) such that any finite coloured linear ordering (with constants for the first and last elements) of length greater than N has some proper $\equiv_{q-1}$ equivalent substructure.*

In the case of the finite coloured orderings $\mathfrak{I}_b$ this means that, up to $\equiv_{q-1}$, only those of lengths up to N need to be taken into account (any substructure of an $\mathfrak{I}_b$ is realised as $\mathfrak{I}_{\hat{b}}$ for suitable $\hat{b}$ by the homogeneity property of $\mathfrak{A}_{q-1}^q$).

But the isomorphism types of (substructures of) $\mathfrak{I}_b$ of size up to N are clearly governed by the $\sim_{N-1}$ type of $\mathfrak{A}_{q-1}^q$, a, whence we get (∗) for $\ell = N - 1$.

Based on this decomposition approach, the following are obtained in [14].

Theorem 6.3.15 ML *is expressively complete for first-order properties that are preserved under bisimulation over the following classes* $\mathcal{C}$ *of partially ordered trees:*

(i) *the class of irreflexive transitive trees.*

(ii) *the class of finite irreflexive transitive trees.*

While the classes of rooted reflexive transitive structures or reflexive transitive trees display a similar behaviour [14], the picture changes if reflexivity is not uniformly prescribed. For transitive tree-like structures in which some nodes *may* be reflexive, a marked difference between finite and not necessarily finite structures becomes important. The first-order formula

$$\varphi(x) = \exists y(Exy \wedge Eyy),$$

expressing accessibility of a reflexive node, is

(a) invariant under bisimulation over the class of *finite* transitive structures, but

(b) not invariant under bisimulation over the class of all transitive structures.

Point (b) is illustrated by the simple example of the infinite irreflexive unfolding of a structure consisting of a single reflexive node. For (a) consider finite transitive structures $\mathfrak{A}, a \sim \mathfrak{A}', a'$ and assume that $\mathfrak{A} \models \varphi[a]$. Consider a play of **I** from a to some reflexive b in $\mathfrak{A}$ followed by a sequence of stationary moves at b (b is reflexive) that is long enough to force the sequence of responses by **II** to visit some node b' twice: as b' is on a cycle, it is reflexive.

[14] shows that an extension of basic modal logic with a modality as suggested by φ above, asserting that there is some *reflexive* successor satisfying ψ, is expressively complete for bisimulation invariant first-order properties over *finite* transitive tree-like structures. For expressive completeness over the wider classes of all finite transitive structures a stronger variant of this new modality is required, which also captures reachability of an E-clique (rather than a single reflexive node) realising several distinct formulae. As indicated above, such extra modalities are necessary in the finite, but not compatible with bisimulation in transitive structures in general. (In fact it is not finiteness, but the absence of infinite strictly forward-directed E-paths, that matters, see [14].)

Over finite transitive structures and some related restricted classes of transitive transition systems, the decomposition based analysis in [14] also extends from first-order to monadic second-order logic.

Among the long open questions in this area remain the finite model theory status of

- the Janin–Walukiewicz result [34] that the modal μ-calculus is expressively complete for monadic second-order properties preserved under bisimulation, and
- expressive completeness of the guarded fragment for the first-order properties preserved under guarded bisimulation, established in the classical setting in [2].

The second issue, concerning guarded bisimulation as a generalisation of modal bisimulation, also leads over to the following section.

6.4 From graphs to hypergraphs

The guarded fragment of FO and, more fundamentally, the concept of guarded bisimulations (compare section 6.2.3) point to a hypergraph structure induced by a relational structure, over and above the graph structure embodied in the Gaifman graph. With the relational τ-structure $\mathfrak{A}$ we can associate the hypergraph of guarded subsets of $\mathfrak{A}$, whose universe is the universe A of $\mathfrak{A}$ and whose hyperedges are the guarded subsets $s \subseteq A$ of $\mathfrak{A}$:

$$H(\mathfrak{A}) = \big(A, \{s \subseteq A \colon s \text{ a guarded subset }\}\big).$$

Generally, with any hypergraph $H = (A, S)$, one also associates the graph over the same universe A whose edge relation is precisely the union of the cliques induced by the hyperedges of H:

$$G(H) = (A, E) \quad \text{where} \quad E = \bigcup_{s \in S} \{(a, b) \colon a, b \in s, a \neq b\}.$$

In the case of the hypergraph $H(\mathfrak{A})$ this just returns the Gaifman graph $G(\mathfrak{A})$.

The graph $G(H)$, however, contains less information, since not every clique in $G(H)$ need be induced by a hyperedge. The complete graph on three elements, K_3, for instance, occurs as $G(H)$ for $H = K_3$ as well as for any hypergraph that has the full set of three elements as one of its hyperedges. In the classical literature on hypergraphs [7], a hypergraph H such that all cliques in $G(H)$ are induced by hyperedges is called *conformal*; conformality plays a role in acyclicity criteria for hypergraphs. In the next section we briefly look at the natural notion of hypergraph bisimulation and discuss corresponding notions of acyclicity and unfoldings.

6.4.1 Hypergraph bisimulation

If we disregard the local relational content in guarded bisimulations, i.e., if we relax the soundness condition on positions in the game from local isomorphism of relational substructures to just local bijections, we obtain a natural notion of hypergraph bisimulation. Guarded bisimulations become a special case of hypergraph bisimulations between the associated hypergraphs of guarded subsets. For questions of acyclicity and of tree decomposability, the actual local relational content does not matter and it makes sense to work with the more fundamental notion of hypergraph bisimulation.

The hypergraph bisimulation game The positions in the bisimulation game on hypergraphs $H = (A, S)$ and $H' = (A', S')$ are local bijections $\rho \colon s \to s'$ between hyperedges $s \in S$ and $s' \in S'$. The challenge/response exchange between players **I** and **II** in a single round, from position $\rho \colon s \to s'$, is played as follows:

- **I** selects either some hyperedge $t \in S$ or some hyperedge $t' \in S'$;
- **II** has to respond with a position $\sigma \colon t \to t'$ (involving the hyperedge proposed by **I** and a match with a hyperedge in the opposite structure) such that ρ agrees with σ on the overlap (between s and t if **I** chose t, or between s' and t' if **I** chose t').

II loses if she has no such response. Otherwise, winning conditions in the q-round game, the finite-round game and the infinite game are as usual. We correspondingly define equivalences in terms of winning positions for **II**.

Definition 6.4.1 For hypergraphs $H = (A, S)$ and $H' = (A', S')$: $H, \mathbf{a} \sim_q H, \mathbf{a}'$ if the bijection $\rho \colon \mathbf{a} \mapsto \mathbf{a}'$ is a winning position in the q-round bisimulation game on the hypergraphs H and H'. Equivalences $H, \mathbf{a} \sim_\omega H', \mathbf{a}'$ and $H, \mathbf{a} \sim H', \mathbf{a}'$ are similarly defined w.r.t. the finite-round and infinite games.

Definition 6.4.2 A *bisimilar cover* of the hypergraph $H = (A, S)$ by the hypergraph $\hat{H} = (\hat{A}, \hat{S})$ is a map $\pi \colon \hat{A} \to A$ such that
(i) π is injective in restriction to every $\hat{s} \in \hat{S}$.
(ii) $S = \{\pi(\hat{s}) \colon \hat{s} \in \hat{S}\}$.
(iii) π comprises a winning strategy for **II** in the infinite bisimulation game in the sense that **II** can maintain positions in which hyperedges are matched through π.

Consider the special case of $H = H(\mathfrak{A})$, the hypergraph of guarded subsets of the τ-structure $\mathfrak{A}$. It is not hard to see that any bisimilar cover $\pi \colon \hat{H} \to H$

by a hypergraph $\hat{H} = (\hat{A}, \hat{S})$ induces a *guarded cover*

$$\pi : \hat{\mathfrak{A}} \to \mathfrak{A},$$

where $\hat{\mathfrak{A}}$ is simply obtained by pulling the relational interpretation on A back to $\hat{A}$ in such a way that every restriction of π to a hyperedge of $\hat{H}$ becomes a local isomorphism. One checks that this leads to a well-defined interpretation of a τ-structure over the universe $\hat{A}$, for which indeed also $H(\hat{\mathfrak{A}}) = \hat{H}$. In particular π now comprises a winning strategy for **II** in the infinite guarded bisimulation game on $\hat{\mathfrak{A}}$ and $\mathfrak{A}$ (compare (iii) above). These simple considerations suggest to view hypergraph bisimulation just as 'guarded bisimulations without relations' – or to view guarded bisimulation as a relational incarnation of a possibly more fundamental notion of hypergraph bisimulation.

6.4.2 Tree-likeness: acyclicity criteria

Full acyclicity (in the hypergraph sense) can be achieved, up to bisimulation, through a process of bisimilar unfolding in close analogy with the tree unfolding of transition systems. We present this basic construction before relating it to the relevant notions of acyclicity and tree-likeness that it exemplifies.

Bisimilar hypergraph unfolding   Consider a hypergraph $H = (A, S)$. We want to find a tree-like hypergraph $\hat{H}$ that provides a bisimilar cover for H; while overlaps between hyperedges have to be reproduced in $\hat{H}$, it should otherwise and in particular globally be as free (free of incidental overlaps) as possible. The construction follows the idea of a tree unfolding of a transition system, but instead of nodes, subsets need to be joined – joined through identifications in overlaps as prescribed in H, compare [25].

With H firstly associate the tree S^* of all finite sequences of hyperedges, with a successor relation linking a sequence $\sigma \in S^*$ to its immediate extensions $\sigma\hat{}s$ for $s \in S$. We obtain the universe $\hat{A}$ of the desired hypergraph $\hat{H}$ as a quotient of the following auxiliary set D, which may be seen as a disjoint union of path-labelled copies of hyperedges $s \in S$:

$$D := \left\{ (\sigma\hat{}s, a) \in S^+ \times A : a \in s \right\} \subseteq S^* \times A.$$

In this set, we want to identify same elements in nodes that are labelled with next-neighbour paths. Let $\doteq$ be the reflexive, symmetric, transitive closure of the relation that links (σ, a) to $(\sigma\hat{}s, a)$ in D. In the following we write $[\sigma, a]$

for the $\doteq$ equivalence class of $(\sigma, a) \in D$. We put

$$\hat{A} := D \,/\, \doteq,$$
$$\hat{S} := \{\hat{s}_\sigma : s \in S, \sigma \in S^*\},$$
$$\text{where } \hat{s}_\sigma = \{[\sigma \hat{\,} s, a] : a \in s\} \text{ for } \sigma \in S^*, s \in S.$$

One checks that $\pi : \hat{H} \to H$, $[\sigma, a] \mapsto a$ is well-defined and a bisimilar hypergraph cover. In line with the above remarks, if the same construction is applied to the hypergraph $H = H(\mathfrak{A})$ associated with the guarded subsets of a τ-structure $\mathfrak{A}$, then the obvious expansion of $\hat{A}$ to a τ-structure yields a guarded bisimilar cover of $\mathfrak{A}$. In both cases, the tree structure of S^* also provides a tree decomposition of the new hypergraph $\hat{H}$, or of the τ-structure $\hat{\mathfrak{A}}$.

Definition 6.4.3 A *tree decomposition* of $H = (A, S)$ consists of a tree T together with a surjective map $\rho : T \to S$ such that for every $a \in A$ the subset $\{t \in T : a \in \rho(t)\} \subseteq T$ is connected in T.

It may be intuitive that the existence of a tree decomposition is an acyclicity condition. Consider a tree decomposition ρ of a finite hypergraph H with finite tree T. One can use ρ to reduce H to the empty hypergraph by repeated application of the following two reduction steps

- removal of an element $a \in A$ that is covered by at most one hyperedge (more precisely, a is removed from A and from the hyperedge covering it).
- removal of a hyperedge s that is contained in some other hyperedge that is retained.

For the claimed reduction, essentially just proceed from the leaves of T: a leaf of T is mapped to a hyperedge that is either contained in the hyperedge at its predecessor node, or it contains some elements not covered by any other hyperedge. Removal of hyperedges or elements based on this procedure is compatible with maintaining a tree decomposition.[14]

If we transfer this notion of a hypergraph tree decomposition to relational structures (cf. Definition 6.5.1 for tree decompositions in that sense), there is an important difference: the usual notion of tree decomposition is more liberal in allowing arbitrary subsets of A to be associated with the nodes of the representation tree, while here we would only admit guarded sets. A cycle (viewed as a hypergraph with size 2 hyperedges) does not admit a hypergraph tree decomposition, but it does admit tree decompositions based on subsets of

---

[14] Note that to deal with infinite hypergraphs, it is necessary to phrase the reduction condition for finite sub-hypergraphs rather than the full graph; e.g., a two-way infinite edge chain is not decomposable as such.

size 3. We return to ordinary tree decompositions of relational structures in section 6.5.1 below.

It follows that every logic invariant under guarded bisimulation (i.e., whose formulae are preserved under guarded bisimulations) has a *bounded treewidth model property* or *generalised tree model property* [21]. This property is of great value in the algorithmic model theory of GF and of its extensions that still are sublogics of GF_∞ like guarded fixpoint logic [24], because it allows a reduction of satisfiability issues to the model theory of trees, via a coding of models in tree representations.

Proposition 6.4.4 (Grädel)   GF *has the following generalised tree model property: any satisfiable* $\varphi \in \mathsf{GF}[\tau]$ *is satisfiable in a model that admits a tree decomposition w.r.t. guarded subsets, and in particular one of treewidth less than the width of* τ.[15]

Returning to hypergraphs, the classical criterion for hypergraph acyclicity is the following. As shown in [6] it coincides (for finite hypergraphs) with hypergraph tree decomposability in the sense of Definition 6.4.3 above, as well as with several other criteria. For classical hypergraph theory compare [7].

Definition 6.4.5   A hypergraph $H = (A, S)$ with associated graph $G(H)$ is called *acyclic* if it satisfies the following two conditions:
  (i) *conformality:* every clique in $G(H)$ is contained in some hyperedge of H.
  (ii) *chordality:* every cycle of length at least 4 in $G(H)$ has a chord: there are two nodes that are not next neighbours along the cycle that are linked (by an edge of $G(H)$/hyperedge of H).

Clearly hypergraph unfoldings are acyclic in this sense, so that every hypergraph admits a bisimilar cover by an acyclic hypergraph. Conformality can be achieved in *finite* conformal bisimilar hypergraph covers according to [31]; this, however, had left open the following for quite some time.

Question 6.4.6   *Does every* finite *hypergraph admit bisimilar covers by* finite *conformal and ℓ-acyclic hypergraphs, for all ℓ?*

Recent progress [43, 44] shows that the answer is *yes* for a natural and seemingly strongest possible notion of bounded acyclicity which forbids short

---

[15] Treewidth is defined to be one less than the maximal size of sets needed in a tree decomposition, here bounded by the width of τ minus 1.

chordless cycles.[16] The following example shows that acyclicity cannot be achieved even locally in finite bisimilar hypergraph covers.

Example Consider a cartwheel hypergraph H_n consisting of an exterior cycle of nodes $a_1, \ldots, a_n, a_1$ plus a central node a, and with hyperedges $\{a, a_i, a_{i+1}\}$ for $i \in \mathbb{Z}/n\mathbb{Z}$. The exterior cycle of length n is without chord, and any bisimilar cover of H_n will still have cycles in the 1-neighbourhood of any node related to a, albeit possibly longer cycles.

The positive resolution to Question 6.4.6 is also the starting point for proving the finite model theory analogue of the classical characterisation theorem [2] for GF in [43]. Previously, only the graph case, or the case of GF[τ] for relational vocabularies of width 2, had been settled positively in [42].

6.4.3 Excursion: extension properties

The basic idea towards the construction of finite conformal bisimilar hypergraph covers in [31] is quite simple – and surprisingly contrary to the intuition of an unfolding. It essentially focuses on the footprints of forbidden cliques in the associated graph $G(\hat{H})$. We illustrate the key idea with a (generic) local example of the task.

Let, for instance, $H = (A, S)$ be a finite hypergraph with a tuple of pairwise distinct nodes $\mathbf{a} = (a_1, \ldots, a_n)$ such that $[\mathbf{a}] = \{a_1, \ldots, a_n\}$ is a clique in $G(H)$ not contained in any hyperedge of H. We want to construct a bisimilar cover $\pi \colon \hat{H} \to H$ by a finite hypergraph $\hat{H} = (\hat{A}, \hat{S})$ such that no lift $\hat{\mathbf{a}} = (\hat{a}_1, \ldots, \hat{a}_n)$ with $\hat{a}_i \in \pi^{-1}(a_i)$ forms a clique in $G(\hat{H})$. Let $A_0 := A \setminus [\mathbf{a}]$ and put

$$\hat{A} := A_0 \cup ([\mathbf{a}] \times \{1, \ldots, n-1\}); \qquad \pi{\restriction}A_0 = \mathrm{id}, \quad \pi(a_j, i) = a_j.$$

We now set $\hat{S}$ to be the set of all subsets $\hat{s} \subseteq \hat{A}$ such that
 (i) $\pi{\restriction}\hat{s}$ is a bijection onto some $s \in S$.
 (ii) for $(a_j, i), (a_{j'}, i') \in \hat{s}$, if $(a_j, i) \neq (a_{j'}, i')$, then $i \neq i'$:
    any two distinct nodes in $\hat{s}$ above $\mathbf{a}$ must have distinct tags in $\{1, \ldots, n-1\}$.

On one hand, one checks that $\pi \colon \hat{H} \to H$ is a bisimilar cover. Crucially, the back-and-forth conditions do not give rise to requirements (of the *back* kind) to produce a hyperedge $\hat{s}$ whose projection to A would cover all of $[\mathbf{a}]$: this is clear, since $[\mathbf{a}]$ is not contained in any hyperedge of H. On the other hand,

---

[16] A weaker notion of acyclicity is used to explore the finite model theory of the guarded and clique guarded fragments in [5].

condition (ii) rules out the possibility of a clique in $G(\hat{H})$ above **a**: if each pair of components in **â** were to be linked by a hyperedge, then they would have to have pairwise distinct tags, which is impossible simply by the pigeon-hole principle.

A uniform application of this idea, for all forbidden cliques simultaneously, yields a finite conformal cover which moreover has useful automorphism properties [31].

An automorphism of a hypergraph is a permutation of its universe that preserves the set of hyperedges. We say that the cover $\pi : \hat{H} \to H$ *lifts automorphisms* of H if for every automorphism ρ of H there is an automorphism of $\hat{H}$ such that $\rho \circ \pi = \pi \circ \hat{\rho}$. The cover is *homogeneous*, if for every pair of hyperedges $\hat{s}_1, \hat{s}_2 \in \hat{S}$ above the same $s \in S$, there is some automorphism σ of $\hat{H}$ mapping $\hat{s}_1$ to $\hat{s}_2$.

Lemma 6.4.7 *Every finite hypergraph $H = (A, S)$ admits a bisimilar cover $\pi : \hat{H} \to H$ by some finite conformal hypergraph $\hat{H} = (\hat{A}, \hat{S})$.*

Every finite relational τ-structure $\mathfrak{A}$ admits a guarded cover by some finite τ-structure $\hat{\mathfrak{A}}$ whose hypergraph of guarded subsets $H(\hat{\mathfrak{A}})$ is conformal.

Moreover, the cover can be chosen homogeneous and such that it lifts all automorphisms of the base structure.

Herwig–Lascar extension theorems, EPPA   A *local automorphism* of a τ-structure $\mathfrak{A}$ is a partial bijection p of A that is an isomorphism between the substructures induced on $\mathrm{dom}(p)$ and $\mathrm{image}(p)$. The following extension theorem for local automorphisms is from [27], also compare [29].

Theorem 6.4.8 (Herwig)   *Let $\mathfrak{A}_0$ be a finite τ-structure. Then there is a finite extension $\mathfrak{A}_1 \supseteq \mathfrak{A}_0$ such that every local automorphism of $\mathfrak{A}_0$ extends to a full automorphism of $\mathfrak{A}_1$. $\mathfrak{A}_1$ can be chosen such that every guarded subset of $\mathfrak{A}_1$ is the image under some automorphism of $\mathfrak{A}_1$ of a guarded subset of $\mathfrak{A}_0$.*

The last condition is in the given situation in fact equivalent to saying that, for every relation $R \in \tau$:

$$R^{\mathfrak{A}_1} = \bigcup_{\rho \in \mathrm{Aut}(\mathfrak{A}_1)} \rho(R^{\mathfrak{A}_0}).$$

If $\mathfrak{A}_1'$ at first only satisfies the automorphism extension property, and $G' = \mathrm{Aut}(\mathfrak{A}_1')$, then replacing $R^{\mathfrak{A}_1'}$ by $\bigcup_{\rho \in G'} \rho(R^{\mathfrak{A}_0})$ preserves the automorphism extension property and yields a structure that also satisfies the additional requirement on guarded subsets. A combination with Lemma 6.4.7 then gives the following strengthening of the theorem [31].

Corollary 6.4.9 *For every finite $\mathfrak{A}_0$ there is a finite extension $\mathfrak{A}_2 \supseteq \mathfrak{A}_0$ such that every local automorphism of $\mathfrak{A}_0$ extends to a full automorphism of $\mathfrak{A}_2$ and such that every clique in $G(\mathfrak{A}_2)$ is the image under some automorphism of $\mathfrak{A}_2$ of some clique in $G(\mathfrak{A}_0)$.*

Proof. Let $\mathfrak{A}_1 \supseteq \mathfrak{A}$ as in Theorem 6.4.8. Let $H_1 = (A_1, S)$ be the hypergraph with hyperedges

$$S = \{\rho(A_0) \colon \rho \in \mathrm{Aut}(\mathfrak{A}_1)\}.$$

We may now apply Lemma 6.4.7 to obtain a conformal bisimilar cover $\pi \colon \hat{H} \to H_1$ with hypergraph $\hat{H} = (\hat{A}, \hat{S})$. The desired τ-structure $\mathfrak{A}_2 = \hat{\mathfrak{A}}$ is obtained by interpreting the relations over the universe $\hat{A}$ such that, for every $\hat{s} \in \hat{S}$, the local bijection $\pi{\restriction}\hat{s} \colon \hat{s} \to s$ becomes a local isomorphism between $\hat{\mathfrak{A}}{\restriction}\hat{s}$ and $\mathfrak{A}_1{\restriction}s$. $\mathfrak{A}_0$ may be isomorphically embedded into this new structure $\hat{\mathfrak{A}}$ by singling out any particular $\hat{s} \in \hat{S}$ above $s = A_0 \in S$. The automorphism properties of the cover as stated in Lemma 6.4.7 guarantee that the local automorphisms of the embedded $\mathfrak{A}_0$ still extend to automorphisms of $\hat{\mathfrak{A}}$. And $G(\hat{\mathfrak{A}})$ does not have any cliques other than those that are unavoidable automorphic copies of cliques already present in the embedded $\mathfrak{A}_0$: this is a consequence of the conformality of $\hat{H}$ and the fact that $G(\hat{\mathfrak{A}})$ consists of the union of the $G(\hat{\mathfrak{A}}){\restriction}\hat{s}$ for $\hat{s} \in \hat{S}$, each of which is an isomorphic copy of $G(\mathfrak{A}_0)$ by construction. □

Further corollaries of this are (simpler proofs of) the extension theorem for local automorphisms within the class of finite triangle-free graphs [27], the class of finite clique-free graphs [28], or the class of finite relational structures with conformal hypergraphs of guarded sets.

The corollary as stated has also been employed in [31] to yield a very transparent proof of the finite model property of the clique guarded fragment, just as Theorem 6.4.8 itself yields a very natural proof of the finite model property for basic GF first given by Grädel [21].

6.5 Locality and special classes of relational structures

6.5.1 Tree-decompositions and treewidth

Bounded treewidth has emerged as one central notion of 'tameness' or 'well-behavedness' in finite relational structures, which is useful both algorithmically and model theoretically. For instance, model checking for first-order or monadic second-order formulae becomes more tractable if the input is restricted to finite

structures of bounded treewidth. But also decidability issues, in particular satisfiability, can often be linked to a priori bounds on the treewidth of target
models – a phenomenon best known, and in its purest form, for logics with the
tree model property, e.g., due to bisimulation invariance. As pointed out above,
the bounded treewidth model property of logics invariant under guarded bisimulation extends this benefit to richer settings. Moreover, bounded treewidth has
featured in recent analogues to classical expressive completeness issues over
finite structures. While bounded treewidth certainly is not the only structural
restriction that helps to overcome well known obstacles in finite model theory,
it seems to occupy a central place in such concerns. We here mainly want to
discuss several such results, especially results concerning expressive completeness for fragments of FO, in the light of connections with techniques stemming
form the fundamental notion of Gaifman locality.

Bounded treewidth is also at the center of Stephan Kreutzer's chapter [39]
in this volume, where the algorithmic impact of bounded treewidth, among
other structural criteria, is treated in depth. There the reader will also find a
much more detailed account of the connections between bounded treewidth and
model checking complexities for first- and monadic second-order logic than
what is sketchily hinted at below.

Relational structures of bounded treewidth   We have already come across
a special form of tree decompositions in section 6.4.2, cf. Definition 6.4.3, and
now briefly review the general notion of a tree decomposition underlying the
definition of treewidth.

Definition 6.5.1   A *tree decomposition* of the finite relational structure $\mathfrak{A}$
consists of a tree T together with a map $\rho \colon T \to \mathcal{P}(A)$ associating subsets of
A with the nodes of T in such a manner that

(i) every relational ground atom of $\mathfrak{A}$ is contained in some $\rho(t)$.

(ii) for all $a \in A$, $\{t \in T : a \in \rho(t)\} \subseteq T$ is connected in T.

The width of the tree decomposition (T, ρ) is $\max_{t \in T} |\rho(t)| - 1$.

The *treewidth of* $\mathfrak{A}$, $\mathrm{tw}(\mathfrak{A})$ is the minimal width among all tree decompositions
of $\mathfrak{A}$.

$\mathcal{C}_k[\tau] := \{\mathfrak{A} : \mathrm{tw}(\mathfrak{A}) \leqslant k\}$ denotes the class of finite τ-structures of treewidth
up to k.

Note that (i) is the same as to say that the subsets used in a tree decomposition
of $\mathfrak{A}$ must cover the guarded subsets.[17] The correction by -1 in the definition

---

[17] That they must not themselves be guarded subsets accounts for the difference in comparison
with Definition 6.4.3; a tree decomposition of $\mathfrak{A}$ is a hypergraph decomposition of some
hypergraph that may be coarser than the hypergraph $H(\mathfrak{A})$ of guarded subsets.

of treewidth is so that trees get treewidth 1 (rather than 2, which is the required patch size).

Model checking complexity   For the complexity of the model checking problem for some logic L over the class $\mathcal{C}$, one distinguishes

- *combined complexity*, where both $\varphi \in$ L and $\mathfrak{A} \in \mathcal{C}$ vary, and the input size is the sum of the input sizes, $|\varphi| + ||\mathfrak{A}||$;[18]
- *data complexity*, where the formula $\varphi \in$ L is fixed, and the variation is in the structure, with input size measure $||\mathfrak{A}||$;
- *expression complexity*, with fixed $\mathfrak{A}$ and varying $\varphi \in$ L.

The following are some well known cornerstones for the model checking complexity of monadic second-order logic MSO, FO and some fragments of FO considered above:

- MSO model checking over $\mathcal{C}_k$ (treewidth k structures) has linear combined complexity due to a fundamental theorem of Courcelle [12], where "linear" refers to a complexity in $\mathcal{O}(||\mathfrak{A}|| \cdot |\varphi|)$. On the class of all finite graphs, on the other hand, MSO clearly captures graph properties at any level of the polynomial hierarchy (this is w.r.t. data complexity).
- FO-formulae have logarithmic data complexity (i.e., poly-logarithmic in $||\mathfrak{A}||$ or $|A|$, but with syntactic parameters of the formula in the exponent) [15, 33].
- The combined complexity of FO model checking is complete for Pspace (this is even true for formula complexity over the fixed naked two-element structure, by a simple reduction of the Pspace complete satisfiability problem for quantified boolean formulae).
- The combined complexity for model checking FO^k, on the other hand, is complete for Ptime for every $k \geqslant 2$, and even linear (in the sense of $\mathcal{O}(||\mathfrak{A}|| \cdot |\varphi|)$) for ML as well as for GF, and still Ptime complete even for ML, [48, 23, 8, 19].

Interestingly, measures of *tree-likeness* improve model checking complexities – both on the side of the structure (e.g., model checking over bounded treewidth structures) and on the side of the formula input (e.g., model checking conjunctive queries with templates of bounded treewidth). We just mention some key results with pointers to the literature, and again refer to [39] for a more thorough treatment of some of these.

---

[18] $||\mathfrak{A}||$ stands for the size of a succinct encoding of the relational structure $\mathfrak{A}$. E.g., for graphs $\mathfrak{A}$ in an adjacency list encoding, $||\mathfrak{A}|| \in \mathcal{O}(n^2)$, but it can be sub-quadratic in the number $n = |A|$ of vertices for graphs with few edges. Finer complexity accounts need to be based on a random access model of computation, so that access to the input structure does not distort the real algorithmic content of formula evaluation.

FO data and combined complexity and local constraints   For FO data complexity, Frick and Grohe [16] establish a linear bound over any class C of structures whose treewidth is *locally bounded*. A class C of structures has locally bounded treewidth if the treewidth of ℓ-neighbourhoods in structures from C is uniformly bounded by some function in the radius ℓ. The underlying model checking algorithm is based on a presentation of the formula in Gaifman form. With this, the checking of 'global' structural properties reduces to local evaluation of FO formulae in ℓ-neighbourhoods and a graph theoretic core algorithm that checks for existence of scattered tuples in the Gaifman graph, vertex-coloured according to the local pre-processing. For generalisations and more recent successes of this approach to first-order model checking complexity in classes tamed by local conditions on graph invariants see Grohe's survey [26] as well as Kreutzer's chapter [39] in this volume, with a view also to the parameterised complexity of the combined model checking problem.

Combined complexity for fragments of FO   The combined complexity of conjunctive query evaluation has been studied intensively, with a natural motivation central to database theory and with interesting connections to constraint satisfaction problems. Also in these investigations tree-likeness (in this case of syntactic features of very special FO formulae) plays a major role. *Conjunctive query evaluation* is the model checking of existential positive prenex FO formulae whose quantifier-free core is just a conjunction of relational atoms, $\varphi = \exists \mathbf{x} \bigwedge_i \alpha_i(\mathbf{x}_i)$ with atomic α_i (in subtuples of variables $\mathbf{x}_i$ of $\mathbf{x}$). The link with homomorphism problems and hence with constraint satisfaction (see for instance [36, 38]) is natural and straightforward. The desired assignment to variables $\mathbf{x}$ over the τ-structure $\mathfrak{A}$ *is* a homomorphism from a τ-structure $\mathfrak{X}_\varphi$ induced by the conjuncts α_i on the set of variables $[\mathbf{x}]$ into $\mathfrak{A}$,

$$\beta : \mathfrak{X}_\varphi \xrightarrow{\text{hom}} \mathfrak{A}.$$

Note that while the data complexity is poly-logarithmic for each individual (first-order) φ or $\mathfrak{X}$, in general one expects an exponential dependency on the number of variables in φ or on the size of $\mathfrak{X}$.

It turns out that the hypergraph $H(\mathfrak{X}_\varphi)$ holds one key to better bounds on the complexity of the associated homomorphism/query evaluation problems. In fact φ is (equivalent to a formula) in GF if $H(\mathfrak{X}_\varphi)$ is acyclic, in which case model checking becomes linear in $|\varphi|$. Indeed, a tree decomposition of $H(\mathfrak{X}_\varphi)$ yields a translation into GF and hence a reduction to the linear model checking of GF. This generalises to φ with a fixed bound on the treewidth of $\mathfrak{X}_\varphi$, where the model checking can be based on the auxiliary acyclic hypergraph of bounded width extracted from the tree decomposition (instead of $H(\mathfrak{X}_\varphi)$ itself).

In these cases, which admit considerable further extensions in terms of weaker notions of bounded widths (e.g., bounded hypertreewidth rather than treewidth [20]), combined model checking remains in Ptime [18, 20].

But also reductions to FO^k can be seen as essential for tractability. For any finite τ-structures $\mathfrak{X}$ and $\mathfrak{A}$, the following are equivalent [13, 38]:

(i) existence of a homomorphism from $\mathfrak{X}$ to $\mathfrak{A}$, $\mathfrak{X} \xrightarrow{\text{hom}} \mathfrak{A}$;

(ii) $\mathfrak{A} \models \exists \mathbf{x} \eta_{\mathfrak{X}}$, where $\eta_{\mathfrak{X}}$ is the positive diagram of $\mathfrak{X}$;

(iii) the *transfer property* $\mathfrak{X} \Rightarrow_{\text{pos}\exists^*} \mathfrak{A}$, meaning that every positive existential sentence true in $\mathfrak{X}$ is also true in $\mathfrak{A}$.

For $\mathfrak{X} = \mathfrak{X}_\varphi$, where φ is a conjunctive query, φ is equivalent to $\exists \mathbf{x} \eta_{\mathfrak{X}}$ (cf. (ii)). For $\mathfrak{X} \in \mathcal{C}_k$, this sentence is expressible in positive existential FO^{k+1} [37, 38], so that (iii) above can be replaced by a transfer condition for all positive existential FO^{k+1} rather than FO. In this context, therefore, the Ptime analysis of winning positions in the (positively restricted, one-sided) $(k+1)$-pebble game [35] on $\mathfrak{X}$ versus $\mathfrak{A}$ decides the homomorphism problem.

6.5.2 Non-classical proofs for (variants of) classical characterisations

With this section we return to expressive completeness issues, related to the existential and the existential positive fragments of FO over classes of finite structures. A first-order sentence $\varphi \in FO[\tau]$ is *preserved under extensions* if in every substructure relationship $\mathfrak{A} \subseteq \mathfrak{B}$ between τ-structures, $\mathfrak{A} \models \varphi$ implies $\mathfrak{B} \models \varphi$. Similarly, φ is *preserved under homomorphisms* if for every homomorphism $\mathfrak{A} \xrightarrow{\text{hom}} \mathfrak{B}$ between τ-structures, $\mathfrak{A} \models \varphi$ implies $\mathfrak{B} \models \varphi$. As an embedding of a substructure is a special case of homomorphism, preservation under homomorphisms implies preservation under extensions. Clearly, existential sentences are preserved under extensions, while existential positive sentences are even preserved under homomorphisms.

The classical results are the following. We explicitly state the more interesting expressive completeness statements.

Theorem 6.5.2 (Łos–Tarski) *The existential fragment of first-order logic is expressively complete for first-order properties that are preserved under extensions.*

Theorem 6.5.3 (Lyndon–Tarski) *The existential positive fragment of first-order logic is expressively complete for first-order properties that are preserved under homomorphisms.*

These are proved classically, e.g. in [11], by means of a compactness argument for the construction of suitable elementary extensions, respectively elementary chain constructions.

Classically, as well as towards possible restrictions of the expressive completeness claim to some class C other than the class of all τ-structures, both essentially amount to finiteness claims for classes of *minimal models* (within C).

We refer to *substructure minimal* models as generators w.r.t. extensions, and, as generators w.r.t. homomorphisms, also to so-called *cores*. In a class closed under homomorphisms, the natural generators are simultaneously minimal w.r.t. the weak substructure relationship and w.r.t. inverse homomorphisms. We review some standard terminology in this connection.

A *weak substructure* relationship between τ-structures, denoted $\mathfrak{A} \subseteq_w \mathfrak{B}$, requires that $A \subseteq B$ and $R^{\mathfrak{A}} \subseteq R^{\mathfrak{B}}$ for every relation R in τ (rather than $R^{\mathfrak{A}} = R^{\mathfrak{B}}{\restriction}A$ as in the substructure relationship $\mathfrak{A} \subseteq \mathfrak{B}$). A *retraction* is a homomorphism h from some structure $\mathfrak{A}$ onto a weak substructure $\mathfrak{A}_0 \subseteq_w \mathfrak{A}$ such that $h{\restriction}A_0 = \text{id}$. It is worth noting that a retraction $h\colon \mathfrak{A} \xrightarrow{\text{ret}} \mathfrak{A}_0$ is accompanied by a trivial inclusion homomorphism back from $\mathfrak{A}_0$ into $\mathfrak{A}$, since $\mathfrak{A}_0 \subseteq_w \mathfrak{A}$. A structure whose only retraction is the identity is called a *core*. Every finite relational structure $\mathfrak{A}$ possesses a retract onto some core and this core is unique up to isomorphism. It is then straightforward to see that a homomorphism closed class of finite structures is generated by its members that are cores; viz., generated as the class of all weak extensions of these. But the subclass of $\subseteq_w$-minimal members generates the same class.

Definition 6.5.4   (a) $\mathfrak{A}$ is a *substructure minimal* ($\subseteq$-minimal) model of φ if
$\mathfrak{A} \models \varphi$ and $\mathfrak{A}' \not\models \varphi$ for all $\mathfrak{A}' \subsetneq \mathfrak{A}$.
(b) $\mathfrak{A}$ is a *weak-substructure minimal* ($\subseteq_w$-minimal) model of φ if $\mathfrak{A} \models \varphi$ and
$\mathfrak{A}' \not\models \varphi$ for all $\mathfrak{A}' \subsetneq_w \mathfrak{A}$.
(c) $\mathfrak{A}$ is a *core* model of φ if $\mathfrak{A} \models \varphi$ and $\mathfrak{A}$ is a core.

Observation 6.5.5   *Let C_0 be a class of finite τ-structures that is closed under extensions. Then the following are equivalent:*
  (i) *C_0 is definable (within the class of finite τ-structures) by an existential first-order sentence.*
  (ii) *C_0 has, up to isomorphism, finitely many substructure minimal members.*

For the crucial direction, (ii) $\Rightarrow$ (i): if $\mathfrak{A}_1, \ldots, \mathfrak{A}_N$ are the isomorphism types of substructure minimal members in C_0, then C_0 is definable by the disjunction over the existentially quantified algebraic diagrams of the $\mathfrak{A}_i$. For (i) $\Rightarrow$ (ii) it suffices to observe that the size of substructure minimal models of an existential prenex sentence φ is bounded by the number of variables.

The above equivalence persists in restriction to any class C of τ-structures that is itself closed under substructures (some such extra condition on the surrounding class C is necessary for (i) $\Rightarrow$ (ii), not for (ii) $\Rightarrow$ (i)).

Similarly one obtains the following, where a disjunction over the existentially quantified *positive* diagrams of $\subseteq_w$-minimal models, which are cores, provides a canonical definition in existential positive FO. We state the equivalence relative to the class of all (finite) τ-structures, but it similarly holds in restriction to any class C of τ-structures that is closed, e.g., under substructures.

Observation 6.5.6 *For any class C_0 of (finite) τ-structures that is closed under homomorphisms, the following are equivalent:*

(i) *C_0 is definable (within the class of finite τ-structures) by a sentence in existential positive FO.*

(ii) *C_0 has, up to isomorphism, finitely many $\subseteq_w$-minimal members.*

(iii) *C_0 has, up to isomorphism, finitely many $\subseteq$-minimal members.*

(iv) *C_0 has, up to isomorphism, finitely many homomorphism minimal core members.*

As we are dealing with finite relational vocabularies τ, a finite bound on the number of isomorphism types of minimal models is equivalent to a bound on the size of minimal models.

It has been known for a long time that the Łos–Tarski theorem (Theorem 6.5.2) fails in the sense of finite model theory (with counterexamples due to Tait and Gurevich, see e.g. [15]).

The status of the Lyndon–Tarski theorem (Theorem 6.5.3) in finite model theory, on the other hand, had been an important open problem for quite some time when it was resolved, positively, by Rossman [47].

Beside the overall finite model theory version, however, one may of course investigate the status of these expressive completeness issues in restriction to various classes of (finite) structures of interest. In the following sections we outline a particular criterion of well-behavedness motivated by considerations of Gaifman locality, which has led to interesting results along these lines.

Wideness criteria

The wideness criteria proposed in [4, 3] couple the existence of large scattered subsets to the size of structures. In the context of the minimal model criteria as in Observations 6.5.5 and 6.5.6 above they can be used to derive upper bounds on the size of minimal models. Models exceeding a certain size cannot be minimal if their richness in scattered sets allows one to extract smaller models on the basis of a Gaifman representation of the first-order property at hand.

Definition 6.5.7   A structure is (ℓ, m)-*wide* if its Gaifman graph contains an ℓ-scattered subset of size m.

A class $\mathcal{C}$ of τ-structures is called *wide* if there is a function $N : \mathbb{N} \times \mathbb{N} \to \mathbb{N}$, such that, for all ℓ and m and $\mathfrak{A} \in \mathcal{C}$, if $|A| \geqslant N(\ell, m)$, then $\mathfrak{A}$ is (ℓ, m)-wide.

$\mathcal{C}$ is called *almost wide* if, for some fixed k, the analogous condition applies after the removal of a suitable subset of at most k elements from the structures $\mathfrak{A}$ at hand.

A typical example of a wide class is the class of graphs of fixed bounded degree. The class of trees, on the other hand, is not wide (there are arbitrarily large trees of diameter 2), but almost wide: a large tree either has long branches or a node of high degree; removal of a single node of high degree also produces a large scattered set. Similarly, in a tree decomposition of fixed bounded width of a sufficiently large graph or relational structure, a large scattered set becomes available at least after the removal of the elements associated with some high degree node in the decomposition tree. A much more profound analysis is necessary to show almost wideness for every class of graphs that excludes a minor [4].

Proposition 6.5.8 (Atserias–Dawar–Kolaitis)   *The class of treewidth k graphs is almost wide. By extension, $\mathcal{C}_k[\tau]$, the class of τ-structures of treewidth up to k, is almost wide.*

More generally, any class of graphs with excluded minor is almost wide, and by extension any class of τ-structures whose Gaifman graphs avoid some minor.

Expressive completeness for extension preservation

The following summarises key results from [3].

Theorem 6.5.9 (Atserias–Dawar–Grohe)   *The size of $\subseteq$-minimal models of a first-order sentence φ that is preserved under extensions can be bounded over the following classes of finite structures:*

  (i) *acyclic relational structures (i.e., directed coloured graphs with acyclic Gaifman graphs);*
 (ii) *wide classes $\mathcal{C}$, like any class of graphs of bounded degree.*
(iii) *$\mathcal{C}_k$, the class of all finite structures of treewidth up to k.*

As a consequence, existential FO is expressively complete for first-order properties preserved under extensions over these classes.

Interestingly, there are almost wide classes over which existential FO is not expressively complete for first-order properties preserved under extensions. A counterexample over the class of planar graphs is given in [3].

The underlying idea in the proof of the theorem is to choose parameters ℓ, q, m from a Gaifman representation of φ, such that φ is preserved under $\equiv_{q,m}^{\ell}$, and then to isolate a proper substructure $\mathfrak{A}_0 \subsetneq \mathfrak{A}$ that at the same time is $\equiv_{q,m}^{\ell}$ equivalent to some extension $\hat{\mathfrak{A}} \supseteq \mathfrak{A}$, in any large enough model $\mathfrak{A}$. The actual argument in [3] involves a sophisticated finite chain construction.

Expressive completeness for homomorphism preservation

The connection between wideness criteria and bounds on the number (or size) of $\subseteq_w$-minimal models, which is crucial according to Observation 6.5.6, is provided by the following theorem. It stems from the analysis of the boundedness problem for Datalog programs (least fixpoint recursion over positive existential FO) over finite structures.

Theorem 6.5.10 (Ajtai–Gurevich)   *Let C be a class of finite τ-structures that is closed under substructures and disjoint unions. If $\varphi \in$ FO is preserved under homomorphisms within C, then there are $\ell, m \in \mathbb{N}$ such that no (ℓ, m)-wide model of φ can be $\subseteq$-minimal.*

The same applies w.r.t. wideness after removal of up to k elements, for fixed k.

Corollary 6.5.11 (Atserias–Dawar–Kolaitis)   *Over any class of finite structures that is almost wide and closed under substructures and disjoint unions, existential positive FO is expressively complete for first-order properties preserved under homomorphisms.*

That minimal models cannot be too wide in the sense of Theorem 6.5.10, comes from a Gaifman representation of φ. We sketch the argument that, for a first-order sentence φ that is preserved under $\equiv_{q,m}^{\ell}$ and under homomorphisms (within C), there are $L, M \in \mathbb{N}$ such that no (L, M)-wide model of φ can be $\subseteq_w$-minimal. More precisely, there are

- M, large enough w.r.t. L, Q, such that within any L-scattered subset of size M in $\mathfrak{A} \models \varphi$ we find some pair of elements $a \neq b$ for which $\mathfrak{A}, a \equiv_{Q,0}^{L} \mathfrak{A}, b$;
- L and Q, large enough w.r.t. ℓ, q, such that $\mathfrak{A}, a \equiv_{Q,0}^{L} \mathfrak{A}, b$ implies the following transfer property for Gaifman rank $(\ell, q, 1)$-assertions:

$$\mathfrak{A} \Rightarrow_{q,1}^{\ell} \mathfrak{B} := \mathfrak{A} {\restriction} (A \setminus \{b\}),$$

meaning that every sentence of the form $\exists x \chi^\ell(x)$ where $\mathrm{qr}(\chi) \leqslant q$ that is true in $\mathfrak{A}$ remains true in $\mathfrak{B}$ ($\mathfrak{A}$ with b removed).[19]

M simply needs to be chosen large w.r.t. the number of quantifier-rank Q types of single elements (in their L-neighbourhood) in order to guarantee the existence of distinct but $\equiv^L_{Q,0}$ equivalent nodes by the pigeon-hole principle.

For such a and b, the desired transfer of $\exists x \chi^\ell(x)$-assertions follows from $\equiv^L_{Q,0}$ equivalence provided $L \geqslant 2\ell$ and Q large enough so that for all $\mathrm{qr}(\chi) \leqslant q$, the assertion

$$\exists x' \big(d(x, x') \leqslant \ell \wedge \chi^\ell(x') \big) \qquad (*)$$

is L-local and of quantifier rank $\leqslant Q$. Compare the diagram below for this proof sketch. In the non-trivial case $\mathfrak{A} \models \chi^\ell[a']$ for some $a' \in N^\ell(b)$, so that after the removal of b, there is no guarantee that still $\mathfrak{B} \models \chi^\ell[a']$. Using $\equiv^L_{Q,0}$ equivalence between a and b, though, $(*)$ is true of a if it is true at b. Hence there is a corresponding $b' \in N^\ell(a)$ such that $\mathfrak{A} \models \chi^\ell[b']$. So $\mathfrak{B} \models \chi^\ell[b']$ follows, since the L-neighbourhood of a is unaffected by the removal of b.

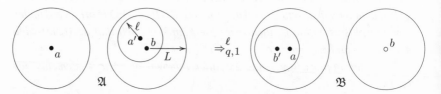

It follows that $\mathfrak{A} \oplus m \cdot \mathfrak{B} \equiv^\ell_{q,m} m \cdot \mathfrak{B}$ (with disjoint sums of m isomorphic copies of $\mathfrak{B}$ plus one copy of $\mathfrak{A}$ on the left-hand side). Therefore, $\mathfrak{B} \models \varphi$ is a smaller model of φ:

$$\mathfrak{A} \xrightarrow{\text{hom}} \mathfrak{A} \oplus m \cdot \mathfrak{B} \equiv^\ell_{q,m} m \cdot \mathfrak{B} \xrightarrow{\text{hom}} \mathfrak{B}.$$

Expressive completeness of the existential positive fragment of FO for homomorphism preservation over the class of all finite relational structures – the finite model theory version of the Lyndon–Tarski Theorem – has been shown by Rossman in [47]. His approach is based on a combinatorial analysis of existential positive types and saturation arguments for these, which can be brought to a sufficient level of closure in a finite chain construction. Leaving aside much of the actual sophistication of the combinatorial analysis, there

---

[19] Note that this is a one-directional transfer rather than an equivalence. E.g., in a graph consisting just of a large cycle, the removal of any single element results in a structure that is inequivalent in the sense of $\equiv^1_{1,1}$.

is one aspect of Rossman's approach that may deserve to be highlighted in connection with the leading themes of this survey. That is the manner in which the new argument is based on explicit model construction (as opposed to an abstract model existence argument), and can be viewed as an upgrading (not of an equivalence, but of a unidirectional transfer relationship) to approximate first-order equivalence, which is orthogonal to the classical argument. This is an interesting parallel with the observations in section 6.3.2. While a traditional proof of the Lyndon-Tarski Theorem can be based on the upgrading indicated in the left-hand diagram, Rossman's proof amounts to the upgrading indicated in the right-hand diagram. In the traditional picture, transfer w.r.t. the full existential positive fragment of FO is upgraded, in a classical saturation argument based on compactness, to yield a homomorphism between structures that are elementarily equivalent to the original ones. In the 'explicit' construction of Rossman's, on the other hand, a specific finite level of transfer (existential positive formulae of quantifier rank up to r) is upgraded to a specific finite level of first-order equivalence that preserves the given sentence φ.

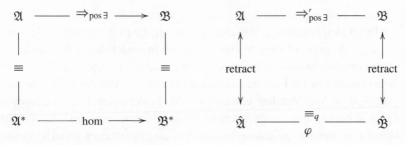

Moreover, Rossman's proof has a classical variant, in which the chain construction is extended to an infinite limit, that yields a completely new, alternative proof of the classical Lyndon–Tarski result with added value. In fact, Rossman shows that existential positive FO is expressively complete for first-order sentences preserved under homomorphisms, level-by-level w.r.t. quantifier-rank. In the classical model theory version of his proof, Rossman realises the above upgrading for $r = q$, while in the finite model theory version there is no elementary bound on r in terms of q.

6.6 Concluding remarks

The focus on a model theory of well-behaved classes of (finite) structures – adapted to specific application areas, or to the study of specific logics, or to other specific model theoretic themes – offers promising perspectives for the development and ramification of finite model theory. Finiteness as the only constraint, which often entails 'negative' results, may not be the best choice for many reasons.

It can be that the class of all finite structure is still not a good match for the natural domain of reasoning for certain application areas; some model theoretic answers – 'positive' or 'negative' – may still be 'too easy' over the class of all finite structures. In modal reasoning, for instance, rootedness or connectivity constraints are arguably essential in the intuitive modelling. More generally, the 'generic finite structure that we mean' may well have more specific structural properties than an 'arbitrary finite structure.'

It can also be that the class of all finite structures is too liberal a setting for structural insights into certain issues. Definability and expressive completeness results, for instance, that fail over the class of all finite structures may not just be recovered but also clarified overall through a better understanding of the structural conditions that support them. In this sense there is not just finite model theory, but there may be many adequate domains of structures for individual issues.

I think both aspects are important from the modelling point of view (i.e., in relation to applications), also clearly from an algorithmic point of view, but also from the point of view of classical issues in model theory. Sophisticated adaptations of classical techniques, like the analysis of types and the use of chain constructions in Rossman's result, enrich finite model theory but also cast fresh light on long-standing classical results. In this context the constructive aspect of explicit model constructions or model transformations – in contrast with smooth abstract existence proofs in classical model theory – is an important methodological contribution.

It seems that the modularity in game-oriented arguments and model con-structions, as illustrated by the power of an analysis in terms of Gaifman locality, has had comparatively little impact on traditional classical model the-ory. The great potential of another aspect of modularity, viz. decomposition techniques, has apparently been realised more fully. The combination of such aspects may lead to a better model theoretic view of more complex hierarchi-cal decompositions in particular for finite structures; and there may be more flavours of structural regularity, smoothness or tameness in finite structures to be discovered.

Bibliography

[1] Alon, N. 1995. Tools from Higher Algebra. Pages 1749–1783 of: Graham, R., Grötschel, M., and Lovasz, L. (eds), *Handbook of Combinatorics*, vol. II. North-Holland.

[2] Andréka, H., van Benthem, J., and Németi, I. 1998. Modal Languages and Bounded Fragments of Predicate Logic. *Journal of Philosophical Logic*, **27**, 217–274.

[3] Atserias, A., Dawar, A., and Grohe, M. 2005. Preservation under extensions on well-behaved finite structures. Pages 1437–1449 of: *Proceedings of 32nd International Colloquium on Automata, Languages and Programming ICALP'05*. LNCS, vol. 3580.

[4] Atserias, A., Dawar, A., and Kolaitis, P. 2006. On preservation under homomorphisms and unions of conjunctive queries. *Journal of the ACM*, **53**, 208–237.

[5] Barany, V., Gottlob, G., and Otto, M. 2010. Querying the guarded fragment. Pages 2–11 of: *Proceedings of 25th Annual IEEE Symposium on Logic in Computer Science LICS'10*.

[6] Beeri, C., Fagin, R., Maier, D., and Yannakakis, M. 1983. On the desirability of acyclic database schemes. *Journal of the ACM*, **30**, 497–513.

[7] Berge, C. 1973. *Graphs and Hypergraphs*. North-Holland.

[8] Berwanger, D., and Grädel, E. 2001. Games and Model Checking for Guarded Logics. Pages 70–84 of: *Proceedings of the 8th International Conference on Logic for Programming and Automated Reasoning LPAR'01*. LNCS, vol. 2250.

[9] Blackburn, P., de Rijke, M., and Venema, Y. 2001. *Modal Logic*. Cambridge Tracts in Theoretical Computer Science. Cambridge University Press.

[10] Blackburn, P., van Benthem, J., and Wolter, F. (eds). 2007. *Handbook of Modal Logic*. Elsevier.

[11] Chang, C. C., and Keisler, H. J. 1990. *Model Theory*. North-Holland.

[12] Courcelle, B. 1990. Graph rewriting: An algebraic and logic approach. Pages 194–242 of: van Leeuwen, J. (ed), *Handbok of Theoretical Computer Science, volume B*. Elsevier.

[13] Dalmau, V., Kolaitis, P., and Vardi, M. 2002. Constraint satisfaction, bounded treewidth, and finite-varaible logics. Pages 310–326 of: *Proceedings of 8th International Conference on Constraint Programming*. LNCS, vol. 2470.

[14] Dawar, A., and Otto, M. 2009. Modal characterisation theorems over special classes of frames. *Annals of Pure and Applied Logic*, **161**, 1–42. Extended journal version of LICS'05 paper.

[15] Ebbinghaus, H.-D., and Flum, J. 1999. *Finite Model Theory*. 2nd edn. Springer.

[16] Frick, M., and Grohe, M. 2001. Deciding first-order properties of locally tree-decomposable structures. *Journal of the ACM*, **48**, 1184–1206.

[17] Goranko, V., and Otto, M. 2007. Model Theory of Modal Logic. Pages 249–329 of: Blackburn, P., van Benthem, J., and Wolter, F. (eds), *Handbook of Modal Logic*. Elsevier.

[18] Gottlob, G., Leone, N., and Scarcello, F. 2001. The complexity of acyclic conjunctive queries. *Journal of the ACM*, **43**, 431–498.

[19] Gottlob, G., Grädel, E., and Veith, H. 2002a. Datalog LITE: A deductive query language with linear time model checking. *ACM Transactions on Computational Logic*, **3**, 1–35.

[20] Gottlob, G., Leone, N., and Scarcello, F. 2002b. Hypertree decompositions and tractable queries. *Journal of Computer and System Sciences*, **64**, 579–627.

[21] Grädel, E. 1999. On the restraining power of guards. *Journal of Symbolic Logic*, **64**, 1719–1742.

[22] Grädel, E. 2007. Finite model theory and descriptive complexity. Pages 125–230 of: *Finite Model Theory and Its Applications*. Springer.

[23] Grädel, E., and Otto, M. 1999. On Logics with Two Variables. *Theoretical Computer Science*, **224**, 73–113.

[24] Grädel, E., and Walukiewicz, I. 1999. Guarded fixed point logic. Pages 45–54 of: *Proceedings of 14th Annual IEEE Symposium on Logic in Computer Science LICS'99*.

[25] Grädel, E., Hirsch, C., and Otto, M. 2002. Back and forth between guarded and modal logics. *ACM Transactions on Computational Logics*, **3**, 418–463.

[26] Grohe, M. 2008. Logic, graphs, and algorithms. Pages 357–422 of: Flum, J., Grädel, E., and Wilke, T. (eds), *Logic and Automata, History and Perspectives*. Amsterdam University Press.

[27] Herwig, B. 1995. Extending partial isomorphisms on finite structures. *Combinatorica*, **15**, 365–371.

[28] Herwig, B. 1998. Extending partial isomorphisms for the small index property of many omega-categorical structures. *Israel Journal of Mathematics*, **107**, 93–124.

[29] Herwig, B., and Lascar, D. 2000. Extending partial isomorphisms and the profinite topology on free groups. *Transactions of the AMS*, **352**, 1985–2021.

[30] Hodges, W. 1993. *Model Theory*. Cambridge University Press.

[31] Hodkinson, I., and Otto, M. 2003. Finite conformal hypergraph covers and Gaifman cliques in finite structures. *Bulletin of Symbolic Logic*, **9**, 387–405.

[32] Hoogland, E., Marx, M., and Otto, M. 1999. Beth definability for the guarded fragment. In: Gebrandy, J., Marx, M., de Rijke, M., and Venema, Y. (eds), *JFAK – Essays Dedicated to Johan van Benthem on the Occasion of his 50th Birthday*. Amsterdam University Press. CD-ROM.

[33] Immerman, N. 1998. *Decsriptive Complexity*. Graduate Texts in Computer Science. Springer.

[34] Janin, D., and Walukiewicz, I. 1996. On the Expressive Completeness of the Propositional mu-Calculus with Respect to Monadic Second Order Logic. Pages

263–277 of: *Proceedings of 7th International Conference on Concurrency Theory CONCUR'96*. LNCS, vol. 1119.

[35] Kolaitis, P. 2007. On the expressive power of logics on finite models. Pages 27–123 of: *Finite Model Theory and Its Applications*. Springer.

[36] Kolaitis, P., and Vardi, M. 2000a. Conjunctive-query containment and constraint satisfaction. *Journal of Computer and System Sciences*, **61**, 302–332.

[37] Kolaitis, P., and Vardi, M. 2000b. A game-theoretic approach to constraint satisfaction. Pages 175–181 of: *Proceedings of Of 17th Conference on Artificial Intelligence AAAI'00*.

[38] Kolaitis, P., and Vardi, M. 2007. A logical approach to constraint satisfaction. Pages 339–370 of: *Finite Model Theory and Its Applications*. Springer.

[39] Kreutzer, S. 2008. Algorithmic Meta-Theorems. In: Esparza, J., Michaux, C., and Steinhorn, C. (eds), *Finite and Algorithmic Model Theory*. CUP. (this volume).

[40] Lorenz, K. 1968. Dialogspiele als semantische Grundlage von Logikkalkülen. *Archiv für Mathematische Logik und Grundlagenforschung*, **11**, 32–55 and 73–100.

[41] Lorenzen, P. 1961. Ein dialogisches Konstruktivitätskriterium. Pages 193–200 of: *Infinitistic Methods, Proceedings of the Symposium on Foundations of Mathematics, Warsaw 1959*. Oxford University Press.

[42] Otto, M. 2004. Modal and guarded characterisation theorems over finite transition systems. *Annals of Pure and Applied Logic*, **130**, 173–205.

[43] Otto, M. 2010a. Highly acyclic groups, hypergraph covers and the guarded fragment. Pages 12–21 of: *Proceedings of 25th Annual IEEE Symposium on Logic in Computer Science LICS'10*.

[44] Otto, M. 2010b. *Highly acyclic groups, hypergraph covers and the guarded fragment*. Draft of extended version of LICS 2010 paper.

[45] Rabin, M. 1969. Decidability of second order theories and automata on infinite trees. *Transactions of the AMS*, **141**, 1–35.

[46] Rosen, E. 1997. Modal logic over finite structures. *Journal of Logic, Language and Information*, **6**, 427–439.

[47] Rossman, B. 2008. Homomorphism preservation theorems. *Journal of the ACM*, **55**.

[48] Vardi, M. 1995. On the complexity of bounded-variable queries. Pages 266–276 of: *Proceedings of 14th Annual ACM Symposium on Principles of Database Systems PODS'95*.

[49] Vardi, M. 1997. Why is modal logic so robustly decidable? Pages 149–184 of: Immerman, N., and Kolaitis, P. (eds), *Descriptive Complexity and Finite Models*. DIMACS Series in Discrete Mathematics and Theoretical Computer Science, vol. 31. AMS.

[50] Vardi, M. 2006. *Games as an algorithmic construct*. Tutorial, Isaac Newton Institute workshop on Games and Verification (Logic and Algorithms programme).

[51] Weinstein, S. 2007. Unifying themes in finite model theory. Pages 1–25 of: *Finite Model Theory and Its Applications*. Springer.